Optical WDM Networks

Optical WDM Networks: From Static to Elastic Networks

Devi Chadha
Indian Institute of Technology, New Delhi, India

This edition first published 2019
© 2019 John Wiley & Sons Ltd

The right of Devi Chadha to be identified as the author of this work has been asserted in accordance with law.

Registered Offices
John Wiley & Sons, Inc., 111 River Street, Hoboken, NJ 07030, USA
John Wiley & Sons Ltd, The Atrium, Southern Gate, Chichester, West Sussex, PO19 8SQ, UK

Editorial Office
The Atrium, Southern Gate, Chichester, West Sussex, PO19 8SQ, UK

For details of our global editorial offices, customer services, and more information about Wiley products visit us at www.wiley.com.

Wiley also publishes its books in a variety of electronic formats and by print-on-demand. Some content that appears in standard print versions of this book may not be available in other formats.

Library of Congress Cataloging-in-Publication Data
Name: Chadha, Devi, author.
Title: Optical WDM networks : From Static to Elastic Networks / Devi Chadha, Indian Institute of Technology, New Delhi, India.
Description: Hoboken, NJ, USA : Wiley, [2019] | Includes bibliographical references and index. |
Identifiers: LCCN 2018061450 (print) | LCCN 2019000068 (ebook) | ISBN 9781119393375 (AdobePDF) | ISBN 9781119393344 (Epub) | ISBN 9781119393269 (hardcover)
Subjects: LCSH: Wavelength division multiplexing.
Classification: LCC TK5103.592.W38 (ebook) | LCC TK5103.592.W38 C53 2019 (print) | DDC 621.39/81–dc23
LC record available at https://lccn.loc.gov/2018061450

Cover Design: Wiley
Cover Image: © Wenjie Dong/Getty Images

Set in 10/12pt WarnockPro by SPi Global, Chennai, India
Printed and bound in Singapore by Markono Print Media Pte Ltd

10 9 8 7 6 5 4 3 2 1

Contents

Preface

The last decade has seen significantly increased use of high bandwidth emerging dynamic applications, such as real-time multimedia streaming, cloud computing, data center networking, etc. These rapid advances require the next-generation optical communication networks to adapt to these changes by becoming more agile and programmable, and to meet the demands of high bandwidth and flexibility with much higher efficiency and reduced cost. It therefore becomes necessary to bring out a comprehensive and up-to-date book of optical networks. This was the driver for this text.

Writing a book on optical networking entails covering materials that span several disciplines, ranging from physics to electrical engineering to computer science and operations research. The treatment of the material requires uncovering the unique strengths and limitations of the appropriate technologies, and then determining how those are exploited in pragmatic network architectures, while compensating for the limitations. The paradigm shift in optical networking which we are seeing with software defined networking requires clear basic concepts to conduct research and development in these newer optical technologies. It is difficult to develop newer sophisticated technology for different applications without understanding its evolutionary process. Thus, the task of writing such a book becomes quite challenging.

Overview of the Book

This book attempts to cover components and networking issues related to second-generation optical networks. The second generation of fiber optic networks exploits the capacity of fiber to achieve hundreds of gigabits per second to terabits per second with dense wavelength division multiplexing (DWDM) by using routing and switching of signals in the optical domain. There is now a matured large bandwidth underlying optical technology available with new tools for network control and management. There is also a recognition of the latest directions of optical network deployment and research. These new directions include cost-effective network architectures tailored to the strengths of current optical transmission and switching equipment, passive optical networks to bring high-speed access to the end user, hybrid optical/electronic architectures supporting the merging of multi-wavelength and Internet technologies, and networks of the future based on all-optical packet switching and software programming. Keeping this view in mind, the book covers the fiber optic wavelength division multiplexing (WDM) networks in ten chapters as detailed below.

Chapter 1 offers an introduction to optical networks with an overview of the fundamentals of network architecture and services provided by it. Chapter 2 gives an overview of the different components needed to build a network, such as transmitters, receivers, amplifiers, multiplexers, switches, optical cross-connects, etc. The next couple of chapters then focus on the different types of networks. The broadcast and select basic static multipoint networks are given in Chapter 3. Chapter 4 describes passive optical network solutions for fiber-to-the-x access network applications, while Chapter 5 covers the metropolitan area networks, basically the ring structures. Chapter 6 describes the wavelength routed wide area networks and how to overlay virtual networks, for example IP or OTN networks over an underlying second-generation optical network. Chapter 7 covers the control plane architecture as it has developed through the recent activities of several standards organizations describing the latest developments in optical network control. It gives detailed discussion of generalized multiprotocol label switching (GMPLS) as it applies to optical networks. Going ahead, Chapters 8–10 are devoted to the advanced techniques used for the design of the upcoming technologies for bringing the expanding capability of the present-day requirements. Chapter 8 covers the effects of signal impairments, survivability, protection, and restoration in optical networks, consistent with the growing importance of optical layer fault management in current networks. As there have been fundamental changes in many aspects of optical networking, Chapters 9 and 10 cover the upcoming technologies of flex-grid and software-defined optical networking.

Exercises are provided for most of the chapters, and many of them suggest avenues for future study. The book is meant to offer several different alternatives for study depending on the interest of the reader, be it understanding the current state of the field, acquiring the analytical tools for network performance evaluation, optimization, and design, or performing research on next-generation networks.

This book has been written primarily as a graduate-level textbook in the field of optical fiber networks. For this reason, the emphasis is on concepts and methodologies that will stand the test of time. Along with this, the advances in the technology are also discussed throughout the text. An attempt is made to include as much recent material as possible so that students are exposed to the many advances in this exciting field. The book can also serve as a reference text for researchers and industry practitioners engaged in the field of optical networks because of the exposure to much of the advancement in the emerging area. Exhaustive reference lists at the end of chapters are provided for finding any further details which could not be included in the book. The listing of recent research papers should be useful for researchers using this book as a reference.

Acknowledgments

A large number of people have contributed to this book either directly or indirectly, and hence it is impossible to mention all of them by name. First, I thank my graduate students who took my course on optical networks year on year and helped improve my class notes through their questions and comments. Much of the book's material is based on research that I have conducted over the years with my graduate students. I would also like to thank students in my classes for developing many of the figures: Gaurav, Devendra, Nitish, Sridhar, Vishwaraj, Shantanu, and a few others; their efforts are highly appreciated.

I am grateful to my Institute, the Indian Institute of Technology, Delhi, India, for providing a cordial and productive atmosphere, and to my colleagues for many useful discussions during the course of writing the book.

This book could not have been published without the help of many people at Wiley International; in particular, Anita Yadav, acquisitions editor, for taking me through the entire process from start to finish, and Steven Fassioms, my project editor, for orchestrating the production of the book.

On the family front, I'd like to acknowledge the invaluable support given by my dear husband, Dev, during this endeavor, and my loving children Manu, Rati, Rashi, and Varun for their understanding when I needed to spend many hours on the book instead of spending time with them. Finally, I thank the Almighty for giving me the strength to embark upon this project and with His blessing conclude it satisfactorily.

1

Introduction to Optical Networks

1.1 Introduction

Any technological development is always driven by the need and demand of the changes in society. The rapid evolution of communication networks from the basic telephone network to the present high-speed large area networks has come with the social need of people to communicate among themselves, with the increasing user demands for new applications, as well as advances in enabling technologies. The fast changes in the present-day telecommunication networks are also driven by the user's need to remain connected anytime and all the time, anywhere and everywhere in the world. The new applications, i.e. multimedia services, video-conferencing, interactive gaming, Internet services, and the World Wide Web, all demand very large bandwidths. Besides this, the user wants the unifying network underneath to be reliable, give the best services, and be cost-effective as well.

What we need today, therefore, is a communication network with high capacity and low cost, that is fast, reliable, and able to provide a wide variety of services from dedicated to best-effort services. The available transmission media best suited to meet most of these requirements is *optical fiber*. Besides having enormous bandwidth in terahertz ($\sim 10^{12}$ Hz), optical fiber has low loss and cost. It is lightweight, with strength and flexibility, and is immune to electromagnetic interference and noise. It is secure and has many more characteristics to make it an ideal high-speed transmission line, therefore optical fiber is most suitable to meet the traffic requirements in today's communication network. The enormous quantity of optical fiber laid throughout the world by the end of twentieth century is the foundation of the information super highway of optical network with huge bandwidth today.

Optical WDM Networks: From Static to Elastic Networks, First Edition. Devi Chadha.
© 2019 John Wiley & Sons Ltd. Published 2019 by John Wiley & Sons Ltd.

1.1.1 Trends in Optical Networking

In order to increase the capacity of point-to-point links, optical fibers first replaced the coaxial and two-wire transmission lines in the existing communication networks. The capacity of the optical links in the network was further increased by using wavelength division multiplexing (WDM) in the laid fiber, thus having several wavelength channels carrying multiple data streams in a single fiber. But all the switching and routing operation in the network still remained in the electronic domain. In the real sense we do not define this network as an *all-optical network*. An all-optical network is the high-capacity telecommunication network which uses optical technology and components not only to provide large capacity optical fiber links for information transmission but also to do all the networking operations, such as *switching and routing* (i.e. facilitation of the correct and suitable path) of the signals on the required path, *grooming* of the low bit rate traffic signals to higher bit rate for better utilization of the enormous capacity of the fiber, and control and restoration functions in the network at the optical level in case of any failure in the network. The operation of the network at the optical or wavelength granularity level has many advantages. As an example, when a single wavelength carries a large number of independent connections and a failure occurs in a fiber cable, it is operationally much simpler to restore services by routing and processing an individual wavelength than to reroute each connection individually. Besides, optical switching functions consume much less power and have lower heat dissipation and footprint compared with their electronic counterparts. With the present optical technology, it is still not possible to achieve all these functions cost-effectively with ease. Hence, both optical and electronic devices are used, which makes the optical network not purely optical. These networks are indeed *hybrid* in nature at present, using both optical and electronic technology.

The communication network, which once supported telephone voice traffic only, now carries more data traffic supporting high-speed multimedia services. At the physical infrastructure level, the available optical components in the optical network can now support traffic at multiple speeds ranging up to tera-bits/sec (Tbps), with each fiber carrying a large number of wavelengths in the WDM systems. Also, along with the optical component infrastructure, the networks are becoming more flexible and agile due to the adoption of intelligent algorithms and protocols for networking, and therefore the network can now respond to new applications and demands with ease. The trend in optical networks is now toward SDON (software defined optical networking) to facilitate programmability of network operations to further increase agility and to provide users with more control over networking functions, thereby resulting in flexibility in the deployment of new services and protocols, better network utilization, QoS (quality of service), higher revenue generation from the flexibility added in the network, and the user managing the network according to his/her requirements. There is a paradigm shift in optical networking now with SDON, which is a fast-evolving technology [1–4].

In the rest of the chapter we will give a brief overview of the optical networks and an introduction to the technologies, terminologies, and parameters involved. In order to have some understanding of the functionalities of different types of networks used, somewhat detailed stratification of a generic network is discussed, hoping for an easier understanding of the networks in subsequent chapters where these will be discussed in detail.

1.1.2 Classification of Optical Networks

On the basis of geographical reach, the telecom network can be subdivided in three categories: the access and local area network (LAN), the regional metropolitan area networks (MANs), and the backbone wide area network (WAN). This is shown in Figure 1.1.

Any two distant users in the access network communicate with each other through the switch/router (or exchange) to the rest of the network infrastructure. The access network can have a span of up to 20 km or so. The *access network* needs to be cost-effective as the cost has to be shared among a smaller number of individual users who are connected through it to the shared communication network underneath. The present-day high-speed optical solution to the access part is with passive optical networks (PONs), with many variants of the Ethernet passive optical network (EPON), gigabit passive optical network (GPON), and WDM PON, etc. The other lower speed access networks are provided by wireless (WiMax, Wi-Fi), cable modems, digital subscriber lines (DSLs), higher speed lines (T1/E1), etc. We discuss the access networks in detail in Chapter 4.

The *metro network* covers distances of a few kilometers to hundreds of kilometers. The metro network aggregates the traffic from the access networks. The technologies used in metro networks are a lot different from those in access networks. Metro networks generally have a ring physical architecture using the legacy Synchronous Optical NETworking (SONET)/synchronous digital hierarchy (SDH), asynchronous transfer mode (ATM), or optical transport network (OTN) networks. A node (router) of the ring or mesh collects or distributes the traffic of the access network connected to the ring while the hub node of the ring on the other end is connected to the edge node of the WAN, as shown in Figure 1.1. The design and management of metro networks are geared toward the different types of traffic streams and services streaming from the access network, and hence the topology can keep changing accordingly. We discuss this in Chapter 5.

The *backbone* segment of the network, which is invariably a mesh physical topology, spans from hundreds to thousands of kilometers, giving nationwide to

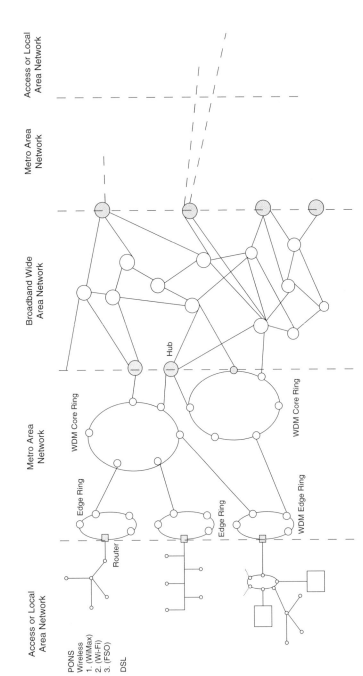

Figure 1.1 Classification of the communication network on the basis of reach.

global coverage, thus connecting the user in the access of LAN to another user or LAN in some other city, country, or continent. Backbone transport networks are shared among millions of users connected through the multiple metro-area networks. The long-haul core networks tend to be more stable in topology, unlike their metro counterpart. In core networks the emphasis is on high-bandwidth pipes and very high bit rate nodes so that more and more data are moved at high speed and efficiently. Traffic in the core network is transported between routers through high-speed gigabit links. The backbone routers use IP-over-SONET or IP-over-ATM-over-SONET technology to route IP (Internet Protocol) or the synchronous stream traffic over the wavelength channel fat pipes. Also, if the end user has large data, it can access the backbone network with a direct WDM channel. This is *optics in the access* with the access data over fiber. Chapter 6 is dedicated to the details of the optical wide area transport networks.

Optical networks could be classified in two generations: *first* and *second*. First-generation optical networks use optical fibers with WDM channels as a replacement for copper cables to increase transmission capacity. However, fiber deployment is mainly for point-to-point transmission. All the routing and switching functions are performed electronically at each node. Optical signals go through the optical to electrical conversion first for performing the switching and processing operation in the electronic domain. Later these electronic signals are reconverted to the optical domain for further transmission at each intermediate node as they propagate along an end-to-end path from one node to another. Figure 1.2a shows one such connection from source node "D" to destination node "B," where the optical signal has to go through the O-E-O conversion at node 1, node 2, and node 3 in the transponders at each connection to the electronic node. The network nodes reach the electronic bottleneck as they are not able to process all the traffic carried by the node, which is in Tbps in WDM systems because of the limitation of the electronic processing speed. The node traffic includes not only the traffic which is intended for it but also the passing-by transit traffic intended for other destination nodes. Examples of first-generation network nodes are the SONET/SDH electronic switch, ATM switch, fiber data distribution interface (FDDI), etc. interconnected with fiber links for transmission. Figure 1.2b shows the schematic of a three-degree first-generation network node which is used in the network where the signal transmission is in the optical domain but at each node the optical signal goes through the optical-to-electronic and then again electronic-to-optical conversion before transmission to the next node.

To overcome the electronic bottleneck for high data speed, the *second-generation* optical networks have optical switching nodes instead of electronic switching nodes. These nodes have optical routing and switching functions and are capable of bypassing those signals that carry traffic not intended for the node directly in the optical domain. Hence, they are also called bypass nodes. With the availability of optical devices, such as optical add-drop multiplexers

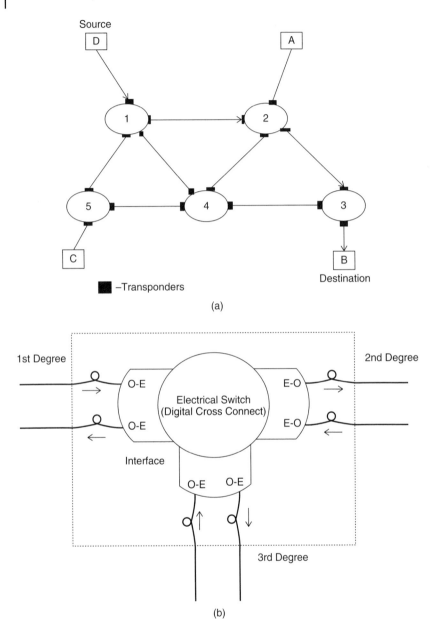

Figure 1.2 (a) First-generation optical network. (b) First-generation 3 degree node.

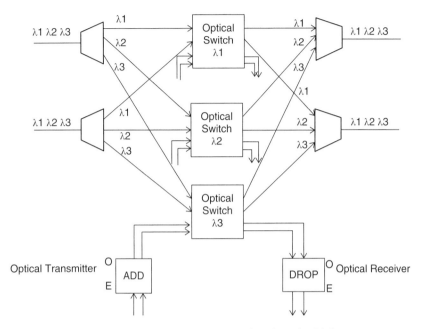

Figure 1.3 Optical bypass second-generation network node with add-drop ports.

(OADMs) and optical cross-connects (OXCs), this has become possible. Figure 1.3 shows the optical bypass capability of the optical node. In this the traffic on the two input ports connected to the node can be bypassed without any O-E-O conversion to the two output ports. Any two nodes in this network can be directly connected together by a *wavelength* (*lightpath*) in a *single-hop* even if they are not connected directly with fiber. WDM optical technology is now widely deployed not only in WANs but also in MANs and LANs.

Two nodes, if not connected directly by a lightpath, can communicate also using a *multi-hop* approach, i.e. by using electronic switching at the intermediate nodes. This electronic switching can be provided by the electronic IP routers, or SONET equipment connected to the optical nodes, leading to an IP-over-WDM or a SONET-over-WDM network, respectively, in the hybrid network. These ports are also shown in Figure 1.3 as the add/drop ports.

1.2 Optical Networks: A Brief Picture

In the previous sections the optical networks were introduced. Next, we give a brief overview of the technologies, parameters, and some other details in order to understand them better in the subsequent chapters where they will be used and discussed in detail.

1.2.1 Multiplexing in Optical Networks

To use the large bandwidth of the fiber efficiently, signals of multiple connections have to be simultaneously transmitted over the fiber; in other words, multiplexing of signals has to be carried out without overlapping in the spectrum. This is employed for all *multi-user* communication in a *multi-access* environment. With the present technology, besides the time division multiplexing (TDM) in the electronic domain (Figure 1.4a), WDM is used in the optical domain on the existing fibers, as shown in Figure 1.4b. With WDM technology, multiple optical signals can be transmitted simultaneously and independently on orthogonal optical wavelength channels over a single fiber. Each of the channels has a rate of many gigabits/sec (Gbps), which significantly increases the usable bandwidth of an optical fiber. WDM systems with 88 WDM channels and bandwidth of $100\,\text{GHz}$ per channel are now available. Thus, on a single fiber with both TDM and WDM schemes, a total traffic of $M(N \times b)$ bits/sec can be carried by wavelength multiplexing M wavelengths, with each wavelength time-multiplexing N data stream each of b bits/sec.

In addition to the increased usable bandwidth of an optical fiber, we exploit WDM for wavelength routing and switching and restoration in the optical domain. The other advantages of WDM are its reduced processing cost compared with electronic processing, data transparency, and efficient failure handling, as mentioned earlier. As a result, WDM has become a technology of choice in the optical networks. Further, within each WDM channel, it is

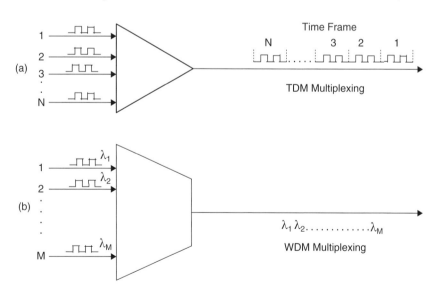

Figure 1.4 (a) Time division multiplexing. (b) Wavelength division multiplexing.

possible to have frequency division multiplexing (FDM) at the electronic level. In this case, the WDM channel bandwidth is subdivided into many radio frequency (RF) channels, each at a different frequency. Each radio frequency channel carries a different information signal. These RF channels are multiplexed to form a composite FDM signal and then the multiplexed signal intensity modulates the optical carrier. This is called *subcarrier multiplexing*. Optical code division multiplexing (OCDM) with code division multiplexing access (CDMA) is another optical multiplexing technique used in the networks. Optical time division multiplexing (OTDM) is also one of the possibilities for multi-user or multi-access technique to enhance the number of connections in the network, and is being actively researched to make it commercially viable.

1.2.2 Services Supported by Optical Networks

Optical networks are required to support services which are diverse in geographical reach, applications, performance, and characteristics. The network may require connectivity in LANs, MANs, or WANs having a global reach. The optical network supports voice, multimedia, and data services. The services offered may vary in performance, with not so demanding email services (best effort) to high-capacity, secure, fast services of VoIP (voice over IP), IPTV (Internet Protocol television), telemedicine, or other highly secure financial services requiring very high performance in terms of having minimum or no errors and delay, and with high reliability of the network. All these different services have different requirements. But all the services are to be handled on a single underlying optical network by adaptively handling these with each special requirement for a large user population.

To handle the variety of traffic demands in a large network, the network functions can be broadly divided into three planes:

- *Transport plane.* The transport plane or data plane has the entire infrastructure, including the electronic switches and routers, the photonic transponders and OXCs, fiber links, etc. associated with the information/data transfer.
- *Control plane.* The control plane takes care of all the activities concerned with the connection provisioning, reconfiguration, performance, and fault management in the network.
- *Management plane.* This is the overarching plane which deals with the many and all operations of administration, maintenance, performance monitoring, fault diagnosis, statistics gathering, etc.

The network with the three functional planes is shown in Figure 1.5. We will be discussing the details of these planes in Chapter 7.

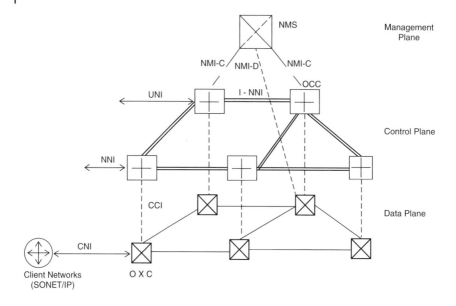

NMS : Network Management System
OCC : Optical Connection Controller
CCI : Connection Control Interface
I-NNI : Internal Network Network Interface
NMI-C : Network Management Interface - Control Plane
NMI-D : Network Management Interface - Data Plane
NNI : Network Network Interface
UNI : User Network Interface
CNI : Client Network Interface

Figure 1.5 Functional architecture of optical network.

1.2.3 WDM Optical Network Architectures

There are broadly three classes of WDM optical network architectures: broadcast-and-select (B&S) networks, wavelength routed networks (WRNs), and linear lightwave networks (LLNs).

1.2.3.1 Broadcast-and-Select Networks

A generic WDM broadcast-and-select network consists of a *passive optical star coupler* network node connecting the user nodes in the network. Each of these user nodes is equipped with one or more fixed or tunable optical transmitters and one or more fixed or tunable optical receivers. Different nodes can transmit messages on different wavelengths simultaneously. The star coupler combines all these messages and then *broadcasts* the combined message to all the nodes. A user node selects a designated wavelength to receive the desired message by tuning its receiver to that wavelength.

The *single-hop* B&S networks are *all-optical*; a message from the network user node once transmitted as light reaches its final destination directly, without being converted to electronic form in between. In order to support high data transmission in these networks we need to have optical transmitters and receivers at the nodes that can tune rapidly. The main networking challenge in these networks is the coordination of transmissions between the various user nodes connected to the central passive optical star coupler. In the absence of coordination or efficient medium access control protocol, collisions occur when two or more nodes transmit/receive on the same wavelength at the same time. To support fast switching efficiently in B&S networks, a *multi-hop* approach can be used, which avoids rapid tuning.

The advantage of B&S networks is in their simplicity and natural *multicasting* capability (ability to transmit a message to multiple destinations). However, they have limitations. First, they require a large number of wavelengths, typically at least as many as there are user network nodes in the network, because there is no *wavelength reuse* possible in the network. Thus, the networks are not scalable beyond the number of supported wavelengths. Second, they cannot span longer distances since the transmitted power is split among various nodes and each node receives only a small fraction of the transmitted power, which becomes smaller as the number of nodes increases. For these reasons, the main application for B&S is for high-speed local area and access networks. The other topologies which are used in these networks are the folded bus and the tree topology. We will be discussing the single-hop B&S networks in Chapter 3.

1.2.3.2 Wavelength Routed Networks

Wavelength routed WDM networks have the potential to avoid the three problems in the broadcast networks: the lack of wavelength reuse, power splitting loss, and scalability to WANs. A WRN consists of routing (or bypass) nodes interconnected by fiber links in an arbitrary mesh topology. The end user node is connected to a network node via a fiber link. Each end node is equipped with a set of optical transmitters and receivers for sending data into the network and receiving data from the network, respectively, both of which may be wavelength-tunable. In a WRN, a message can be sent from the source node to the destination node using a wavelength continuous route called a *lightpath* on one of the WDM channels, without requiring any O-E-O conversion and buffering at the intermediate nodes. This process is known as *wavelength routing*. The destination node of the lightpath accesses the network using receivers that are tuned to the wavelength on which the lightpath operates. A lightpath is an *all-optical communication path* between two nodes, established by allocating the same wavelength throughout the route of the transmitted data. Thus, it is a high-bandwidth pipe carrying data up to several gigabits per second, and is uniquely identified by a physical path and a wavelength. The requirement that the same wavelength must be used on all the links along the selected route is known as the *wavelength continuity constraint*. Two lightpaths cannot be assigned the same wavelength on any single fiber. This requirement is known as *distinct wavelength assignment constraint*. However, two lightpaths can reuse the same wavelength if they use disjoint sets of links. This property is known as *wavelength reuse*. A WRN is illustrated in Figure 1.6.

The simultaneous transmission of messages on the same wavelength over fiber link disjoint paths or *wavelength reuse* property in WRNs makes them more scalable than B&S networks. Another important characteristic which enables WRNs to span long distances is that the transmitted power invested in the lightpath is not split and is sent to the relevant destination only. Given a WDM network, the problem of routing and assigning wavelengths to lightpaths is of paramount importance in these networks. Good algorithms are needed in order to ensure that functions of routing and wavelength assignment are performed using a minimum number of wavelengths. Connections in WRNs can be supported by using either a single-hop or a multi-hop approach. In the multi-hop approach, a packet from one node may have to be routed through some intermediate nodes with an O-E-O conversion before reaching its final destination. At these intermediate nodes, the packet is converted to electronic form and retransmitted on another wavelength. These attractive features – wavelength reuse, protocol transparency, and reliability – make WRNs suitable for WANs.

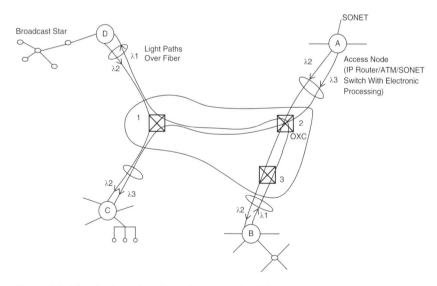

Figure 1.6 All-optical wavelength routing network architecture.

1.2.3.3 Linear Lightwave Networks

In the case of WRNs we multiplex several wavelengths and each of these orthogonal wavelengths can form a *lightpath* to route a connection in the network. In LLNs, meanwhile, we use *waveband*, which is a collection of a number of orthogonal wavelengths to route over the network as a unit. In these networks several wavebands are multiplexed on a fiber and several wavelengths are multiplexed in each waveband. The routing nodes de-multiplex and multiplex wavebands but not wavelengths within a waveband. Like the wavelength, the waveband-routed networks have the property that each waveband channel can be recognized in the optical network nodes (ONNs) and routed individually and hence network capacity can be improved with the spectrum reuse. Since an LLN node does not distinguish between wavelengths within a waveband, individual wavelengths within a waveband are separated from each other at the end node's optical receiver demultiplexer. The two constraints of *waveband continuity* and *distinct waveband assignment* as in WRN apply to LLNs. Further, there are two routing constraints unique to LLNs: *inseparability*, that is, channels belonging to the same waveband when combined on a single fiber cannot be separated within the network; and *distinct source combining*, that is, on any fiber, only signals from distinct sources may be combined. We will be limiting our discussion to wavelength routing networks and not considering the LLNs any further.

1.2.4 Services Types

The optical networks are used for different applications, varying from voice and data to other multimedia transmission. In general, network services for all the applications provided can be classified as follows:

I) *Connection-oriented services.* These services are possible on circuit-switched networks, where the connection over the network is established before the traffic can be sent and is maintained until the connection is dropped. Examples of such services/protocols are the legacy telephone services, ATM, TCP (transmission control protocol), MPLS (multiple protocol label switching) services. All nodes have to maintain state information once the handshake procedure has been executed. Thus, QoS is maintained as decided by the service level agreement (SLA).

The connection-oriented services can have two types of connections:

- *Dedicated connections.* In the circuit-switched networks where the path is dedicated for a connection over long periods of time, possibly for months, days, or hours, it is said to be a dedicated connection. The user has a service agreement with the service provider for the connection. The connection is disrupted only when fault occurs and is then quickly put on another path.

- *Demand-oriented connections.* The demand-oriented connections are also circuit switched and provided to the user when demanded. As the network resources are shared, the services are provided on available free resources, otherwise the connection is blocked or dropped. Once the demand is accepted, the path is reserved and kept connected as demanded by the client. The demand may be over several seconds or milliseconds.

II) *Connectionless services.* Connectionless service does not require the establishment of any connection prior to sending data. The sender starts transmitting data to the destination according to the access control in use. Therefore, the connectionless services are less reliable than connection-oriented services. These are packet switched services; the connection is made as the data flows over the network, with IP providing the best effort (BE) services. They are much more economical but with reduced QoS compared with connection-oriented services.

1.2.5 Types of Traffic

- *Static and dynamic.* The traffic can be either fixed or changing dynamically with time, both in its source destination pair connection and in terms of demand size. In the case of static traffic demand, connection requests are known beforehand or a-priori with the estimation of long-term traffic demands of the network. This is expressed as the traffic matrix, which is static. In the case of dynamic traffic, the connection requests change with time in the network in random fashion and so does the traffic matrix.
- *Stream or synchronous traffic.* The stream traffic flow is uniform and constant and with the same clock period for every switch in the network – for example, SONET/SDH with fixed rate.
- *Asynchronous or random traffic.* The random traffic is dynamic with random arrival rate and length. It can take some probability distribution, i.e. uniform, Poisson, etc., or it may be bursty in nature, for example Internet traffic.

1.2.6 Switching Granularities

In optical networks we talk of different types of granularities, such as granularity of traffic, of switching speed, or of connection capacity.

Depending on the traffic demands of the various clients, optical networks need to be able to provide connections of different *traffic demand granularity*, such as 1 Gbps, 40 Gbps, etc. Also, the granularity of the connection's switching speed can be either fast or slow according to the required application and performance, QoS, etc. In the optical switching network, the switching can also be on the basis of granularity of connection capacity, such as whether it is space switching from one fiber to another, waveband switching, wavelength switching, or time-slot switching. The connection capacity in space switching is of the order of Tbps, i.e. the total capacity of all the WDM channels in a fiber, while in the case of time-slot switching it is a single user connection, so the order of Mbps may be involved. The connections provided by optical switching networks must be able to adapt to time-varying network conditions and traffic requirements dynamically in order to optimize network performance and utilization of network resources.

1.2.6.1 Optical Circuit Switching

For the connection-oriented services in circuit switching we can perform switching at the following granularity:

1) *Space or fiber switching.* Any connection from one incoming fiber can be transferred to another outgoing fiber by switching in space. This is the coarsest level of switching in terms of granularity, as a large number of connections carrying a huge amount of data are switched in this case.
2) *Waveband switching.* This is the next level of lower switching in which a waveband is switched at the node. A waveband consists of a number of wavelengths. These wavebands are switched independently from each other – that is, network nodes are able to switch individual wavebands arriving on the same incoming fiber to a different waveband in the outgoing fiber.
3) *Wavelength switching.* This is a further finer granularity in which individual wavelengths in a waveband or a fiber itself are first de-multiplexed to the level of individual wavelengths, and then any of these wavelengths can be independently switched in the input fiber to the other wavelength in the output fiber.
4) *Sub-wavelength.* Sub-wavelength switching is also possible when OTDM is possible. Each slot in the TDM time frame carries single client data which can be switched to another slot. But this is still at the research stage. TDM in the electronic domain is done to multiplex lower rate data connections to accommodate low rate client data, as in SONET/SDH systems.

1.2.6.2 Packet Switching for Bursty Traffic

Circuit switching in optical networks is not economical in case of random traffic load. If the random bursty traffic is provided to a dedicated connection, then large bandwidth is wasted when low traffic density is greater than its statistical average. Also, on the contrary, the performance will deteriorate considerably when optical burst with high packet density arrives. In such situations switching is recommended for the random traffic by individual packet as it arrives at the switch. Packet switching is done at two granularity levels:

1) *Individual packet switching.* In the case of packet switching, each packet has a header with source-destination address which is recognized at the nodes.

After reading the address, a node forwards the packet to the next node on a wavelength channel. Thus, individual packets are switched independently from one to another node.

2) *Burst switching*. Unlike in the case of packet switching, packets are aggregated at the ingress node and sent as bursts across the network. For each burst a reservation control signal is sent on a dedicated control wavelength channel prior to sending the burst on one of the data wavelength channels after a pre-specified offset time.

1.3 Optical Network Layered Architecture

Until the early twentieth century before the fiber was well laid, only limited data traffic was carried over the existing telephone network which used to carry voice services. With optical fiber now the main transmission medium, the networks have become more data-centric rather than voice-centric, and digital transmission standards of higher bit rate, such as SDH/SONET, ATM, and IP networks, are well developed for this purpose. Internet Protocol is being used for transmitting all types of traffic, be it voice, video, or data, with all types of multimedia services on the optical fiber network. In other words, the services supported by the underlying optical network now are vast, extremely diverse in terms of connectivity, bandwidth, performance, reach, cost, and with many other features. Therefore, the network has to adapt to different special features for each type of service.

With the ubiquitous optical network to provide the variety of services mentioned above, the complexity of the network increases. Different components in the network have to perform a variety of functions. Therefore, as the OSI (open systems interconnection) model in computer networks, we can conceptually divide the functionalities of any node in the network in the constituent layers with client-server relations with the neighboring layers. The layered architecture allows flexibility not only in designing the equipment but also in design and implementation of the network. By employing a layered model, designers are left with enough leverage to add functions at each layer and to concentrate only on certain functions in the network provided by the specific layer while maintaining interoperability with other layers. The layered model helps in understanding and designing the optical network architecture, which is flexible, robust, and scalable.

In an optical network, when a user makes a request for connection, the source node makes a connection to its destination end by passing through a number of intermediate network elements or nodes. A variety of functions has to be carried out at each network element, such as framing of the packets, grooming, multiplexing, routing, and switching. In the layered architecture each layer carries out specific functions as required by the layer above it, then the layer after performing the required functions, instructs the layer below to perform the necessary specified function. Thus the data flows between each layer. Each intermediate network element, between the source and the destination node along the path, has a certain number of layers depending on the network element's function, starting from the lowest layer up to a certain higher layer in the hierarchy.

In the seven-layered ISO (International Organization for Standardization) structure there are two types of layers: host and media. The upper four layers – the *transport, session, presentation,* and *applications* layers – are the *host layers* and the lower *media layer* has three sub-layers. These sub-layers are concerned with the functionalities of the data flow over the common network media. The lowest layer in the hierarchy is the *physical* layer, followed by the *data link* layer and the *network* layer. The physical layer provides the pipe with a certain bandwidth to the above layer. The pipe can be an optical fiber, a coaxial line, or any physical channel. The data link layer provides multiplexing, de-multiplexing of channels, framing of the data to be sent over the physical layer, etc. The network layer above the data link layer provides the end-to-end service to the message. In the present text we will be concentrating on these lowermost media layers in the optical networks.

The functional layered structure of a multilayered WDM optical network is shown in Figure 1.7. In the case of optical networks we have three layers for the functional layer abstraction: *client or user network, logical,* and *physical.* Each adjacent layer has a client server relationship with the lower layer serving the adjoining upper layer. The bottom layer, the physical layer, serves the adjoining

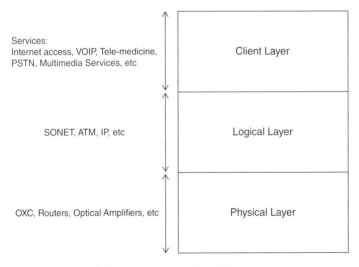

Figure 1.7 The multilayer optical network model.

second layer by providing the required basic optical channel bandwidth service to *logical network nodes* in the logical layer. As the physical layer is a fiber pipe, it therefore encompasses optical multiplexing, transport, and switching in the optical devices based on optical technology. The *logical switching nodes* (LSNs) of the logical layer, including IP routers, Ethernet switches, ATM switches, SONET/SDH switches, and OTN switches, are all based on electronic technology. The protocols associated with each of these switches do the formatting and partitioning of the data and move the data from source to destination over the optical physical layer by specific methods. Each of these LSNs organizes the raw offered capacity by the physical layer to the needs of the clients in the *user layer*, which is also called the *service* or *application layer*. The application layer includes all types of services, such as voice, video, and data. Thus, the several logical networks in the logical layer can provide a variety of specialized services to a connection demanded by the upper client layer.

For example, the SONET switches in the logical layer will electronically multiplex the low-speed data streams of the client layer into higher-speed streams, which in turn use optical wavelength channels provided by the physical layer to transmit the multiplexed data as an optical signal. The SONET channels can support a wide variety of services, such as telephone services, data, video, etc. In another case Internet services can be provided to some users through the IP layer sitting on the ATM over the SONET when reliability of ATM services is required before multiplexing them in the SONET layer. For certain IP we may have the IP layer directly over the optical physical layer, and yet in another case when the application layer connection requires a very high bandwidth, a raw *wavelength service* may directly be over the physical layer bypassing the logical layer, thus providing totally transparent optical network functions. Finally, in the physical layer, multiple wavelengths or lightpaths are optically multiplexed, switched, and routed in the WDM fiber physical layer, providing large bandwidth connectivity to the upper layers.

Next, we first give a simplistic view of how connection is made in an optical network before explaining in detail the functionalities of each layer. A typical optical network picture is given in Figure 1.8. The ONNs, which are OXC or wavelength or space switches, are interconnected with fibers in the physical optical layer. These ONNs are responsible for connecting different fibers as per the requirements of the logical connection demanded by the LSNs. These optical nodes are connected to the LSNs through the optical line terminal (OLT) or the network access terminal (NAT). NAT is a photonic device through which the electronic signal from the LSN is converted to the optical signal for onward transmission to the ONN, or vice versa. They are basically the transceivers. The LSN which can be either IP routers or the SONET/OTN switches are electronic. They are connected to the user equipment on one side and to the ONN on the other side through the NAT, thus they connect clients to the physical layer using a wavelength channel in the fiber.

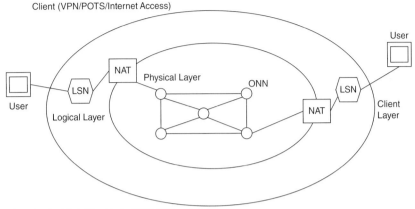

LSN - Logical Switching Node (SONET/IP router)
ONN - Optical Network Node
OLT - Optical Line Terminal

Figure 1.8 Schematic of an optical network.

The physical layer containing optical components, i.e. fiber, optical amplifiers, optical routers, etc., executes linear operations and is transparent to signal protocol, format, and bit rate, etc. The LSN in the logical layer has electronic components which can execute nonlinear operations on electronic signals and thus are dependent on the signal format, bit rate, etc. As optical technology advances in signal processing, it will be possible for the logical layer's nonlinear operational functions to be carried out in optical devices. At present, the transparent physical layer is considered to be almost linear and the logical layer, which is opaque, carries out all the nonlinear operations with electronic processing.

1.3.1 Layers and Sub-layers

Figure 1.9 illustrates details of the multilayered optical network functional layered structure, which gives the different functions of the physical and logical layers shown in Figure 1.8.

Physical layer. The functionality of the physical layer in an ONN, unlike from other communication networks, is divided into two sub layers: (a) the optical layer, and (b) the fiber layer.

(a) *Optical layer.* The function of the optical layer is to provide lightpaths or λ-channel to carry the traffic of logical nodes in the upper logical layer, such as the SONET/SDH switches, IP/MPLS routers, as well as the electronic layer of the OTN. Thus, it can be viewed as the server layer to the upper logical layer that makes use of the services provided by the optical

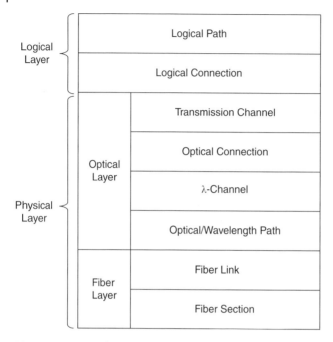

Figure 1.9 Layer/sub-layer structure of optical network connection.

layer to make a logical connection between the source-destination pair. The main features of such lightpath service of the optical layer are:

- The lightpath is set up and taken down as required by the logical layer.
- Typically the client layer specifies the bandwidth required to carry its connection traffic on the lightpath. The optical layer provisions the lightpath bandwidths as negotiated between the logical layer and the optical layer.
- An adaptation function may be required at the input and output interface of the optical and logical layer to convert user signals to signals that are compatible with the optical layer. The optical layer supports a specific range of signal types, bit rates, and protocols; these need to be established between the client and the optical layer.
- The lightpaths need to provide a guaranteed level of performance: bit error rate (BER), jitter, delay, etc.
- Multiple levels of services need to be supported for different connections: secured, unprotected, best-effort service, extensive fault management service connections, etc.
- Multicasting and broadcasting functions need to be supported.

Enabling the delivery of these services requires a *control and management* interface for the data plane supporting both ONNs and logical nodes. This interface allows the LSNs to specify the set of *lightpaths* to be set up or taken down and sets the service parameters associated with the logical paths. Also it enables the optical layer to provide performance and fault management information to the logical layer. The management system communicates with the optical layer element management system (EMS) and the EMS in turn manages the optical layer. There is a hierarchical system: central network management system (NMS) and the distributed EMS. Signals from the NMS and EMS are transmitted through the distributed control network (DCN) to all the network elements. The details of these will be discussed further in Chapter 7.

Thus, to summarize, the functionalities of the optical layer cover all the aspects of optical connection, i.e. optical signal generation and transmission, optical multiplexing and routing, survivability, etc. The electro-optic network terminal (i.e. the NAT) provides the required interface between electronic end systems, the LSNs in the logical layer, and the ONNs in the optical layer through the access fiber. Each of these interface units is a transceiver with electro-optic devices, the laser and the photo-detector.

As the optical layer performs multiple functions, the functionalities of this layer can be further sub-divided in the following sub-layers, as shown in Figure 1.9:

- *Transmission channel.* The transmission channel in the optical layer carries out the adaptation function of converting the logical signal to transmission signal. The electro-optic conversion is carried out in the transmission processor, and on the receiver side the reverse operation of converting the optical signal to an electrical signal is done in the receive processor. The other functions carried out in these processors are the modulation/de-modulation, multiplexing/de-multiplexing, coding/decoding, and other functions to adapt well to the optical topological connections and vice versa before the signals are converted to optical/electronic format by the optical transmitter/optical receiver, respectively, in the network interface unit at the source and destination.

- *Optical connection.* The function of the optical connection sub-layer is to provide the end-to-end optical communication channel between the two terminating transceiver pairs at the access interface unit in the optical layer. The terminating interface units tune their transceivers to the assigned wavelength of the λ-channel. The λ-channel is the basic information unit carrier in the physical layer and is independently routed and switched by the ONNs to establish the optical connection.

- *λ-channel.* In a WDM system, λ-channels are the smallest granularity in the optical spectrum partitioning, each having a distinct wavelength to carry the information. Several lower bit rate signals of the logical channel are time division multiplexed on a single λ-channel. The λ-channels can be switched at the nodes (OXC) to different paths at the nodes. Thus, λ-channels can be multiplexed (WDM), access (WDMA – wavelength division multiple access), and switched as an identity at the nodes.

- *Optical path*: This sub-layer manages the optical connection from source to destination through sequences of optical nodes (OXC) on a selected wavelength λ-channel.

(b) *Fiber layer*. The fiber layer takes care of the optical infrastructure aspects. Its functionality is divided into two sub-layers:

- *Fiber link*. The fiber links or inter-nodal hop between the ONNs along the optical path are the fiber link layer.
- *Fiber section*. The fiber sections between the amplification sections are managed by the fiber section sub-layer.

Logical layer. The logical layer above the physical layer has its functionality in providing the virtual unidirectional connections between the pair of end systems in the client layer external ports. This connection is provided with the help of a pair of source and destination NATs, discussed in the optical layer, at the interface of the physical and logical layers. The logical layer functionalities can be further divided into two sub-layers:

- *Logical connection*. The logical connection carries electronic signals in a particular format as required by the end system – for example, it may be an ATM cell, IP packet, analog video, or SONET digital bit stream. These logical connections are carried by an optical connection with the resources of the physical layer. The logical connection is carried over the optical connection through the *transmission channels*. The transmission channel has the adaptation function of converting the logical signal to an optical transmission signal.
- *Logical path layer*. The virtual connection provides the end-to-end connection between the pair of end systems. It is carried on the logical path layer with the help of one or multiple logical hops or connections made with the LSNs.

The multi-layered architecture thus breaks down the different functions, such as multiplexing, switching, grooming, O-E-O conversion, routing, of the complex optical network into separate layers and thus these can then be handled with ease. The number of sub layers in different networks and their devices will differ from one to another depending on the size, topology, and functions involved for making connections. For a long link length in a WAN, all the sub layers will be present, while in the case of link between any two nodes in the MAN or LAN, the sub layer structure will be somewhat simpler because of lesser functions which have to be performed by the nodes of the link in these cases.

1.4 Organization of the Book

In Chapter 1, optical networks were summarized in order to give an overview of the different categories of the optical networks, their working and parameters, and the terminologies which are generally used. Chapter 2 deals with building blocks of the optical networks. With an overview of the enabling infrastructure in the physical layer of the optical network, the chapter give details of the network photonic and optical switching devices used in the current WDM networks. For more comprehensive details of fibers and their characteristics, lasers, photo-detectors, and the different optical amplifiers used in the optical networks, readers are encouraged to turn to the dedicated literature on these topics [5–15]. Chapter 3, on single-hop B&S networks used in access networks and for LANs, looks in detail at topologies, media access control (MAC) protocols, and other specifics. Also discussed in this chapter are basic concepts of traffic types in optical network topologies which are more generic. Chapter 4 examines the various access networks, their limitations, and the available high-bandwidth PONs with their newer avatars.

The metro area networks discussed in Chapter 5 examine the SONET and OTN protocols, etc. The wavelength routed WANs are dealt with in Chapter 6. This chapter examines the wavelength routing and assignment (RWA) problem, the logical topology design issues, and the concept of integer linear programming (ILP) and heuristic solutions of these networks.

With an overview about the NMS in Chapter 1, Chapter 7 discusses in detail the very important topic of network management, control, and signaling to make the network work with details of GMPLS (generalized multi-protocol label switching). In Chapter 8, impairment awareness and survivability, protection, and restoration are covered in detail, consistent with the growing importance of optical layer fault management in networks. The chapter considers the very important issue of optical network impairments and survivability and how they are taken into account while designing networks.

There is a need for the optical layer to have low cost, flexibility, and reconfigurability instead of the present fixed nature in order to be geared up to accommodate the varying traffic demands and services. The present dense wavelength division multiplexing (DWDM) networks based on ITU-T split the useful fiber spectrum into fixed 50 GHz and 100 GHz spectrum slots. The optical devices and switches in use are still not fully flexible and therefore the network does not have the flexibility to adjust itself to the traffic demands. Addressing the need for the optical networks to evolve in this direction, the last two chapters of the book, Chapters 9 and 10, are devoted to the topics of elastic optical networks (EONs) and SDON.

1.5 Summary

In this introductory chapter we first talked of the present-day demand of communication traffic and the need for an optical fiber network to meet the high-bandwidth, large area network to support all types of new and flexible services.

The networks were classified in terms of geographical reach: the LAN, MAN, and the WAN, and the two generations. The first-generation WDM networks use optical fiber as a replacement for copper cable to get higher capacities, and the second-generation networks use the other characteristics of the fiber to provide circuit-switched lightpaths by routing and switching wavelengths inside the network.

The chapter gives a brief overview of the second-generation optical networks in order to make the subject easier to understand in subsequent chapters where these topics will be discussed in detail. The different network architectures, traffic types, connections, services, types of switching, and the network layered abstractions are introduced.

Problems

1.1 (a) What factors limit the scalability and throughput of the B&S and the WRN networks?

(b) Figure 1.9 gives the layered structure for a multi-wavelength hybrid optical network connection. Draw a similar layered structure for a transparent multi-wavelength B&S and a transparent all-optical WRN connection.

(c) What is the bandwidth in THz of the signal at 1300 nm wavelength for a 150 nm spectrum?

(d) What do you understand by "electronic bottleneck"? What is the way to eliminate it?

(e) What is the difference between "FDM" and "WDM"?

1.2 In a 640 km 1st generation link 40 Gbps data is transmitted using 16 separate fiber pairs. Calculate the number of regenerators required if they are placed every 120 km. If, instead, a 16-channel WDM system is used for the above link with one fiber pair only, how many optical amplifiers with a gain of 30 dB each will be required if the amplifiers are placed every 80 km and a regenerator required only after 320 km. Compare the regenerators required in the two cases.

1.3 For the six-node mesh wavelength-routed optical WDM network given in Figure 1.10, connections 1–4 and 2–6 are on wavelength λ_1, and on

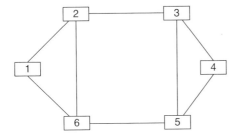

wavelength λ_2, respectively. Establish the connections 2–4 and 3–6 using the minimum number of wavelengths with no wavelength conversion available at any node.

1.4 If the six-node network was connected in a passive star topology, how will the above connections be made with minimum number of wavelengths?

1.5 What type of service; dedicated or demand connection-oriented *circuit switched* or connectionless *packet switched*, should be provided to the following services:

(a) VoIP
(b) IPTV
(c) telemedicine
(d) highly secure financial service
(e) social networking

1.6 All the above services are provided by the same underlying core fiber network. Explain how it is possible to give such diverse services using one network.

References

1 Goransson, P. and Black, C. (2014). *Software Defined Networks: A Comprehensive Approach*. Morgan Kaufmann.
2 Lopez, V. and Velasco, L. (2016). *Elastic Optical Networks: Architectures, Technologies, and Control*. Springer International Publishing.
3 Kreutz, D., Ramos, F.M.V., Verissimo, P. et al. (2015). Software-defined networking: a comprehensive survey,. *Proc. IEEE* 103 (1): 14–76.
4 Thyagaturu, A.S., Mercian, A., McGarry, M.P. et al. (2016). Software defined optical networks (SDONs): a comprehensive survey. *IEEE Commun. Surv. Tutorials* 18 (4): 2738–2786.

5 Okamoto, K. (2006). *Fundamentals of Optical Waveguides*, 2e. New York: Academic.

6 Buck, J.A. (2004). *Fundamentals of Optical Fibers*, 2e. Hoboken, NJ: Wiley.

7 Fukuda, M. (1999). *Optical Semiconductor Devices*. Hoboken, NJ: Wiley.

8 Neamen, D.A. (2006). *An Introduction to Semiconductor Devices*. New York: McGraw-Hill.

9 Personick, S.D. (1973). Receiver design for digital fiber communication systems. *Bell Syst. Technol. J.* 52: 843–886.

10 Schneider, K. and Zimmermann, H.K. (2006). *Highly Sensitive Optical Receivers*. New York: Springer.

11 Desurvire, E. (2002). *Erbium-Doped Fiber Amplifiers: Principles and Applications*. New York: Wiley.

12 Yariv, A. and Yeh, P. (2006). *Photonics: Optical Electronics in Modern Communications*. New York: Oxford University Press.

13 Islam, M.N. (ed.) (2004). *Raman Amplifiers for Telecommunications 1 – Physical Principles*. New York: Springer.

14 Islam, M.N. (ed.) (2004). *Raman Amplifiers for Telecommunications 2 – Sub-Systems and Systems*. New York: Springer.

15 O'Mahoney, M.J. (1988). Semiconductor laser optical amplifiers for use in future fiber systems. *J. Lightwave Technol.* 6: 531–544.

2

Network Elements

2.1 Introduction

In the last chapter we had an overview of optical networks regarding their architecture and working. In this chapter we will discuss the hardware or the infrastructure used in the networks to provide the physical optical connection between any two end systems. We appreciate the fact that network performance is ultimately inhibited by the limitations and non-ideal characteristics of the physical layer infrastructure, which includes fiber, transmitter, receiver, switches, cross-connects, and other components and devices used. Performance depends on the specific technology used in fabricating these devices and on the architecture of the network. There has been rapid and continuous growth at the material and fabrication levels of these devices and, hence, the elements of the optical network infrastructure are seen to be continuously changing and improving. In this chapter we therefore concentrate basically on the generic characteristics of the network elements in order to understand how they function without giving too much detail about their specific technology or architecture.

In order to know the different network modules used in the optical network, Figure 2.1 shows an arbitrary connection between a source-destination (s-d) pair link in a WDM network. The end systems make a virtual connection among themselves over the optical network through the electronic logical network nodes, such as the IP router or Synchronous Optical NETwork (SONET) switch. The logical network nodes access the physical optical layer through the optical line terminal (OLT), also known as the *network access terminal (NAT)*. Each logical connection signal which is an aggregation of multiple source-destination pair signals modulates a laser at a particular wavelength. Several other such logical signals modulate different wavelength laser sources which are then multiplexed together to be sent over the fiber. The signal is transmitted over the fiber through the intermediate optical network nodes (ONNs), optical amplifiers (OAs), etc. to reach the destined NAT

Optical WDM Networks: From Static to Elastic Networks, First Edition. Devi Chadha.
© 2019 John Wiley & Sons Ltd. Published 2019 by John Wiley & Sons Ltd.

on the receiver end. The NAT receiver does the reciprocal of the transmitter end, i.e. de-multiplexes the multiple wavelengths, followed by O-E conversion, and after processing the received signals in the electronic domain sends them on the other end of the link to the logical node. The in-line optical amplifiers are periodically deployed in long links after every 80 km or so. The optical add-drop multiplexers (OADMs) are used at locations where any wavelength is to be added or dropped at a station while the rest of the *lightpaths* (LPs) are bypassed or routed on the link. The optical nodes or the optical cross-connects (OXC) besides carrying out the add/drop function do many other operations. It is much more complex and has many more ports and wavelengths to handle compared with the OADMs.

The network elements supported in an optical network as shown in Figure 2.1 can be divided into two categories:

(i) *Optical/photonic network resources.* Optical fiber, NAT which comprise of the laser transmitters and optical receivers (ORs), OADM, optical amplifiers, OXC, optical wavelength converters, etc. in the physical layer.

(ii) *Electrical network resources.* The logical switching nodes (LSNs), such as SONET/asynchronous transfer mode (ATM) switches, IP routers, digital cross-connects (DXCs), optical transport network (OTN) switches in the logical layer.

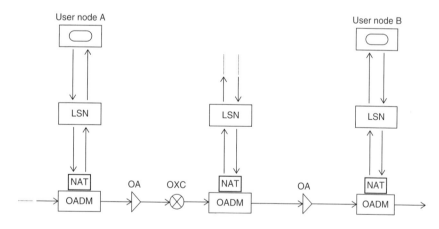

LSN: Logical switching node
NAT: Network access terminal
OXC: Optical cross-connect
OA: Optical amplifier
OADM: Optical add/drop multiplexer

Figure 2.1 Network modules in a long link between a source-destination pair.

In this chapter we discuss the essential optical devices and systems used in the implementation of second-generation optical networks. The emphasis is on conveying the understanding of functioning and performance of current devices in network applications. The chapter starts with an overview of optical fiber, laser diode transmitters, optical detector receivers, and optical amplifiers. Dedicated texts [1–7] on optical fiber transmission systems discuss fibers, lasers, p-type, intrinsic and n-type (PIN) and avalanche photodiode (APD) receiver structures, and optical amplifiers in much greater detail; we will therefore concentrate more on the network elements, such as optical couplers, filters, optical switches, wavelength converters, and other network functional modules. In the following sections we will be describing in detail the optical and photonics components in the physical layer. In Sections 2.2–2.5 we give few details of the optical fiber, laser transmitter, PIN and APD receivers, and optical amplifiers, explaining their characteristics and functionalities with respect to communication networks. Section 2.6 covers the basic optical elements which are commonly used in the buildup of network devices and functional blocks, such as optical couplers, switches, filters, gratings, etc. In Section 2.7 we discuss the various types of multiplexers (MUXs) and de-multiplexers (DMUXs), followed by details of different types of wavelength routers in Section 2.8. In Section 2.9 various large port switching fabrics are discussed, and finally the wavelength converters followed by the functions blocks, such as the NAT, reconfigurable OADMs and OXC are discussed in Sections 2.10 and 2.11.

2.2 Optical Fiber

Theoretically, a single-mode silica fiber has a very high bandwidth and can allow potentially nearly 50 Tbps of bit rate, which is about four orders of magnitude higher than the currently achievable electronic processing speed of tens of Gbps. Apart from the huge bandwidth capacity, optical fibers have a number of other significant characteristics, such as low attenuation (about $0.2\,dB\,km^{-1}$), low signal distortion, low power requirement, low space require-ment, low cost, and very low dispersion. However, because of the limit of the electronic processing speed, it is unlikely that all the bandwidth of the optical fiber can be exploited for transmission. For this reason, effective multiplexing technology is required to efficiently exploit the huge bandwidth capacity of the optical fibers. WDM technology has provided an easier practical solution for meeting this requirement. With WDM technology, we divide the large spec-trum of the fiber in the low attenuation and low dispersion bands of 1550 and 1310 nm, respectively (Figure 2.2) into smaller orthogonal channels of fixed spacing and separation. Multiple optical signals can be transmitted simultane-ously and independently in these different optical channels over a single fiber,

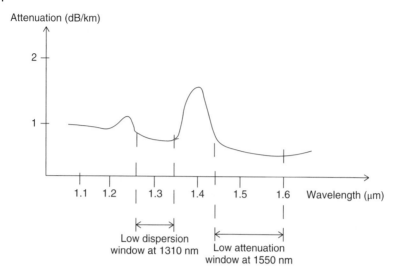

Figure 2.2 Attenuation characteristics of silica fiber with low dispersion window at 1310 nm and low attenuation window at 1550 nm.

each at a rate of tens to hundreds of Gbps without crossing the electronics processing speed barrier. This significantly increases the usable bandwidth of an optical fiber. In recent times, commercial optical systems of 44–100 number of WDM channels per fiber, with each having up to 100 Gbps of speed, have been put in the field. Furthermore, 400 Gbps and 1 Tbps rates flex-grid technology is likely to be deployed in coming few years in the *elastic optical networks* [8, 9]. In addition to the increased usable bandwidth of an optical fiber, WDM has a number of other advantages, such as reduced processing cost, data transparency, and efficient failure handling. As a result, WDM has become a technology of choice for meeting the tremendous bandwidth demand in the telecommunication infrastructure. Optical networks employing fixed and flex-grid WDM technology are therefore considered as promising network infrastructure for next-generation telecommunications networks and optical Internet.

2.2.1 Loss and Bandwidth Windows

The signal power in the fiber attenuates exponentially with distance. The two main loss mechanisms in an optical fiber are *material absorption*, and *Rayleigh and Mei scattering* [10]. The material absorption of pure silica is negligible in the entire 1.3–1.6 μm band that is used for optical communication systems. Figure 2.2 shows the attenuation loss in silica as a function of wavelength. We see that the loss has local minima at 1310 nm and 1550 nm wavelength bands with typical losses of 0.4 dB km^{-1} and 0.1 dB km^{-1}, respectively.

The wavelength bands around 1.3 μm and 1.5 μm are the primary bands used for optical communication as the low dispersion and low attenuation bands, respectively. The usable bandwidth of optical fiber in these bands is approximately 14 THz at 1.3 μm and 15 THz at 1.55 μm. The low-loss dense wavelength division multiplexing (DWDM) band at 1.55 μm is divided into three regions; S- or the short band is from 1495 nm to 1530 nm; the middle band from 1530 nm to 1565 nm is the conventional or C-band; and the band from 1565 nm to 1625 nm is called the long band or the L-band. The optical amplifiers used in the systems have a bandwidth of approximately 35 nm. The conventional erbium-doped fiber amplifiers (EDFAs) operate in the C-band. The high-capacity WDM systems band are being used from 1530 nm to 1625 nm in the C- and the L-bands today.

In the fixed WDM system, the parameter of interest is the channel spacing, which is the spacing between two wavelengths or frequencies in a WDM system. The channel spacing can be measured in units of frequencies as the laser tuning is done at a fixed frequency instead of wavelength. The relationship between the frequency spacing Δf and the wavelength spacing $\Delta \lambda$ centered at λ_0 can be expressed as:

$$\Delta f = -\left(c \Big/ \lambda_0^2 \right) \Delta \lambda \qquad (2.1)$$

From Eq. (2.1) we observe that if multiple channels are spaced apart equally in frequency, they are not spaced apart exactly equally in wavelength. WDM systems today primarily use the 1.55 μm wavelength region. The reason for using the 1.55 μm wavelength is the inherent lowest loss in that region and excellent optical amplifiers available in that region to increase the reach.

There is a strong need for standardization of WDM systems so that WDM components and equipment from different vendors can interoperate. The wavelengths and frequencies used in WDM systems have been standardized on a frequency grid by the International Telecommunications Union (ITU) [11]. It is an infinite grid centered at 193.1 THz. The ITU standardizes the grid in the frequency domain based on equal channel spacing of 25 GHz, 50 GHz, or 100 GHz, which corresponds approximately to 0.2 nm, 0.4 nm, and 0.8 nm wavelength channel spacing around the center frequency, respectively. A few channels of this grid are shown in Table 2.1. The 50 GHz grid is obtained by adding a channel exactly half way between two adjacent channels of the 100 GHz grid. Continuing this process, a 25 GHz grid can also be defined and it can support 600 wavelengths. This standardization helps in the deployment of WDM systems because component vendors can build wavelength-selective parts to a specific grid, which helps significantly in inventory management and manufacturing.

Table 2.1 ITU-T G.694.1 selected DWDM grid with 100-GHz and 50-GHz spacing in the L-, C-, and S-bands.

Channel no.	L-band		C-band		S-band	
	THz (100 GHz)	THz (50 GHz)	THz (100 GHz)	THz (50 GHz)	THz (100 GHz)	THz (50 GHz)
1	186.00	186.05	191.00	191.05	196.00	196.05
2	186.10	186.15	191.10	191.15	196.10	196.15
11	187.00	187.05	192.00	192.05	197.00	197.05
12	187.10	187.15	192.10	192.15	197.10	197.15
21	188.00	188.05	193.00	193.05	198.00	198.05
22	188.10	188.15	193.10	193.15	198.10	198.15
31	189.00	189.05	194.00	194.05	199.00	199.05
32	189.10	189.15	194.10	194.15	199.10	199.15
41	190.00	190.05	195.00	195.05	200.00	200.05
42	190.10	190.15	195.10	195.15	200.10	200.15
49	190.80	190.85	195.80	195.85	200.80	200.85
50	190.90	190.95	195.90	195.95	200.90	200.95

2.2.2 Linear and Nonlinear Effects

The signal propagating through the fiber is affected by its linear and nonlinear characteristics [10, 12]. The linear characteristics are the attenuation, Rayleigh scattering, Mei scattering, and linear dispersion effects due to material, chromatic, and polarization dispersion. Linear characteristics affect the signals when the system is operated at moderate power up to a few milliwatts and at bit rates of about $2.5\,\mathrm{Gb\,s^{-1}}$ or less. However, at higher bit rates such as $10\,\mathrm{Gb\,s^{-1}}$ and above, and/or at higher transmitted powers, it is important to consider the effect of nonlinearities in the fiber. In the case of WDM systems, nonlinear effects can become important even at moderate powers and bit rates.

There are two categories of nonlinear effects. The first category is due to the dependence of the refractive index on the *intensity* of the propagating signal through the fiber. The most important nonlinear effects in this category are *self-phase modulation* (SPM), *cross-phase modulation* (XPM), and *four-wave mixing* (FWM). In the case of SPM, the transmitted pulses undergo chirping. This induced chirp factor becomes significant at high power levels and causes the pulse-broadening effects of chromatic dispersion. In a WDM system with multiple channels, chirp in one channel depends on the variation of the refractive index with the intensity of the other adjacent channels. This effect is the cross-phase modulation. The other effect due to refractive index variation with the applied electric field is that of FWM. If the WDM system consists of W wavelength channels of frequency $f_1, ..., f_W$, then they give rise to additional new signals at frequencies, such as $2f_i - f_j$ and $f_i + f_j - f_k$ due to nonlinearity of the third order in the fiber. These signals cause *cross-talk* to the existing signals in the system due to their proximity to the transmitted WDM signals. These cross-talk effects are particularly severe when the channel spacing is close. Reduced chromatic dispersion enhances the cross-talk induced by FWM. Thus, systems using dispersion-shifted fibers are much more affected by FWM effects than systems using standard single-mode fiber.

The second nonlinear effect of the fiber is stimulated scattering, which can be of two types: *stimulated Brillouin scattering* (SBS) and *stimulated Raman scattering* (SRS). In scattering effects, energy gets transferred from one light wave to another wave at a longer wavelength and the lost energy is absorbed by the molecular vibrations or phonons in the medium. This generated longer wavelength is called the *Stokes wave*. The first wave, therefore, is called a *pump* wavelength because it causes amplification of the Stokes wave. As the pump propagates in the fiber, it loses power to the Stokes wave as it gains power. In the case of SBS, the pump wave is the signal wave and the Stokes wave is the unwanted wave that is generated due to the scattering process. In the case of SRS, the pump wave is a high-power wave and the Stokes wave is the signal wave that gets amplified at the expense of the pump wave. As the Stokes wave in SRS is in the forward direction, it can be used as the means to form the *Raman amplifier*.

Further discussion on linear and nonlinear dispersion effects can be found in Chapter 8.

2.3 Laser Transmitters

In optical networks we use *semiconductor diode laser* transmitters to generate the optical carrier signals. A p-n junction semiconductor laser diode is forward biased and has polished edges perpendicular to the junction forming the resonance cavity. By using different types of semiconductor materials, light with various ranges of frequencies may be released. The actual frequency of light emitted by the laser is determined by the length of the cavity. There are several types of laser with different structures and performance parameters. Multiple-quantum-well (MQW) lasers have much improved performance compared with the bulk cavity laser diode mentioned earlier. The quantum wells are placed in the region of the p-n junction. By confining the possible states of the electrons and holes by the multiple quantum wells, it is possible to achieve much smaller linewidth of lasers.

2.3.1 Laser Characteristics

The important characteristics of lasers are the laser linewidth, frequency stability, and the number of longitudinal modes [13, 14].The finite linewidth decides the *spectral width* of the laser. The increase in linewidth causes increased dispersion when light propagates along a fiber, which in turn effectively leads to increasing the spacing of the WDM channels to reduce crosstalk. *Frequency instabilities* in lasers are due to mode hopping, mode shifts, and wavelength chirping in finite linewidth lasers. Mode hopping is caused by variation in the injection current above a given threshold. Mode shifts are due to temperature changes, and the wavelength chirp is due to variations in injection current of the laser diode. Therefore, due to frequency instabilities, the spacing between the channels has to be increased, which leads to reduced numbers of channels that can be placed in the selected optical bandwidth of the fiber. Frequency instability in the laser transmitters can be avoided with temperature-controlled power supply, which controls the bias current and the threshold. Lasers with limited or single longitudinal mode are preferred for WDM systems as the number of longitudinal modes decides the total number of modes generated by the lasers. The modal dispersion will increase with the increase in number of modes.

2.3.2 Tunable Lasers

Transmitters in WDM networks frequently use tunable lasers. Besides the other characteristics mentioned above, tuning range and the tuning time are important characteristics for tunable lasers. The optical network systems can have either continuously or discretely tunable lasers. The tunable lasers can be mechanically tuned, acousto-optically, electro-optically, or injection current tuned.

Among the mechanically tuned lasers, the most common is the Fabry-Perot (FP) external cavity resonator which can have a very large tuning range, but due to the mechanical nature of the tuning and the length of the cavity the tuning time is high, of the order of milliseconds. Meanwhile, the MQW types of F-P lasers have advantages of lower threshold current, lower noise, better linearity, and high temperature stability, but as they are more susceptible to back reflections, it causes instability in the laser output light. The external acousto-optic tunable filter (AOTF) lasers have a moderate tuning range and a tuning time of around 10-µs or so. In the case of external electro-optically tunable filter lasers, the tuning times are in the order of some tens of nanoseconds. Both these lasers do not allow continuous tuning over a range of wavelengths but have discrete tuning over few wavelengths.

The injection-current tuned distributed feedback (DFB) laser with low linewidth uses a diffraction grating placed in the lasing medium for the selection of wavelength. The injection-current tuned distributed Bragg reflector (DBR) laser has the grating outside of the lasing medium. The DBR-based laser can provide a broad range of biasing conditions and therefore has a large tunability range. But at high biasing, current mode-hopping occurs in DBR lasers.

Instead of a tunable laser one can use an array of fixed-tuned lasers integrated on a chip, with each laser operating at a different wavelength. This laser array has the advantage of providing multiple wavelength transmissions simultaneously. The micro-electro-mechanical switches (MEMS)-based vertical-cavity surface emitting laser (VCSEL) array lasers are important in this category. VCSELs are MQW devices, and Bragg reflectors with mirror layers form the laser reflectors. The advantages of VCSELs are a low threshold current of less than 1 mA, high stability with temperature variation, and fast tuning time of 1–10 ps. Besides this they cost less than their tunable laser counterparts [15].

2.3.3 Modulation Techniques

A typical laser transmitter is given in Figure 2.3 showing how the data signal modulates the laser for onward transmission on the link. The lasers can have either *internal* (direct) or *external* modulation for modulating the laser light with data signal. Internal modulation is possible up to a certain bit rate, above which it can lead to *chirping*, or variations in the laser's amplitude and frequency when the laser is turned on. Above 10 Gbps, external modulation is preferred. Both analog and digital techniques are used to modulate the laser light before transmission over the fiber. Analog modulation schemes used are the amplitude modulation, frequency, and phase modulation. The commonly used digital techniques include amplitude-shift keying (ASK), frequency-shift keying (FSK), and phase-shift keying (PSK). The binary ASK or the on-off keying (OOK) is commonly used because of its simplicity. In OOK modulation, the laser is turned on and off, but this being a directly modulating scheme, it cannot be used for high bit rates. In long-haul, high-capacity WDM systems, advanced coherent modulation formats [16] are used in order to manage signal impairments arising from fiber nonlinear effects, dispersion, and amplified

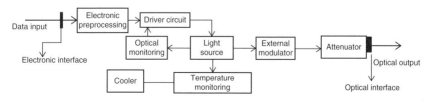

Figure 2.3 Typical laser transmitter.

spontaneous emission (ASE) noise. The modulation scheme for long-haul high-speed WDM transmission links not only should have a simple and cost-effective configuration for signal generation but must also be power and spectrally efficient and should be less susceptible to fiber nonlinearities and dispersion.

2.4 Optical Receivers

Optical receivers are one of the critical units in any optical link. The received signal is very weak and hence receiver sensitivity is a major design issue to take into consideration. The detection devices for optical networks are the PIN semiconductor photodiode and APD [17, 18]. The receivers can have either *direct* (incoherent) or *coherent* detection. The performance of coherent detection receivers is much better than direct detection as they have higher receiver sensitivity, allowing reception of weak signals from a noisy background. Coherent detection-based receivers have a more complex circuitry than direct detection, requiring phase information of the optical signal for coherent detection. They normally use a monochromatic laser as a local oscillator in the receiver.

In the case of intensity modulated direct-detection receivers, the photodetector is followed by a front-end amplifier, also called the *preamplifier*, and often incorporates an equalizer. The amplified signal passes through the detection circuit in the case of digital transmission. The photodetector converts the optical signal to a photo current and the pre-amplifier raises the power of the photocurrent to a level sufficient for further electronic processing. In the digital case, this processing is primarily clock recovery, sampling, and threshold detection of the received data signal to extract the digital bit stream from the received signal. Figure 2.4 shows the receiver structure with direct detection, which beside the detection circuit includes the electronic pre- and post-amplifiers followed by the data processing circuit.

Two types of pre-amplifiers are typically used following the photodetector: a *high-input impedance* voltage amplifier and a *trans impedance* amplifier, which is actually a current to voltage converter. The trans impedance amplifier is most frequently used in current systems, and provides large bandwidth and low noise. The amplified and filtered signal is then sampled and threshold

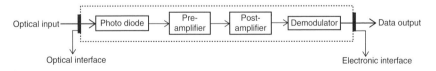

Figure 2.4 Optical direct-detection receiver structure.

detected. The clock recovery unit processes the output signal to determine at what point in time the signal should be sampled. Using this information, the received signal is sampled periodically at a point within the bit period to provide a high probability of correct detection. The signal samples are then compared with a decision threshold.

2.5 Optical Amplifiers

In an optical communication system, the optical signals are attenuated by the optical fiber as they propagate through it. All the optical components, such as multiplexers, filters, couplers, etc., in the link also add loss as the signal passes through them. After a certain distance, the cumulative loss in the fiber causes the signal to become too weak to be detected. Therefore, for long-distance links, as the signal gets attenuated, optical amplifiers are required to boost the signal level, typically after every 80 km. The transparent optical amplifiers only do amplification or have 1R (reamplification), but are not capable of restoring the shape or timing of the signal. An optical amplifier provides total data transparency as it is independent of the signal data rate, modulation format, etc. But over very long links, as signal travels through multiple optical amplifications, noise accumulates and it gets corrupted both in phase and shape, hence needs complete regeneration after around 300–400 km. The optical signal is first converted into an electronic data signal and retransmitted optically after undergoing reamplification, reshaping, and retiming electronically. This is 3R regeneration. The reshaping of the signal reproduces the original pulse shape by removing the spread of the digitally-modulated signals and removing the noise in the signal. Retiming synchronizes the signal to its original bit rate and bit timing pattern. 2R regeneration is also possible in which the optical signal is amplified and reshaped and then it directly modulates a laser without retiming.

The advantage of using optical amplifiers in a WDM system over optoelectronic regeneration is that it eliminates the need to separate all the wavelengths before amplifying. In 3R, each WDM signal first has to be de-multiplexed into individual wavelengths before amplifying electronically, and then multiplexed again before being retransmitted. Nevertheless, optical amplifiers have fairly large gain bandwidths, so as a consequence, a single amplifier can simultaneously amplify several WDM signals. In contrast, we would need a regenerator for each wavelength in the case of 3R. Thus, optical amplifiers have become essential components in high-performance long-distance optical communication systems.

The amplifier introduces spontaneous emission noise, therefore the optical noise as well as the signal will be amplified as they travel through the link. Other optical amplifier parameters of interest besides the gain bandwidth and noise are the amplifier gain, gain saturation, gain efficiency, and polarization sensitivity. With the *gain bandwidth* specifying the spectral bandwidth of the amplifier, in a WDM network gain bandwidth will limit the number of channels possible for a given channel spacing. The *gain saturation* is the maximum gain of the amplifier after which the output power no longer increases with an increase in the input power. The saturation power is typically defined as the output power at which there is a 3-dB reduction in the ratio of output power to the input as compared with the small-signal gain. *Gain-efficiency* measures the gain as a function of pump power in dB/mW. The dominant noise sources in optical amplifiers are the spontaneous emission of photons in the active region of the amplifier and the beat noise due to several noise frequency components. The noise due to spontaneous photon emission is called the ASE and depends on the amplifier gain spectrum, the noise bandwidth, and the *population inversion parameter* [19, 20]. In long-distance links, ASE noise is of major concern when multiple amplifiers are cascaded; the ASE noise generated by previous amplifiers gets amplified by each subsequent amplifier, reducing the received signal to noise ratio (SNR) at the end of the link.

2.5.1 Types of Optical Amplifiers

Optical amplifiers can be divided into two basic classes: semiconductor optical amplifiers (SOAs) and fiber amplifiers (FAs), which will be discussed briefly in the following section.

2.5.1.1 Semiconductor Optical Amplifier

The structure and functioning of an SOA (Figure 2.5a) is similar to the injection semiconductor laser with minimal facet reflectivity to reduce the positive feedback of the signal required for lasing. The signal to be amplified is sent through the active region of the semiconductor and is amplified by the stimulated emission in the active region with a stronger signal emitted from the semiconductor as it passes through the active region [21].

A traveling wave amplifier (TWA) type of SOA is used in WDM systems. The TWA MQW-type SOA can provide higher bandwidth with higher gain saturation than the bulk devices, and with a single pass it can provide gain in the order of 20–25 dB. The SOA amplifies over a much wider band, a 60–65 nm range, than an EDFA. It provides faster on-off switching times in the order of few pico-seconds, with the advantage to integrate with other components on a chip. Due to faster switching of the order of pico-seconds SOAs are preferred as active switches but cannot be used as in-line amplifiers where fiber amplifiers are preferred.

The carrier lifetime of an SOA in the high-energy state is very short. As indicated earlier, this means that signal fluctuations at Gbps rates cause fluctuations in the gain of the amplifier at those rates, producing cross-talk effects

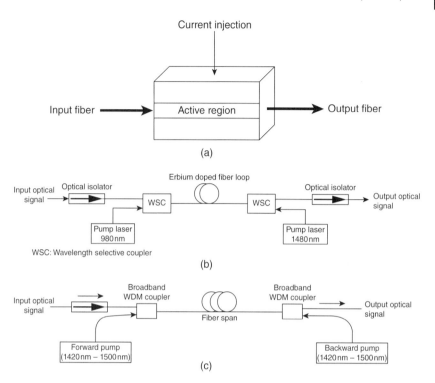

Figure 2.5 (a) Semiconductor optical amplifier. (b) Erbium-doped fiber amplifier. (c) Raman amplifier.

among the signals in the WDM channels. Second, because of its asymmetrical geometry, the SOA is polarization dependent. Third, the coupling losses between the fibers and the semiconductor chip reduce the usable gain and output power.

2.5.1.2 Fiber Amplifiers

Doped-Fiber Amplifiers EDFA is the most common fiber amplifier used in optical networks [1, 20]. It provides gain for wavelengths between 1525 nm and 1560 nm. An optical pump source sends high power along the length of the fiber at a lower wavelength than the modulated incoming laser light. The pump signal excites the electrons of the doped fiber into a higher energy level. The input signal stimulates these excited electrons and thus in turn the input light is amplified. The pump source for EDFA is either 980 nm or 1480 nm optical amplifier. The 980 nm pump has higher gain efficiency than the 1480 nm pump. The 3 dB gain bandwidth for the EDFA is around 35 nm, a gain of 25–30 dB, saturation power is around 20 dB, and gain-efficiency is in the order of $10\,\mathrm{dB}\,\mathrm{mW}^{-1}$. Also, EDFA has a longer lifetime of a few milliseconds. Therefore, it introduces no cross-talk when amplifying WDM signals, until the bit rate drops into the 10 Kbps range because of the longer lifetime. These characteristics make EDFA the choice for the in-line amplifier in the WDM system. The schematic of EDFA is shown in Figure 2.5b.

Raman Amplifier   A Raman amplifier is the other amplifier in the category of fiber amplifiers. The schematic diagram of a Raman amplifier is shown in Figure 2.5c. The high-power pump at an appropriate wavelength produces Raman scattering in the fiber [22]. An input optical signal at a larger wavelength then leads to SRS. In this process, pump and input optical signal are coherently coupled producing Raman gain.

The Raman amplifier is relatively a broad-band amplifier with a band width of more than 5 THz, and has reasonably flat gain over a wide wavelength range. As Raman gain exists in all types of optical fiber, it is being used in long-haul silica fiber-optic transmission systems, providing a cost-effective means of amplifying the signal transmitting through it. However, Raman gain requires much larger pump power (tens of milliwatts per dB of gain) compared with EDFA which requires less than 1 milliwatt per dB.

2.6   Optical Network Components

There are variety of optical components used in the network. These components are made by different materials as well as technologies. We first explain the basic optical components used as the building blocks of the different network modules. These optical components can be both *passive* and *active* types, making them either static or reconfigurable according to the state of the network.

2.6.1   Passive Coupler Devices

Under the generic name of *coupler* there are many devices which can combine light or split the light between different fibers or integrated optical waveguides. The basic structure of the coupler is a *directional coupler*. A directional coupler is a four port device used for combining or splitting the optical signal in the network with different splitting ratios.

Passive coupler devices can be made by fusing fibers or with integrated optic waveguides. The *2 × 2 fiber coupler* fabricated by fusing together two single mode fibers over a uniform section of length, called the coupling length, is shown in Figure 2.6. Each input and output fiber has a long tapered section of length to avoid scattering. With negligible reflected power, the input power from the input port is divided in the two output ports depending on the

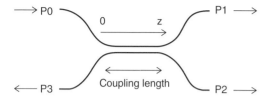

Figure 2.6 Fused-fiber coupler.

coupling length having a diameter equal to the single mode fiber. Coupled optical power in the output ports can be varied by changing the axial length of the coupling region or by the size of the reduced radius in the coupling region.

Using the coupled mode theory [2, 23], the electric field in the fibers can be obtained by quantitative analysis. Assuming no power loss in the coupler for the ideal case, with reference to Figure 2.6, the power coupled from one input fiber to the other output fiber port over an axial distance z can be obtained as:

$$P_2 = P_0 \sin^2(\kappa z) \tag{2.2}$$

where κ is the *coupling coefficient*. By conservation of power, for identical-core fibers we have the throughput power P_1, in the direct output port as the difference between the input power and the coupled power P_2, in the coupled port. The phase of the direct port power always leads the coupled port power by 90° and is obtained as:

$$P_1 = P_0 \cos^2(\kappa z) \tag{2.3}$$

By virtue of the above relation, at a distance when $\kappa z = \pi/2$, all of the power gets transferred from direct port fiber to the coupled port fiber in the output. Now the coupled port fiber becomes the driving fiber, so that for $\pi/2 < \kappa z < \pi$ the phase in direct port lags that in the coupled port, and so on.

2.6.1.1 Coupler Parameters

The coupler is characterized by the following four parameters:

- *Splitting ratio.* With reference to Figure 2.6, with P_0 being the input power and P_1 and P_2 the output powers, the splitting ratio is defined as:

$$\frac{P_2}{P_1 + P_2} \times 100\% \tag{2.4}$$

If the power is divided evenly between the two outputs ports, with half of the input power going to each output, then it is a *3-dB coupler*.

- *Excess loss.* Excess loss of a coupler is defined as the ratio of the input power to the total output power. In a practical device due to inherent losses, the total output power from the ports is always less than the input power. In decibels, the excess loss for a 2×2 coupler in dB is given by:

$$10\log\frac{P_0}{P_i + P_2} \tag{2.5}$$

- *Insertion loss*. The *insertion loss* refers to the loss for a particular port-to-port path. For the path from input port i to output port j, the insertion is given as:

$$10\log\frac{P_i}{P_j} dB \tag{2.6}$$

- *Cross-talk or return loss*. Cross-talk *measures* the degree of isolation between the input at one port and the optical power scattered back into the other input port. It is a measure of the optical power level P_3 shown in Figure 2.6, and is expressed as:

$$10\log\frac{P_3}{P_0} dB \tag{2.7}$$

Example 2.1 A 2×2 coupler with a 10/90 splitting ratio has the insertion losses as 11 and 0.5 dB, respectively, for the 10% and 90% channels. If the input power $P_o = 200\,\mu W$, find P_1 and P_2.

Solution

IL for P1 = 11 dB; IL for P2 = 0.5 dB

$$\frac{P_o}{P_1} = 10^{0.05}$$

Therefore,

$$P_1 = \frac{P_o}{10^{0.05}} = \frac{200}{10^{0.05}}\mu W = 178.25\mu W$$

and

$$P_2 = \frac{P_o}{10^{1.1}} = \frac{200}{10^{1.1}} = 15.89\mu W$$

2.6.1.2 Scattering Matrix Formulation of the 2×2 Coupler

Scattering matrix formulation of an individual cascaded element in a system can be useful in determining the transmission response and design of a larger

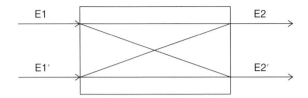

Figure 2.7 Functional block diagram of a 2×2 coupler.

system. Here we obtain the scattering for a generic 2×2 coupler used in many devices such as multiplexer/de-multiplexer along with other elements. Figure 2.7 shows the functional block diagram of the 2×2 coupler for the calculation of the scattering matrix. For an ideal symmetrical passive and lossless coupler, if the input field strengths E_1 and E_1', entering the two input ports are from distinct optical sources so that they are incoherent, the output field strength, E_2 and E_2' are related by the scattering matrix given as:

$$\left[E_2\right]=\left[S\right]\left[E_1\right] \tag{2.8}$$

where E_2 is the column matrix of the output field strength and E_1 the column matrix for the input field strengths. The scattering matrix, S is obtained as [7]:

$$S=\begin{bmatrix} \sqrt{1-\alpha} & j\sqrt{\alpha} \\ j\sqrt{\alpha} & \sqrt{1-\alpha} \end{bmatrix} 0 \le \alpha \le 1 \tag{2.9}$$

where α is the power coupling coefficient with values between 0 and 1. The coupling coefficient, α, gives the power coupled to the adjoining line. If the parameter α is fixed we call it a *static* (passive/non-configurable) device. If α can be varied through external control (i.e. thermal, electronic, or mechanical) then the device is *dynamic* or *controllable*. We are considering the device to be ideal with no excess loss, but in practice the device will have excess losses due to device imperfections.

There are many functions which are possible with the generic 2×2 couplers. When power enters only from one arm of the input coupler, it acts as a *power divider* or *splitter*. When the port at E_2' is terminated with an absorbing load and output is only in the port at E_2, then it works as a *combiner* for two inputs from ports E_1 and E_1'. There will be a combining loss in the dummy load even when the device is lossless. When α = 0.5 the device is called a *3 dB coupler*. For the dynamic case, by continually varying α it can be used as a *variable attenuator*. It can act as a switch with bar-state of α = 0 and cross-state with α = 1. We discuss this in detail in the following sections.

2.6.2   Switching Elements

Switching elements are one of the very important and commonly used essential components of any optical network. Switches are used for various functions: path or wavelength switching, protection switching, etc. As the optical transmission network is developing in the direction of ultra-high speed and large capacity, a large number of switching elements are used for path switching in the optical nodes. Also network protection switching and recovery become critical issues where the high-speed switches play an important role in switching to the redundant path or node. With increasing network size, the size of the switching matrix of the optical switches will continue to grow and large-capacity, high-speed, low-loss optical switches will be needed in future networks and they will play an important role in the development of optical networks.

The switching devices are classified into two basic classes: *logic* switching and *relational* switching. When the incident information-carrying signal of the device controls the state of the switch in such a way that some Boolean functions are performed on the inputs, it is called *logic switching*. Logic switching is *opaque* in nature as it is primarily done in the electronic domain. At present, most optical networks employ electronic processing, switching, and processing of data by converting an optical signal to electronic form to perform the logical functions. These switches have the limitation of data processing speed due to electronics and also the O-E-O conversion at an intermediate node in the network can introduce extra delay and cost. In few networks, optical logic switching is also now made available.

As the name indicates, in *relational switching* there is a relation between the inputs and the output of the switch. This relation is a function of the control signal applied to the switch but it is independent of the input information signal. The control of the switching function is performed electronically, but the optical stream is transparently routed from a given input of the switch to a given output. Such transparent switching allows the switch to be independent of the data rate and format of the optical input signals. This allows signals at high bit rates to pass through. In the following sections, we review a number of different optical switching elements and architectures used in optical networks.

In an optical network, switching is done in both the optical and the electronic domain. There can be wavelength or lightpath switching, packet switching or protection switching, as well as time-slot switching. Switches can perform various functions and are of different varieties. The classification takes into account their size or where they are used or the switching speed. The switching speed can vary from a few milliseconds in the case of protection or wavelength switching to a few nanoseconds in the case of packet switching. The switches can be made of different materials: $LiNbO_3$ switch, fiber switch,

MEM technology silicon switch [2, 24], and semiconductor material switch [25]. The control of the switching state is done electronically, mechanically, acoustically, etc. The characteristics that determine the performance of the switches are switching speed, delay, loss uniformity, polarization sensitivity, cross-talk, insertion loss, extinction ratio, reliability, complexity, and cost. The switches can be classified also as opaque and transparent, as discussed above.

Optical switches play a very important role in the optical network and have several applications in the network:

1. *Lightpaths switching.* These are the fixed and reconfigurable spatial switches made with 2×2 switches used to switch the incoming lightpath from the input fiber port to the required output port of the OXC. The switching time ranges over a few milliseconds.
2. *Space switching for path protection.* Protection switches are used for switching from the main fiber to the protection fiber/device in case of failure. These can be low/high-duty switches with high switching speed of tens of milliseconds and giving switchover back to the main fiber within 2–100 ms.
3. *Cell or packet switching.* As the name indicates, in circuits with packet switching in the optical layer these are used in the optical nodes for providing transmission of packets. These are very fast switches, with switching speed varying from 1 ns to 100 ns depending on packet length.
4. *External modulation.* They are used for turning on/off the data source with switching speeds of a small fraction of bit duration in the order of a few pico-seconds or so depending on the data rate.

Figure 2.8 shows the basic 2×2 switching element which is a building block of almost all of the switching fabrics in the optical nodes to be discussed. The 2×2 cross-point element routes optical signals from two input ports (A&B) to two output ports (A'&B') and has two states, *cross* and *bar*, controlled by the electrical signal applied to the device control ports. In the cross-state, the signal from the upper input port is routed to the lower output port with a transmission coefficient of α_{12}, and the signal from the lower input port is routed to the upper output port with a transmission coefficient of α_{21}. In the bar state, the signal from the upper input port is routed to the upper output port, and the signal from the lower input port is routed to the lower output port with a transmission coefficient of α_{11} and α_{22}, respectively, as shown in Figure 2.8. This switch is not wavelength selective, hence the input optical signal is switched to one or the other port incoherently. The 2×2 fiber cross-connects are simple and cheap devices, and it is possible to construct larger switching fabrics with large port counts by using many of them.

The dynamic switches are broadly of two types: (i) the generic directive switch, in which light is physically directed to one of two different outputs, and (ii) the gate switch, in which optical amplifier gates are used to select and filter input signals to specific output ports.

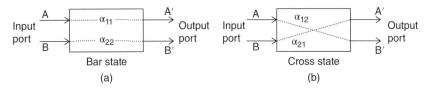

Figure 2.8 2×2 cross-connection switch: (a) bar state, with α_{11} and α_{22} as transmission coefficients, (b) cross state, with α_{12} and α_{21} as transmission coefficients.

2.6.2.1 Directive Switches

The common structures of directive switches are shown in Figure 2.9a–c. The controllable directional coupler (Figure 2.9a) consists of a pair of optical channel waveguides that are parallel and in close proximity over a finite interaction length as in the case of the static coupler. Light input to one of the waveguides couples to the second waveguide because of overlapping of fields in the two guides. Depending on the length of the interaction region, the coupling can be partial or complete. This is much like the static fiber coupler discussed earlier. For the device to act as a switch, the electrodes are placed over the waveguides. By the voltage difference applied across the two electrodes, the propagation constants in the waveguides can be changed. When the applied voltage is sufficiently increased so that no light couples between the two waveguides, it is the bar state of the switch. In the cross-state which corresponds to zero applied voltage the light is totally coupled in the second guide. There are some other types of directional switches, such as the M-Z interferometric switch (Figure 2.9b) [1, 2], and the Y-branch waveguide switch shown in Figure 2.9c [26] with a few improved features.

The waveguide optical switches discussed above are commonly used in optical networks. In these transparent relational switches, the control of switching is electro-optic, acousto-optic, thermal-optic, and magneto-optic effects.

2.6.2.2 Gate Switches

The switching devices discussed above were passive in nature and had finite insertion loss. The gate switches are powered devices with active devices. In the $N \times N$ gate switch, each input signal first passes through a $1 \times N$ splitter. The signals then pass through an array of N^2 gate elements and are recombined in $N \times 1$ combiners and sent to the N outputs. The gate elements are implemented using SOAs, which can be turned on or off to pass only selected signals to the outputs. The amplifier gains can compensate for coupling losses and losses incurred at the splitters and combiners. A monolithic structure uses a combination of passive waveguides for combining/splitting and active waveguides with the SOAs. Made as a PIC (photonics integrated circuit), the gate switching array with active elements provides gain as well as switching control

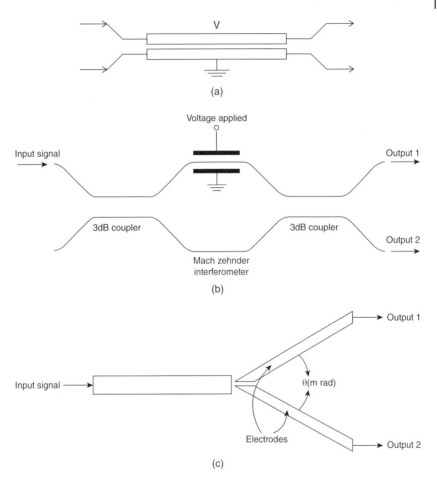

Figure 2.9 Schematic of optical directive switches. (a) Controllable directional coupler switch. (b) MZ interferometric switch. (c) Y-switch.

[27]. A 2×2 amplifier gate switch is illustrated in Figure 2.10. The disadvantage of gate switch is that the splitting and combining losses limit the size of the switch.

2.6.2.3 Micro-Electro Mechanical Switches

Optical MEMS are fabricated using micro-machining techniques. MEMS optical switches are typically fabricated in polysilicon on a silica substrate. The fabricated micro-machined mirrors are free rotating or torsion hinged and are actuated electrostatically. MEMS-based switches are of two types, i.e. 2D and 3D types. The 2D switches have a planar architecture, with an $N \times N$ switch

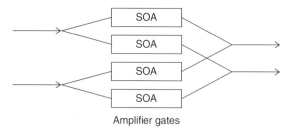

Amplifier gates

Figure 2.10 2×2 amplifier gate switch.

consisting of an integrated array of N^2 moving mirrors. The connections between the N inputs and N output fibers are made with free-space optical beam in cross-bar fashion. The 3D optical MEMS are more commonly used because they are compact and can make large port count switching fabrics, overcoming the scalability limitation in the 2D approach. 3D MEMS can achieve input and output port counts of over 1000 and are used in large-scale OXCs in which spatial parallelism is utilized. They have low and uniform insertion loss, very low cross-talk, are polarization insensitive, and have high scalability [24, 28] with wavelength insensitivity under various operating conditions, which provides high application flexibility in network design.

Figure 2.11 shows the basic configuration of the 3D MEMS optical switch. Light beams from an array of input fibers are focussed on an array of tilting mirrors, with one mirror dedicated to each input fiber and a similar arrangement used for the output fibers. The optical signals are switched independently by the MEMS mirrors with two-axis tilt control and then focused onto the optical fibers at the output ports. In the switch, any connection between input and output fibers can be made by controlling the tilt angle of each mirror independently. The 3D MEMS can be built in sizes up to (1000×1000) bidirectional ports. The combination of thousands of ports and bit rate independence results in a very high scalability, leading to compact and stable OXC switches.

2.6.2.4 Liquid Crystal Optical Switch

Liquid crystal (LC) materials have an anisotropic refractive index property which can be modulated with an applied electric field. With this they have special polarizing properties in the presence of varying applied control voltage [29]. The working principle of the liquid crystal optical switch is based on this polarization control, i.e. in one path light is reflected by the polarization, while in the other path light passes through. Because the electro-optic coefficient of liquid crystal is high, that makes it the most effective photoelectric material. Two types of LC materials that are commonly used in communications applications are the twisted nematic crystals and smectic crystals. The smectic crystals exhibit ferroelectric polarization behavior, acting as polarizers whose axis of polarization can be switched between two states by the

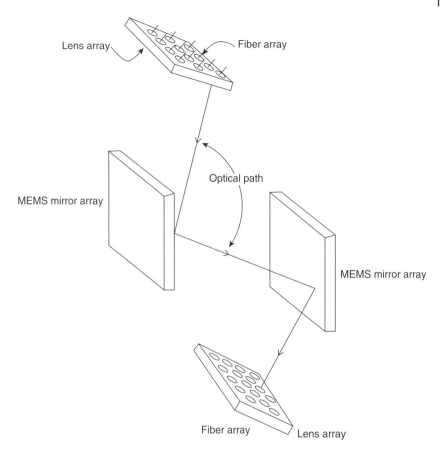

Figure 2.11 3D MEMS configuration.

application of an electric field. They are relatively fast, with switching times in the order of $10\,\mu s$. Liquid crystals are most commonly used in spatial light modulators (SLMs) where a slab of LC material sandwiched between other elements is arranged to produce an array of pixels controlled through a grid of electrodes to produce prescribed light patterns. Though LC switches have high insertion loss, they have very low switching time, low cost, and no moving parts, unlike MEMS.

2.6.3 $N \times N$ Star Coupler

The static $N \times N$ broadcast star coupler is one of the common passive devices used in optical networks to distribute incoherently the input optical signal coming from any port to all the output ports equally. Power is divided uniformly to all the output ports. It is an important device and large port count $N \times N$ *star couplers* are built by interconnecting the (2×2) couplers in different connecting fabrics matrices used in many switching devices.

The 4×4 broadcast star model in the WDM network can be explained by Figure 2.12a where four orthogonal wavelengths enter the four input ports and they are equally divided in the four output ports by the coupler. In this model, the transmitters on the end nodes transmit the signals on distinct wavelengths to the network, and the coupler combines these signals and distributes the aggregate signal to the receivers. The $N \times N$ coupler with larger port counts is fabricated with integrated optic devices while for lower port counts of up to 32 or so it can be fabricated with fusing and pulling the fibers. The principal role of star couplers is to combine the powers from N inputs and divide them equally among N output ports.

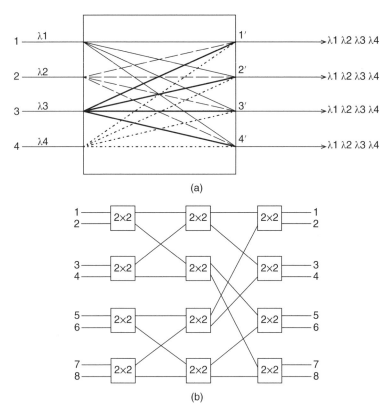

Figure 2.12 (a) Broadcast star. (b) Broadcast star using 2×2 couplers.

In a star coupler, the *splitting loss* is given by:

$$10\log\left(\frac{1}{N}\right) \tag{2.10}$$

For input power P_{in} in one port and with N output ports, the *excess loss* is given by:

$$10\log\left(\frac{P_{in}}{\sum\limits_{i=1}^{N}P_{out,i}}\right) \tag{2.11}$$

A large port count $N \times N$ star coupler can be made also with 2×2 3-dB coupler elements, for $N = 2^n$ with $n > 1$. Figure 2.12b shows an 8×8 star coupler device formed by twelve 3 dB couplers. Any input signal entering from an input port gets equally divided in all the eight output ports transparently.

In general, for a $N \times N$ coupler, there are $N/2$ elements in the vertical direction and $\log_2 N$ elements along the horizontal direction, requiring a total number of 3 dB couplers to construct an $N \times N$ star of:

$$N_c = \left(\frac{N}{2}\right)\log_2 N \tag{2.12}$$

The *excess loss* in decibels in an $N \times N$ coupler is:

$$-10\log\left[P_T^{\log_2 N}\right] \tag{2.13}$$

where the fraction of power loss in each 3 dB coupler element is P_T with $0 < P_T < 1$. As the signal passes through the $\log_2 N$ stages of the $N \times N$ star and gets divided into N outputs, the total loss, which is the sum of the splitting and the excess loss, is:

$$10\log\left[\frac{P_T^{\log_2 N}}{N}\right] = 10\log\left(1 - 3.22\log P_T\right)\log N \tag{2.14}$$

Hence, the loss increases logarithmically with N.

Example 2.2 2×2 couplers with loss of 0.5 dB each are used to make an optical $N \times N$ star coupler for distribution of optical signals between the connected nodes. If each receiver at the node requires a minimum of 100 nW and each transmitter at the node is capable of emitting 0.5 mW, calculate the maximum number of nodes served by this $N \times N$ coupler.

Solution

For an $N \times N$ star coupler made with 2×2 coupler devices, there are N/2 rows and $\log_2 N$ columns in the star of 2×2 coupler devices.

The power budget $= P_{TX} - P_{RX} = -3\,\text{dBm} - \left(-40\,\text{dBm}\right) = 37\,\text{dBm}$,

and

Total power loss in the 2×2 devices from an input port to the output port:

$= IL + \text{Splitting Loss}$

$= \left[0.5\left(\log_{10} N\right) / \left(\log_{10} 2\right) \right] + 10 \log N = 37$

Or N = 1490

As N = 2^n, taking n = 8; N = 1024.

2.6.4 Gratings

Gratings are periodic structures widely used in optical devices for separating the input light into its constituent wavelengths. They are available in both fiber and integrated optic devices. There are two types of gratings available: the transmission type and reflective types. As the name suggests, in the transmission type it is transmitted light which gets separated in its constituent wavelengths and in the case of the reflected type it is the reflective light which gets separated when the light is incident on the device. Further, gratings are also classified as *short-period* and *long-period* gratings, based on the period of the grating. Short-period gratings are also called *Bragg gratings* and have periods that are comparable to the wavelength, whereas long-period gratings have periods that are much higher than the wavelengths, ranging from a few hundred to a few mm. Fiber Bragg gratings (FBGs) are used to make optical filters, multiplexer, de-multiplexers, dispersion compensators, etc. Long fiber gratings are primarily used as band-reject filters.

Figure 2.13 shows the various parameters for a reflection grating. Here, θ_i is the incident angle of the light, θ_d is the diffracted angle, and Λ is the *period* of the grating. Constructive interference at a wavelength λ occurs in the imaging plane when the rays diffracted at the angle θ_d satisfy the grating equation given by:

$$L = \sin\theta_i - \sin\theta_d = m\lambda \tag{2.15}$$

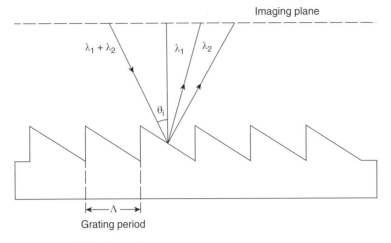

Figure 2.13 Reflective grating.

Here, m is called the *order* of the grating. A grating can separate individual wavelengths since the grating equation is satisfied at different points in the imaging plane as shown in Figure 2.13 for different wavelengths.

2.6.4.1 Fiber Bragg Gratings

In FBG the grating in the fiber is formed by the periodic variation of high and low refractive index in the core of the fiber by directly photo imprinting the grating pattern in the core of an optical fiber. The maximum reflectivity occurs when the Bragg condition holds, i.e.:

$$\lambda_{Bragg} = 2\Lambda n_{eff} \tag{2.16}$$

where n_{eff} is the mode effective index of the core of the fiber. At this Bragg wavelength, the peak reflectivity, R_{max} for the grating of length L and coupling coefficient κ is given by:

$$R_{max} = \tanh^2(\kappa L) \tag{2.17}$$

The full bandwidth $\Delta\lambda$ over which R_{max} holds is:

$$\Delta\lambda = \left(\frac{\lambda_{Bragg}^2}{\pi n_{eff} L}\right)\left[(\kappa L)^2 + \pi^2\right]^{1/2} \tag{2.18}$$

The full width half maximum (FWHM) bandwidth of the grating with approximation can be expressed as:

$$\Delta\lambda_{FWHM} = \lambda_{Bragg} S \left[\left(\frac{\delta n}{2n_{core}} \right)^2 + \left(\frac{\Lambda}{L} \right)^2 \right]^{1/2} \tag{2.19}$$

where $S = 1$ for strong gratings with near 100% reflectivity, $S = 0.5$ for weak gratings, and δn is the index difference of the periodic high and low index of the fiber to form the grating. For a uniform sinusoidal modulation of the fiber core index, the coupling coefficient is given by

$$\kappa = \frac{\pi \delta n \eta}{\lambda_{Bragg}} \tag{2.20}$$

with η being the fraction of optical power contained in the fiber core, which can be approximated by:

$$\eta \approx 1 - V^{-2} \tag{2.21}$$

where V is the V-number of the fiber. Equation (2.21) is valid under the assumption that the grating is uniform in the core. A Bragg grating when used as filter will reflect the specific wavelength of light back to the source while passing the rest of the wavelengths. Two primary characteristics of a Bragg grating are the spectral bandwidth and the reflectivity. Typical spectral bandwidth is of the order of 0.1 nm, while reflectivity in excess of 99% is achievable. Fiber gratings have low insertion loss and they find application for the implementation of multiplexers, de-multiplexers, and tunable filters.

2.6.4.2 Arrayed Waveguide Grating

Arrayed waveguide grating (AWG) is an integrated device and can be fabricated at low cost [30, 31]. It is made of three blocks: two multiport star couplers interconnected with an array of waveguides with differential lengths. The input and output waveguides, multiport couplers, and the arrayed waveguides are all fabricated on a single substrate, usually of silicon, and the waveguides are of silica or Ge-doped silica.

An AWG can function as a multiplexer, a demultiplexer, and also as a router. When only a single input port is used with N output ports, the device acts as a de-multiplexer, and if operated in the reverse direction it can act as a multiplexer. Typically an AWG, as shown in Figure 2.14, has a phased array of multiple waveguides with path length differences between neighboring guides to form a wavelength router. In an $N \times N$ system, it creates a static routing pattern determined by the geometry of the device. An AWG provides a fixed routing of an optical signal from a given input port to a given output port based on the

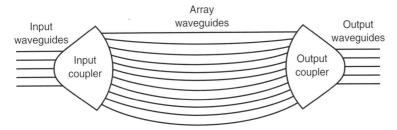

Figure 2.14 The arrayed waveguide grating router.

wavelength of the signal. A multiplexed signal of different wavelengths coming from an input port will be routed to a different output port. Being a static device, AWG router has a fixed routing matrix which cannot be reconfigured.

2.6.5 Optical Filters

There is a large variety of optical filters available made with fibers or integrated optics, such as *Fabry-Perot, multilayer dielectric thin-film, Mach-Zehnder interferometers* or *acoustic-optic* filters [32]. Fabry-Perot cavity-type filters and multilayer dielectric thin-film cavity filters are narrow band filters, while Mach-Zehnder interferometer (MZI) type filters made with integrated optics are wide-band filters. For large multiport devices *AWGs* are preferred over the MZI.

The characteristics of optical filters include insertion loss, selectivity, polarization sensitivity, etc. The filters may be *fixed* or *tunable*. Tunable optical filters are also characterized by their *tuning range* and *tuning time*. Larger tuning range specifies that the filter can be tuned to a larger range of wavelengths. The *tuning time* of a filter specifies the time required to tune it from one wavelength to another. There are other two parameters, *free spectral range* (FSR) and *finesse*, which characterize resonant multiband filters:

- *Free spectral range*. The FSR of the filter is the spectral width between two successive passbands in a multiband filter. The transfer function of the filter repeats itself periodically. The period of such filters is referred to as the FSR. With reference to Figure 2.15, the peaks of the various spacing occur at the wavelengths that satisfy the condition $N\lambda = 2nD$, where n is the refractive index of the dielectric in the etalon, D is the length of the cavity, and N is an integer. The distance between the adjacent peaks is the FSR, expressed as:

$$FSR = \frac{\lambda^2}{2nD} \tag{2.22}$$

The FSR of the filter depends on cavity lengths or waveguide lengths.

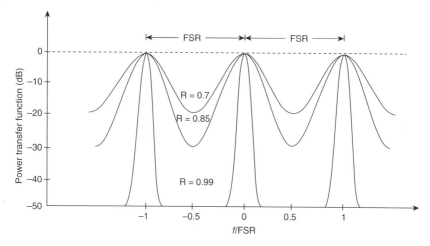

Figure 2.15 Free spectral range of a tunable filter.

- The *finesse* of a filter is related to the width of the passband. It is the ratio of FSR to 3 dB filter bandwidth. The FWHM is defined as the width of each passband where the transfer function is half of its maximum value. The *finesse*, F of the filter with reflectivity, R of the cavity mirror is given by:

$$F = \frac{\pi \sqrt{R}}{1 - R} \tag{2.23}$$

The number of channels in an optical filter is limited by the FSR and finesse. All channels within a FSR can be tuned and high finesse results in more channels being able to fit into one FSR. With low finesse, as the filter passband is broad the channels need to be spaced further apart to avoid cross-talk, resulting in the filter being able to tune to fewer channels. Optical filters along with the gratings are used in many devices, such as routers, multiplexers, de-multiplexers, add-drop multiplexers, etc., which we will be discussing in the following paragraphs.

2.6.5.1 Fabry-Perot Filter

In the *Fabry-Perot (F-P)* filter etalon when monochromatic light ray travels back and forth between two mirrors as shown in Figure 2.16a, and the distance between mirrors equals an integral number of wavelengths, then the light passes through the etalon. The transmission, T of an ideal etalon in which there is no light absorption by the mirrors is an *Airy* function, expressed as:

$$T = \left[1 + \frac{4R}{(1-R)^2} \sin^2 \left(\frac{\phi}{2} \right) \right]^{-1} \tag{2.24}$$

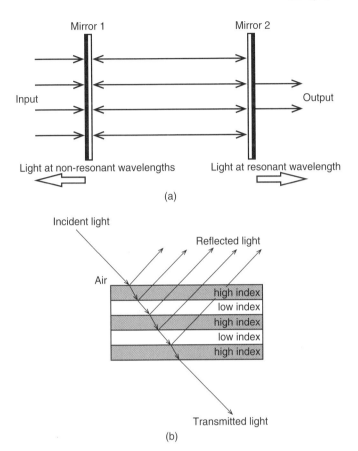

Figure 2.16 (a) Fabry-Perot filter. (b) Multilayer dielectric thin-film filter.

where R is the *reflectivity* of the mirrors. Ignoring any phase change at the mirrors, the phase change in the output wave at λ is:

$$\varphi = \left(\frac{2\pi}{\lambda}\right) 2nD\cos\theta \qquad (2.25)$$

where n is the refractive index, D is the distance between mirrors, and θ is the angle to the normal of the incoming light beam. As discussed earlier, the peaks of the spacing, called the *passband*, occur at those wavelengths that satisfy condition $N = 2nD\cos\theta$, where N is an integer. The ratio FSR/FWHM gives an approximation of the number of wavelengths that a filter can accommodate. This is the *finesse F* of the filter given in Eq. (2.23).

The F-P etalon has a tuning time of the order of tens of milliseconds due to mechanical tuning of the mirrors. When used for WDM communication systems, the FSR of the Fabry-Perot etalon corresponds to the typical channel spacing of 100 GHz or 50 GHz, thus requiring cavity lengths of about 1mm or so. Furthermore, the Fabry-Perot filter can cover a whole communication band, which is typically 30-40 nanometers large.

2.6.5.2 Multi-Layer Dielectric Thin-Film Filter

Similar to the F-P filter, the multi-layer dielectric thin-film resonant cavity filters work on the same approach for filtering out one or more wavelengths from a number of wavelengths. *Thin-film filters (TFFs)* [30] are fabricated by depositing alternating layers of low-index and high-index materials onto a substrate layer as shown in Figure 2.16b. These devices act as a bandpass filter by passing through a particular wavelength and reflecting all the other wavelengths. The cavity length determines the wavelength of the pass-through signal. TFF normally consists of more than one cavity to have flat filter bandwidth with sharper roll-off. TFF technology has high thermal stability, low insertion loss, and is polarization insensitive.

2.6.5.3 Acousto-Optic Filter

AOTFs [33] are made of a piezo-electric crystal transducer, such as $LiNbO_3$ or TeO_2 crystal. They have a very small tuning time of about 10 ps and are limited by the flight time of the *surface acoustic wave*. The tuning range for acousto-optic filters covers the entire 1300–1560 nm spectrum. The gratings in the crystal are formed when the radio frequency (RF) waves are passed through the crystal transducer. These RF waves convert the sound waves to mechanical movement, changing the refractive index of the crystal material and in turn enabling the crystal to act as grating. When light is incident upon the transducer it is diffracted at an angle which depends on the angle of incidence and the wavelength of the light. A single optical wavelength can be chosen and filtered through the material by changing the RF waves while the rest of the wavelengths destructively interfere. More than one wavelength can also be filtered out if more RF waves are passed through the grating simultaneously. Thus, the filter can be tuned to several channels at the same time, but in this case the received signal is the superposition of all of the received wavelengths, therefore if more than one of those channels is active, cross-talk will occur. An acousto-optic filter is used as a tunable filter for wavelength tunable lasers because of its wide tuning range and fast switching speed of several microseconds.

2.7 Optical Multiplexer and De-Multiplexer

A *multiplexer* (MUX) is used to combine several WDM optical channels into a serial spectrum of closely spaced wavelength signals and couple them onto a single fiber. A *de-multiplexer* (DMUX) performs the complementary function to separate the optical signals into appropriate WDM wavelength channels. The multiplexed optical signals at different wavelengths have no significant optical power outside the designated channel spectral width; one, due to sufficient separation between the channels with the guard bands between them, and two, due to narrow line widths of the DFB/DBR laser used. On the de-multiplexer end they have very narrow optical filters with sharp wavelength cut-offs at the input of the detector. These MUXs/DEMUXs are made with several techniques, i.e. arrayed waveguides and other frequency-sensitive devices as discussed in the previous sections. We discuss here one type of optical multiplexer/de-multiplexer to explain the design and analysis.

2.7.1 Mach-Zehnder Interferometer (MZI) Multiplexer

Figure 2.17 shows a 2×2 MZI multiplexer/de-multiplexer which consists of three stages: a 3 dB directional coupler, a central section where one of the waveguides is longer by ΔL to give a wavelength-dependent phase shift between the two arms, and another 3 dB coupler which recombines the signals at the output. The two input signals at two WDM wavelengths enter the 3 dB 2×2 coupler. The output signals from the coupler undergo a differential phase shift due to ΔL difference in the two lines; when recombined in the second coupler the two signals get combined to give the multiplexed signal in one arm while no signal is there in the second arm of the coupler.

The function of this arrangement is that by splitting the input beam and introducing a phase shift in one of the paths, the recombined signals will interfere constructively at one output and destructively at the other. The multiplexed signals then finally emerge from only one output port.

We can obtain the desired multiplexed output in one arm of the device by multiplying the propagation/scattering matrix of the three stages: 3 dB coupler,

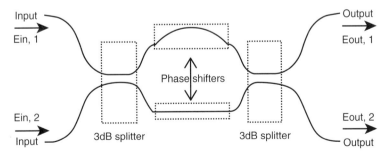

Figure 2.17 Layout of 2×2 Mach-Zehnder interferometer multiplexer.

the phase shifter with ΔL difference between the two arms, followed by the 3 dB combiner, as:

$$S = S_{coupler}S_{\Delta\varphi}S_{coupler} = j\begin{bmatrix} \sin\left(\dfrac{k\Delta L}{2}\right) & \cos\left(\dfrac{k\Delta L}{2}\right) \\ \cos\left(\dfrac{k\Delta L}{2}\right) & -\sin\left(\dfrac{k\Delta L}{2}\right) \end{bmatrix} \tag{2.26}$$

From Figure 2.17, if inputs to the MZI are at different wavelengths – $E_{in,1}$ at λ_1 and $E_{in,2}$ at λ_2 – then the output fields $E_{out,1}$ and $E_{out,2}$ can be obtained when the signal is passed through the propagation matrix of (2.26) (refer to Eq. (2.8)) as:

$$E_{out,1} = j\left[E_{in,1}(\lambda_1)\sin\left(\frac{k_1\Delta L}{2}\right) + E_{in,2}(\lambda_2)\cos\left(\frac{k_2\Delta L}{2}\right) \right] \tag{2.27}$$

$$E_{out,2} = j\left[E_{in,1}(\lambda_1)\cos\left(\frac{k_1\Delta L}{2}\right) - E_{in,2}(\lambda_2)\sin\left(\frac{k_2\Delta L}{2}\right) \right] \tag{2.28}$$

where, $k_m = 2\pi n_{eff}/\lambda_m$ with m = 1, 2 for two wavelengths. The output powers are found from the light intensity, which is the square of the field strengths. By ignoring the higher frequency terms which will be filtered out, the power terms obtained are:

$$P_{out,1} = E_{out,1}.E_{out,1}^* = \sin^2\left(\frac{k_1\Delta L}{2}\right)P_{in,1} + \cos^2\left(\frac{k_2\Delta L}{2}\right)P_{in,2} \tag{2.29}$$

$$P_{out,2} = E_{out,2}.E_{out,2}^* = \cos^2\left(\frac{k_1\Delta L}{2}\right)P_{in,1} + \sin^2\left(\frac{k_2\Delta L}{2}\right)P_{in,2} \tag{2.30}$$

where $P_{in,m} = |E_{in,m}|^2$. From Eqs. (2.29) and (2.30), we see that if we want all the power from both inputs to leave the same output port (say port 2), we need to have

$$\frac{k_1\Delta L}{2} = \pi \text{ and } \frac{k_2\Delta L}{2} = \pi/2, \text{or}\left(k_1 - k_2\right)\Delta L = 2\pi n_{eff}\left[\left(\frac{1}{\lambda_1}\right) - \left(\frac{1}{\lambda_2}\right)\right]\Delta L \tag{2.31}$$

Hence, the length difference in the interferometer arms should be

$$\Delta L = \left[2n_{eff}\left\{\frac{1}{\lambda_1} - \frac{1}{\lambda_2}\right\}\right]^{-1} = \left[c/(2n_{eff}\Delta f)\right] \tag{2.32}$$

where Δf is the frequency separation of the two wavelengths. This being a passive device with reciprocity, on sending the multiplexed signal from the reverse direction it will work as a de-multiplexer separating the two wavelengths.

Higher port count devices can be obtained by adding more stages. For example, in the case of an $(N \times 1)$ MZI multiplexer, where $N = 2^n$ with $n \ngeq 1$, n being the number of multiplexer stages, the number of MZIs in a stage j will be $2^{(n-j)}$ and the path difference in an interferometer element of stage j is then given as:

$$\Delta L_{stagej} = \frac{c}{2\left(2^{n-j} n_{eff} \Delta f\right)} \tag{2.33}$$

The $(N \times 1)$ MZI multiplexer can be used as a $(1 \times N)$ de-multiplexer by reversing the light-propagation direction. These devices are not scalable as the number of MZI required increases, making it not viable. When large port count static multiplexer/de-multiplexers are required, AWG devices are preferred.

Example 2.3 Design a 8×1 multiplexer using 2×2 MZI that can handle a channel separation of 50 GHz. Assuming the shortest λ be 1550 nm, determine the differential length in each stage with ($n_{eff} = 1.5$).

Solution

For N = 8, n = 3, the number of multiplexer stages are: 3.

The first stage will have four identical MZI, the second stage will have two identical MZI, and the third stage has one MZI.

From Eq. (2.33), in the first stage all MZI have differential length of:

(i) $\dfrac{c}{8n_{eff}\Delta f} = 1.666x10^{-12}c = \Delta L = 500\, \text{\textcrm}$

(ii) In the second stage the two MZI will have a differential length of:

$\dfrac{c}{4n_{eff}\Delta f} = 3.33x10^{-12}c = 2\Delta L = 1000\, \text{\textcrm}$

(iii) The last stage has MZI of differential length:

$\dfrac{c}{2n_{eff}\Delta f} = 6.6x10^{-12}c = 4\Delta L = 2000\, \text{\textcrm}$

2.8 Routers

Routers are used in optical networks for routing the traffic, be it a *lightpath* or packet traffic, from the source node to the destination node passing through the intermediate network nodes. The routers used are of different types in both structure and technology. In the following we give the types of routers commonly used in networks.

2.8.1 Static Wavelength Router

In the wavelength switched routing networks the *lightpaths* have to be routed over the fiber between the source-destination nodes. Signals from different input ports of the router are sent to different output ports based on the wavelengths to route the lightpath to their destination.

The $(N \times N)$ wavelength router requires two blocks to build: first, N number of $(1 \times N)$ wavelength DMUX, and second, N number of $(N \times 1)$ wavelength multiplexer. For a $(N \times N)$ *static or non-reconfigurable* wavelength router, each of the N input fibers carries a set of $\{\lambda_1, \lambda_2 \dots \lambda_N\}$ wavelengths and is connected to a $(1 \times N)$ DMUX. The DMUX spatially separates the wavelengths and send them on N output ports. The output stage has again N number $(N \times 1)$ MUXs, each of which is identical to DMUX but used in the opposite direction. Each of the N number *MUX* combines the different λs connections coming from the N number *DMUX* and connects them to the output fiber. A static wavelength router is also known as a static wavelength *cross-connect* (WXC) in the wavelength-routed networks. In the static WXC, the cross-connection pattern is fixed at the time when the device is made and cannot be changed dynamically. Figure 2.18 illustrates a 4×4 wavelength static router as an example.

The connection between the two devices is such that different λs from different ports are combined by the MUX. The N^2 fibers between the input and output stages are connected in a way that prevents identical wavelengths from different input ports from being combined on the same output port, thus avoiding interference among these channels. Hence, the path taken by any signal through the node is determined uniquely by its wavelength and port number. The particular input port and its wavelength get routed to the particular output port depending on a *routing matrix* characterizing the router; this matrix is determined by the internal connections between the DMUX and the MUX.

An AWG can be made to work as a static wavelength router as PLC (Planar lightwave circuit) with silica on silicon technology.

2.8.2 Reconfigurable Wavelength Router

In order to have dynamic traffic connectivity in a wavelength routing network it is essential to have reconfigurability. As discussed in the last sub-section,

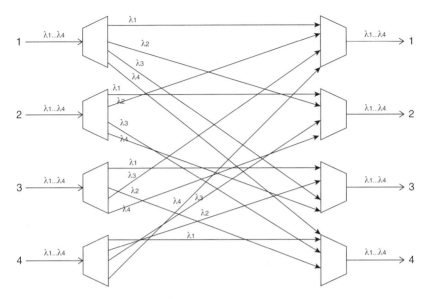

Figure 2.18 4×4 Fixed wavelength router.

the routing for a static router cannot be changed once it is designed. We can get reconfigurability in the static wavelength router by adding a space switching stage between the DMUX and MUX columns to form the *dynamic* wavelength-routing router. This is also referred to as a wavelength selective cross-connect (WSXC). The switching stage can reconfigure the connectivity between the DMUX and MUX by electronically controlling the switching wavelength.

Figure 2.19 gives the functional block diagram of the $N \times N$ reconfigurable router [34]. Each incoming fiber connected to the DMUX carries N number WDM wavelengths. The de-multiplexed N wavelengths are directed to an array of N, $N \times N$ optical switch fabric between the DMUX and MUX. Each $(N \times N)$ switch fabric is for a particular wavelength; hence all signals on a particular wavelength are directed to the same switch fabric. The switched signals are then directed to the MUXs which are associated with the output ports. The optical routing space switches in their simplest form may be made of (2×2) two state switches with no wavelength selectivity. The 2×2 cross-point switches may be reconfigured to adapt to changing traffic requirements by the control signaling. Large space switch matrices/fabrics are formed by the interconnection of (2×2) or low port count space switches. As the reconfigurable wavelength routers provide additional control in setting up connections, they are more flexible than static routers. The routing is a function of both the wavelength chosen at the source node and the configuration of the switches in the network nodes.

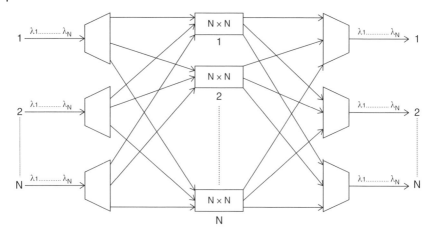

Figure 2.19 $N \times N$ reconfigurable wavelength-routing switch with N wavelengths.

2.8.3 Optical Packet Routing Switches

Lightpath routing, which was done in static and reconfigurable routers, is used for circuit switching in which the routing of traffic is done after the path has been established. In the case of packet switching the packet itself carries the path ID. The packets have header bits with source-destination address, and each wavelength channel carries time division multiplexed information packets. The packets, which are typically of random variable length, arrive at variable random rates from different sources and are not synchronized in time. In electronic packet switching, the header bits are read in the electronic processor once there is an O-E conversion at the node. But in the case of optical packet switching, the packets contending for transmission on a given output line at a node may require optical buffering to resolve their contention. Unlike the electronic storage, the available optical buffer/storage is limited. It can be provided either with an optical delay line or by deflection routing or rerouting. In most optical packet switches (OPSs), the header processing and switching functions are carried out electronically and the routing of the payload is done in the optical domain. Several packet switches are proposed which have the data part fully optical but the control of switching operation is performed electronically [34–37].

2.9 Optical Switching Fabrics

In Section 2.6.2 we read about different types of switches with lower port counts of 2×2 or 4×4, etc. In present-day high-traffic networks, $N \times N$ switching fabrics are made of large port counts ranging from a few hundred to a few thousand ports, with each fiber port carrying a large number of wavelength channels. The number of port counts increases with increased traffic load and number of users on the network. The complexity and cost of the switching fabric also increase with the increase in ports. There is a variety of these switching fabrics with different performance parameters.

2.9.1 Classification of Switching Fabrics

The switching fabrics can be classified according to their input/output port connections and accordingly they can be classified broadly in three categories.

2.9.1.1 Permutation Switching Fabric

The interconnection pattern between the different input ports to the output ports of the switching fabric is represented by a connection matrix. The interconnection between the switches in the case of $N \times N$ permutation switching fabric is such that there can be only point-to-point (one input port to another output port) unicast connection. It cannot have broadcasting ($1 \times N$) or multicasting ($N \times 1$) connections. The permutation fabric matrix, therefore, will have one cross-point closing in each row and column. A connection matrix for the 4×4 permutation fabric is shown in Figure 2.20a. The switching matrix is square in this case as there can be only permutation connections possible. The total possible input/output connection patterns or states are N! in a permutation switch fabric.

2.9.1.2 Generalized Switching Fabric

In the case of generalized switching fabric one can make all types of input/output connections, i.e. point-to-point, point-multipoint ($1 \times N$), or multicasting ($N \times 1$). In general, it is an ($N \times R$) matrix with 2^{NR} connecting states. Though the permutation switch has only one point closing in a row or column of the connection matrix, for generalized fabric switch there can be multiple points closing in a row or column depending on whether it has broadcasting or multicasting connections. Unlike the permutation switch, which has only the control for bar and cross state, the generalized matrix has additional controls to specify the number of connections, as shown in Figure 2.20b for a 4×4 matrix.

2.9.1.3 Linear Divider and Combiner Switching Fabric

In the generalized switching fabric when we have the combining and splitting ratios also defined for multicasting and broadcasting, respectively, they are known as *linear divider and combiner* (LDC) switching. As the combiner/splitting ratios can take any value between 0 and 1, therefore, LDC can have a continuum of connection states.

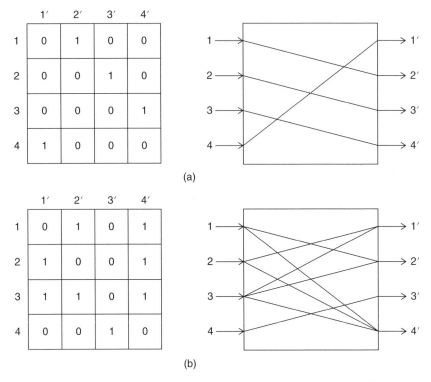

Figure 2.20 (a) 4×4 permutation switch connection matrix. (b) Generalized switching connection matrix.

2.9.2 Classification According to Blocking Characteristics

In any network a very large number of switches cannot be made available to provide dedicated connections to all possible connections required. This is done in order to provide better utilization of the resources but without compromising on the performance of the services. Therefore, with the prior estimate of the traffic on the network, they are designed that they have no or least little blocking to the admission of requests with the available number of switches. So the switches we consider here are non-blocking switches and are classified according to their blocking characteristics for randomly made requests to the network for the connections between any input and any output port. According to the blocking of the randomly appearing input requests, the switch fabrics can be of the following types.

(i) *Strict sense non-blocking switch fabrics.* In the strict sense non-blocking switch fabrics, any new incoming connection request is allowed to use any free path in the switch. The new connection is made without any rearrangement of the already existing connections. This is possible with the large number of switches being made available. The routing algorithms in this case are very simple but the network requires much more hardware in order to have no connection disruption. Strictly non-blocking switch configurations have good cross-talk properties and low latency but typically have more switching devices than necessary.

(ii) *Wide sense non-blocking switch fabrics.* These fabrics can realize any new connection without rearranging the already existing active connection, provided the correct rule is used for routing each new connection. Therefore, the routing algorithm has to be intelligent and more complex but requires lesser hardware as compared with strict sense fabrics. The active connections are also not interrupted as in the case of strict sense.

(iii) *Rearrangeable non-blocking switch fabrics.* These types of fabrics are much simpler in hardware as compared with the strict sense and wide sense cases but have two problems. One, active connections may have to be interrupted momentarily while rerouting is taking place to establish the new connection, and second, a complex computation or algorithm is required to determine the device setting to route the new connection. But these switches require minimum hardware and are therefore less expensive.

2.9.3 Types of Space Switching Fabrics

There are large numbers of switch fabrics ranging in variety of materials and fabrication technology. The switching control can be electromechanical, electronic, and photonic, such as integrated optics, micro-mirrors or MEMS, liquid crystals, etc. While selecting a particular switching fabric, besides the blocking characteristics discussed above there are other considerations, such as cost, size, number of ports, loss uniformity among the different input and output connections. Insertion loss is introduced by each switch element in the fabric. Hence, the loss between any two input and output port connections depends on the number of switch elements in its optical path. The measure of loss uniformity depends on the difference between the number of switches between the longest and the shortest path connection. Another important characteristic affecting the performance of the switch fabric is the number of cross-overs between any two connections. These cross-overs introduce cross-talk and loss. Also, fabrication by integrated optics becomes more complex in the presence of cross-overs. Large size switching fabrics are fabricated using silica on silicon, $LiNbO_3$, PLC using directional couplers, and some other technologies.

Among the many switching fabrics, cross-bar, Benes, Clos, etc. are common in optical networks. We discuss a few of these in the following paragraphs.

2.9.3.1 Cross-Bar Switching Fabric

A cross-bar fabric can be used both as an $(N \times N)$ permutation switch and as a $(R \times N)$ generalized switch. In the case of a permutation $(N \times N)$ switch fabric, it consists of N input lines, N output lines, and N^2 cross-points. The cross-points are implemented by controllable (2×2) optical switching elements. An $(R \times N)$ fabric will require RN (2×2) switches. There are many ways of the implementing cross-bar concept in the optical domain.

Two possible realizations of an optical (4×4) cross-bar are shown in Figures 2.21a,b. In Figure 2.21a, if the switch at the cross-point is OFF, light from an input guide continues on its horizontal path through the cross-point. If the switch is ON, light is deflected from the horizontal guide downward on the vertical guide at the cross-points. Therefore, when all the switches are in

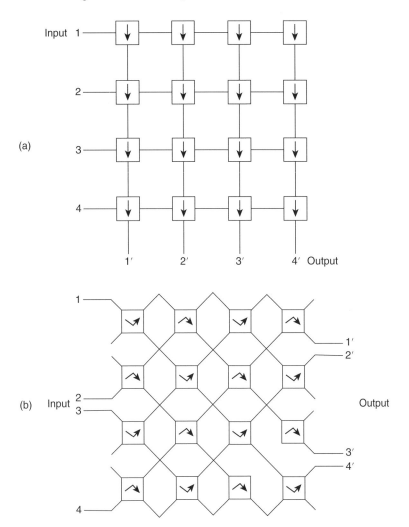

Figure 2.21 (a) Cross-bar (4×4) switch. (b) Cross-bar (4×4) switch.

the OFF state, there are no active connections through the switch. Insertion loss in this case is better but the losses are not uniform because of the inequality of path length, which also produces unequal levels of cross-talk on each path. In contrast, all paths in Figure 2.21b are the same length and traverse the same number of switches from left to right, so that it has path-independent insertion loss. Light remains in its guide when passing through a cross-point with OFF switch, but light is diverted from an input guide to an output guide when the switch is in the ON state. Each path traverses four switches. The cross-bar switch is wide-sense non-blocking.

To summarize, the cross-bar switches have no cross-over and hence lesser cross-talk. They have larger fabric structure size as the switching elements number grow with N^2 as N increases.

2.9.3.2 Clos Switch Fabric

Unlike the single-stage cross-bar switching fabric, Clos architecture [38] is a multi-stage non-blocking switching fabric. It can be constructed as a strict sense or wide sense non-blocking switch and is widely used in practice to build large port count switches. A three-stage ($N \times N$) port Clos switch is shown in Figure 2.22. Three design parameters, *M, K*, and *P*, with $N = PK$ are used in the

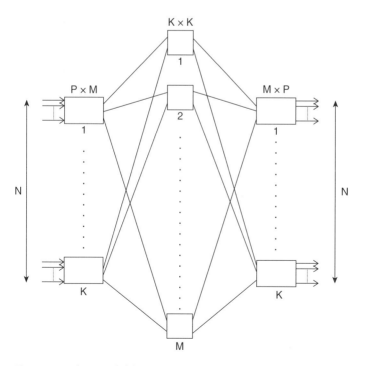

Figure 2.22 Clos switch fabric.

design of a ($N \times N$) Clos switch fabric. The first and third stages consist of K number ($P \times M$) switches. The middle stage consists of M number ($K \times K$) switches. Each of the K switches in the first stage is connected to all the switches in the middle stage. Likewise, each of the K switches in the third stage is connected to all the switches in the middle stage.

The necessary and sufficient condition to make the switch strict-sense non-blocking is to have $M = (2P - 1)$. Usually, the individual switches in each stage are designed using cross-bar switches. Thus, each of the [$P \times (2P - 1)$] switches in the first and third columns requires ($P \cdot (2P - 1)$) elementary (2×2) switch elements, and each of the ($K \times K$) switches in the middle stage requires K^2 elementary switch elements. Using ($K = N/P$), the number of elementary switch elements is minimized when ($P \approx (N/2)^{1/2}$). Using this value for P, the number of switch elements required for minimum cost configuration is approximately ($4\sqrt{2}N^{3/2} - 4N$), which is significantly lower than the N^2 required for a cross-bar.

The Clos architecture has several advantages. The loss uniformity between different input-output combinations is better than a cross-bar, and the number of switch elements required is significantly smaller than for a cross-bar.

2.9.3.3 Spanke Switch Fabric

The Spanke architecture, shown in Figure 2.23, is a strict-sense non-blocking switch and is one of the popular architectures for building large switches. An ($N \times N$) switching fabric has $N \cdot (1 \times N)$ number splitters in the input and $N \cdot (N \times 1)$ number of combiners in the output, unlike the other switching fabrics which use (2×2) switching elements. Spanke uses ($1 \times N$) and ($N \times 1$) optical switches,

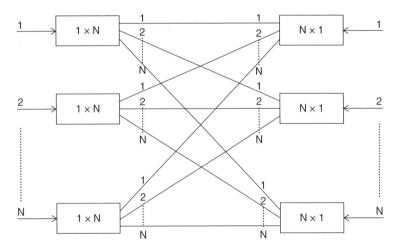

Figure 2.23 Spanke switch fabric.

which can be easily made with MEMS technology. Each connection passes through two switch elements, which is significantly smaller than the number of switch elements in the path for other multi-stage designs and therefore has much lower insertion loss than other multi-stage designs. It has a total of *2N* elements and has uniform path loss, the cross-overs are minimal. and the cost increases linearly with *N*, which is significantly better than in other switch architectures.

2.9.3.4 Benes Switch Fabric

We have seen that constructing large switch fabrics is more economical through a multi-stage fabric. Multi-stage configurations usually can be realized with far fewer cross-points than the cross-bar, especially for large-size switches. One such multi-stage switch fabric is the Benes fabric.

The Benes switch fabric is basically a non-blocking rearrangeable multi-stage permutation switch. Hence, it is more economical with respect to strict-sense or wide-sense switching fabrics, but usually at the cost of increased cross-talk and latency. Figure 2.24 shows a Benes (8×8) fabric with five stages using (2×2) switches. Comparing the cross-bar switch numbers with the number of elements required in the Benes switch, we find that for large value of *N*, the Benes fabric is very close to optimal as it uses a minimum of hardware. For example, the rearrangeable non-blocking (8×8) switch, shown in Figure 2.24, uses only 20 number (2×2) switches, while in comparison a (8×8) cross-bar switch will require 64 numbers of the basic (2×2) switching elements.

In general, a $(N \times N)$ Benes switch has $(2 log_2 N - 1)$ stages [39] with $N/2$ number of (2×2) switches per stage. They have uniform loss but the numbers of cross-overs are more, making it difficult to fabricate in integrated optics. Each path for a connection has $(2 log_2 N - 1)$ switches, therefore total of $(N/2)$ $(2 log_2 N - 1)$ number (2×2) switches are used, with *N* being a power of two. The loss is equal through every path in the switch as each connection is made with the same number of switches.

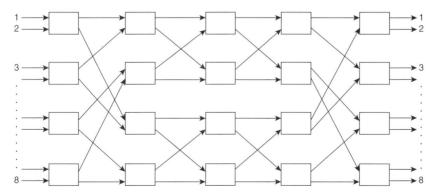

Figure 2.24 (8×8) Benes switch.

Example 2.4

Refer to Figure 2.24 for the (8×8) Benes and an (8×8) cross-bar switch for the type given in Figure 2.21a. What type of blocking characteristics do they have? If each (2×2) switch has an insertion loss of 1 dB and connection has to be established between input port 1 and output port 6 in both cases, how will their performances compare?

- Benes is the rearrangeable type with insertion loss of 5 dB.
- Crossbar is a wide-sense non-blocking with insertion loss of 13 dB.

2.9.3.5 Spanke-Benes Switch Fabric

Another rearrangeable non-blocking switching fabric is the Spanke-Benes [39] architecture shown in Figure 2.25, which is an (8×8) switch using 28 (2×2) switches and without any waveguide crossovers. This switch architecture is called the *N-stage planar architecture* since it requires N stages to realize an $N \times N$ switch. The main feature is that it has no cross-over and hence is called an N-stage planar fabric with total of $N(N – 1)/2$ number 2×2 switches. The loss is not uniform though, with the longest path length N and the shortest $N/2$.

2.10 Wavelength Converter

If the nodes in the network do not have provision for changing the wavelength of the lightpath or λ-*channel* from the input to the output port of the node, the nodes are said to have *wavelength-continuity constraint*, which restricts their performance. In order to improve the throughput of the wavelength routed networks, wavelength conversion can be added along the path in the nodes. This decreases blocking in case of dynamic traffic, increasing overall load carrying ability, and making protection and restoration easier. The wavelength-continuity constraint in the network suffers higher blocking and call drops.

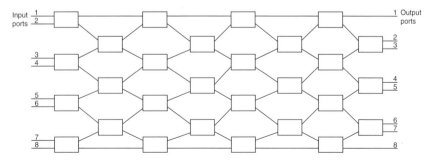

Figure 2.25 Spanke-Benes switching fabric.

The technique to convert the data arriving on one wavelength along a link to another wavelength at an intermediate node and forwarding it along the next link is referred to as wavelength conversion. Wavelength converters are also used for converting the logical signal at 1310–1550 nm to transmit in the fiber layer, or in general, in the network wherever data is to be transmitted on a different wavelength than the incoming wavelength [40, 41].

The characteristics desired for wavelength converters are that they should be fast, have a broad spectrum, should be polarization insensitive, and have transparency of bit rates and signal formats. Wavelength converters can be broadly classified as opto-electronic wavelength converters and all-optical wavelength converters. The opto-electronic wavelength converters are opaque in which the optical signal must first be converted into an electronic signal, and in all-optical wavelength converters the signal remains in the optical domain. Hence, we can have both transparent and non-transparent wavelength converters.

2.10.1 Opto-Electronic Wavelength Converters

Opto-electronic wavelength converters are the most commonly used in optical networks. The O-E-O architecture used in these converters is shown conceptually in Figure 2.26. At the input, it will accept a modulated signal at a wavelength (λ_1) which lies within the range of the detector. The detected output once amplified modulates the continuous wave (CW) laser diode signal in the external modulator and obtains the modulated data signal at the desired wavelength, independent of the input wavelength. This is also known as the 3R approach, which is when the reshaping, retiming, and amplification of the signal are done besides the wavelength conversion. Reshaping, retiming, and reamplification are done in the electronic domain before reconverting it to optical form at wavelength λ_2.

The drawback of the electronic approach is the speed limitation, high power, and higher cost with the increasing data rate. In this 3R converter, signal transparency is lost, but it has the inherent advantage of regeneration of the signal which does amplification, reshaping, and also retiming of the signal.

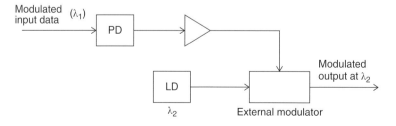

Figure 2.26 Opto-electronic wavelength converter.

2.10.2 All-Optical Wavelength Converters

All-optical wavelength conversion can be of two types: transparent and opaque. The transparent conversion method is typically based on wave-mixing properties. Wave-mixing arises from the nonlinear optical response of a medium when more than one frequency waves are passing through it. It results in the generation of another wave whose intensity ideally is proportional to the product of the interacting wave intensities. Wave-mixing preserves, phase and amplitude information, have fast conversion time, of the order of sub-picoseconds, but the conversion efficiency is poor. The opaque conversion typically depends on electro-optic phenomena in materials, such as strong nonlinearities in semiconductors. In this case the output generated wavelength is independent of the input wavelength. The conversion time is of the order of tens of picoseconds but it consumes less power than the transparent converters.

2.10.2.1 Transparent All-Optical Wavelength Converters

Transparent devices for wavelength conversion are based on FWM [10] or *difference frequency conversion* (DFC), which are produced by nonlinear effects in fibers and optical waveguides and are considered as transparent and coherent as they maintain the phase and amplitude information of the input signal.

FWM is a third-order nonlinearity in silica fibers, which causes three optical waves of frequencies, $f_i, f_j,$ and f_k $(k \neq i, j)$, to interact in a multichannel WDM system to generate a fourth wave of frequency given by $f_{ijk} = (f_i \pm f_j \pm f_k)$. Specifically, when two signals at neighboring optical frequencies ω_1 and ω_2 are passed through a fiber with a third-order nonlinearity, a third frequency $(2\omega_1 - \omega_2)$ will appear very close to the signal frequencies passing through the fiber. This is known as FWM. If we let ω_1 be the pump frequency, ω_p and ω_2 be the signal frequency, ω_s in the fiber, then the converted FWM frequency, ω_c, is equal to $(2\omega_p - \omega_s)$. FWM is also achievable in other passive waveguides and in an active medium, such as an SOA. FWM is a promising technique for wavelength conversion in optical networks owing to its ultra-fast response and transparency to bit rate and modulation format. It can have a 3 dB conversion range of around 50 nm at 1550 nm wavelength with a conversion efficiency of around (−15 dB).

DFC is based on a square law nonlinearity of the fiber. It is also bidirectional and fast, but it suffers from much lower efficiency and has higher polarization sensitivity.

2.10.2.2 Opaque All-Optical Wavelength Converter

The SOA-based gate wavelength converter is one of the most popular opaque systems for wavelength conversion. The main technique using this principle is cross-gain modulation (XGM), using a nonlinear effect in an SOA [10, 42]. Much like with cross-gain modulation , one can use the cross-phase modulation (XPM) approach.

The block diagram of an SOA-based gating device wavelength converter is shown in Figure 2.27a. The data signal at λ_1 is to be converted to λ_2; both are applied at the input of the SOA gate. The wavelength λ_2 is a CW signal, also called the probe signal. The principle behind XGM is that when the high-intensity input data signal passes through the SOA, the carrier density of the amplifier changes. This in turn changes the gain of the amplifier according to λ_1. Hence, the CW probe signal at the wavelength λ_2 gets modulated by the gain variation and carries the same information as the original input data signal. The data pulses at high intensity are able to drive the SOA into saturation and intensity of the λ_2 signal appearing at the output is low. When the input data pulse is low at λ_1, the probe signal λ_2 is amplified by the SOA. Thus, a replica of the data signal appears at the output with inverted polarity at wavelength λ_2. The filter at the output removes the data signal and leaves the gated probe at λ_2 carrying the inverted signal of the data signal.

In the XPM approach, an interferometric gating device is used to convert phase modulation produced by the nonlinear effect in SOA into intensity modulation for wavelength conversion.

Figure 2.27b shows one such wavelength converter using SOA as the nonlinear device with MZI configuration. The input data signal is applied to the SOA in one of the arms of the MZI at wavelength λ_1 and the probe CW signal at λ_2 is applied at the input of the MZI. As a result of XPM, the refractive index of the SOA material changes with the signal strength, which in turn will cause a phase change in the propagating probe signal accordingly. The couplers at the two ends of the MZI have an asymmetrical coupling ratio. Because of the

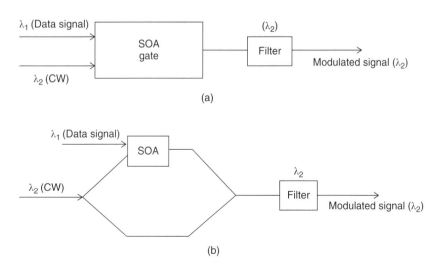

Figure 2.27 (a) SOA wavelength converter based on XGM. (b) SOA wavelength converter based on XPM.

asymmetrical coupling ratio, the phase change induced by the amplifier on the probe signal is different due to the power differential. The MZI translates this relative phase difference between the two arms into an intensity-modulated probe signal at the output.

Though wavelength conversion with FWM has performance advantages in WDM networks because of transparency and hence can handle multiple wavelengths simultaneously, SOAs using XGM/XPM have the advantages of conversion gain and a simple structure.

2.11   Optical Network Functional Blocks

Having discussed some basic optical components, in the following sections we give the details of the different functional blocks or nodes in the WDM optical network architecture. The basic functional blocks in the optical network are the NAT/network access station (NAS)/OLT, OADM, reconfigurable add/drop multiplexer (ROADM), and the OXC.

2.11.1   Network Access Terminal

NATs are deployed at the two endpoints of a fiber connection to provide access to the optical network underneath. The services of the *optical path* layer are used by the NAT to provide the different functions to the end systems attached to the external ports of the NAT.

The block diagram of the NAT is given in Figure 2.28. At the input or on the client side of the NAT there are ports to interface with IP routers or SONET/synchronous digital hierarchy (SDH) switches or other client node, and on the output side it has interfaces with the line ports of the network. Hence, the basic functionality of the NAT is providing adaptability of the

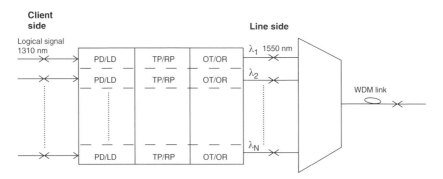

Figure 2.28 Schematic of network access terminal.

client side signals to the line side interface besides providing the traffic grooming, multiplexing/de-multiplexing of the WDM channels, adding overheads for purposes of network management and monitoring, etc. The building blocks of the NAT are the WDM transponders and wavelength multiplexers and de-multiplexers. On the transmission side, the WDM transponder consists of an optical receiver or photodiode (PD), an electrical transmission processor (TP), followed by the optical transmitter (OT), which converts the electrical transmission signals to optical signals at 1550 nm on the line side. Each optical transmitter in the NAT is then connected by the fiber through the MUX to multiplex the wavelength channels of the different logical connections.

On receiving the signal from the line side, after de-multiplexing each individual wavelength passes through the WDM transponder consisting of the optical receiver (OR), followed by the electrical receiver processor (RP). The receiver takes the optical signal from the line side and converts it to a client node compatible signal. The wavelengths generated by the transponder typically conform to standards set by the ITU in the 1550 nm wavelength window, while the incoming signal from the client side may be a 1310 nm signal. The RP conducts all the reverse functions done by the TP and obtains the client side signal to be carried by the logical connections. The NAT also has an optical supervisory channel to monitor the performance of optical line amplifiers along the link, as well as for a variety of other management functions discussed in Chapter 7. Besides this, the transponder typically also monitors the bit error rate of the signal at the ingress and egress points in the network.

In the following paragraphs we explain the functionality of each block of the NAT with reference to Figure 2.28.

Transmission processor (TP): The input ports of the TP are connected through short-reach (SR) interface or very-short-reach (VSR) interfaces to the unidirectional *logical connections* (LC) of the client network nodes and output ports are connected to the optical transmitter. The LCs are optical signal at 1310 nm. These are first converted to the electronic domain for carrying out various processing functions. The TP converts an LC signal to a transmission signal suitable for modulating a laser in OT; for example, the laser diode is modulated by a unipolar signal, therefore TP converts the input electronic bipolar signal to unipolar OOK or to a unipolar subcarrier signal to modulate the laser. These are known as adaptation functions. It may also perform coding and multiplexing functions for electronic grooming. The TP may add additional overheads for network management and monitoring purposes in the network. All the above functions of TP are done in the electronic domain before being converted to the optical domain at 1550 nm. The TP design may vary as per the LSN to which it is connected.

Optical transmitters (OT): The OT may have tuned lasers or array of fixed lasers at different ITU standard wavelengths in the 1550 nm window. These lasers are modulated by the logical *transmission channels* and generate the corresponding λ-channels. The different λ-channels are eventually multiplexed and coupled onto the access fiber as the *optical connection*.

Optical receiver (OR): The OR on the receiving side does the O-E conversion process. The input access pair from the ONN carrying the multiplexed λ-*channels* is de-multiplexed to an individual wavelength channel and each wavelength is passed onto the OR, which converts the channels to the corresponding electrical *transmission channels*.

Receiving processor (RP): RPs carry out the inverse functions of TP. After reshaping and retiming the weak received noisy and corrupted electrical signals, it does the error correction, de-multiplexing, and electrical demodulation and gets back the electrical logical signals. These signals in turn are made client network node compatible at 1310 nm at the interface. The interface between the client and the transponder may vary depending on the client node, bit rate, and distance and/or loss between the client and the transponder.

One thing to note is that like all other functional blocks in the network, such as the OADMs and the OXC, the design and complexity of the NAT are vendor specific but the interfaces provided at the input and output have to be ITU standardized, providing overall compatibility with other vendors' products. Transponders [43] typically constitute the bulk of the cost, footprint, and power consumption in a NAT. Also, considering the time division multiplexing (TDM) and WDM multiplexing schemes in the WDM networks, the number of connections in the network can be enhanced by both means: by increasing the number of channels in the WDM and by number of slots in TDM. This capacity built up in the optical layer is increased with the optical infrastructure, which besides the optical bandwidth of the fiber depends on the number of transponders, tuning of the lasers, and speed of electronics in the NAT or line cards as well.

2.11.2 Optical Network Node

An ONN performs different functions in the optical path layer, such as establishing optical path, routing and bypassing connections, and releasing the connection after the demand is over. Their structure will vary from a simple star coupler to a much complex OXC depending on whether they are used in the access, metro, or large area network segment. The complexity of the nodes also depends on the flexibility they provide to the network, whether they can handle static or dynamic load, have high or low capacity, carry out fault tolerance and monitoring functions, or have other network control and management features. In the case of broadcast & select access networks, the network nodes are all-optical, while in the case of wavelength routing networks they are hybrid, with all signaling and few other operations in the electrical domain and lightpath transmission and switching in the optical domain.

The network nodes can be classified as static and dynamic nodes. The nodes are said to be static or dynamic if the routing and switching of the node can be changed with the change of traffic matrix either in source-destination pairs or with changes in the traffic load. This change is carried out by the switching elements in the node. For the static node, as the connections are permanent and cannot change with traffic demands there is no switching. A typical example of a static node is a passive $(N \times N)$ broadcast star. In the case of static nodes with wavelength selectivity we have to have a wavelength selective $(N \times N)$ wavelength router which has a $(1 \times N)$ wavelength DMUX at the input port, followed by fixed wavelength connectivity to the N input ports of a $(N \times 1)$ wavelength MUX at the output port, as given in Section 2.8.

A dynamic node basically comprises dynamic wavelength space switches (*WSSs*). The WSSs, as given in Section 2.8.2, also known as λ-space switches, have wavelength selectivity added in the basic switch. They can be of a variety of types, such as transparent or translucent, or can be opaque. The basic $(N \times N)$ WSS has a three-stage configuration. The first and third stages have N number $(1 \times N)$ de-multiplexers and N number $(N \times 1)$ multiplexers, respectively. The center stage can be of many variants. A simpler one (given in Figure 2.19) has N number $(N \times N)$ dynamic switching planes. Each plane operates for a particular wavelength $\lambda_1, \lambda_2 \dots \lambda_N$ from the N wavelengths on each input fiber. Thus, each plane is wavelength insensitive. These switches are one of the previously discussed switching fabrics in Section 2.9. After the 1st stage de-multiplexes the individual wavelengths and sent to the input of the switches which in turn switches the different wavelengths to the MUX on different fibers. These types of WSS are used when wavelength continuity is required for routing the *lightpath*, as there is no wavelength converter available in the WSS.

2.11.2.1   Optical Add-Drop Multiplexers

OADMs are ONNs that provide capability to add and drop traffic in the wavelength routing network. They are located at sites in the network link where it drops and adds a number of wavelength channels at destination and source nodes, respectively, while enabling the rest of the lightpaths or the λ-channels a through pass at the intermediate nodes, thus reducing the number of unnecessary opto-electronic conversions without affecting the traffic that is transmitted transparently through the intermediate node. OADMs can thus increase capacity rate and extend the optical reach in a backbone network. An OADM has fewer line-port counts and can be used in linear, ring, or mesh network architectures. Most practical OADMs use FBGs, dielectric TFF, or AWGs for its operation.

The architecture of an OADM will depend on many attributes [44, 45], such as number of ports $(N \times N)$, number of wavelengths supported, number of wavelengths that can be added or dropped, constraints on the wavelength that can be added or dropped, i.e. if the add drop of the wavelength is service hit or not, modular architecture so that you pay as you grow, complexity of physical path design, i.e. how the adding/dropping of channels changes the characteristics of the OADM, and whether it is reconfigurable or not.

Fixed Optical Add-Drop Multiplexers   The OADM can operate in either fixed or reconfigurable mode. In fixed OADMs, the add/drop and through channels are fixed or predetermined and cannot be rearranged after installation, except by changing the connections between ports manually. In reconfigurable OADMs both through and added/dropped channels of the node can be dynamically reconfigured and connections can be set up on the fly. Besides the add/drop ports with transponders, the OADM has basically a WSS structure.

The possible architectures of fixed *series* and *parallel* OADM are illustrated in Figure 2.29. In parallel architecture, predefined fixed wavelengths are dropped in one stage, and in series architectures the drop of individual wavelength is in multiple stages.

There are certain advantages and disadvantages in each case. With parallel OADMs, the through loss is fixed and is not dependent on the number of channels added or dropped, unlike in the serial architecture. But the design is cost-effective only when large numbers of channels have to be dropped; otherwise, one has to de-multiplex all the channels even if they are not to be dropped but have to be bypassed. This adds more losses due to extra multiplexer and de-multiplexer. The serial design, meanwhile, can take care of a few of these shortcomings but at the cost of reduced flexibility. A single channel is dropped and added from an incoming set of channels. The adding and dropping of any channel will affect existing channels. It is useful where a small number of

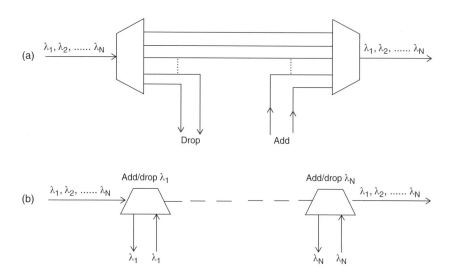

Figure 2.29 Fixed OADM architecture: (a) parallel, (b) serial.

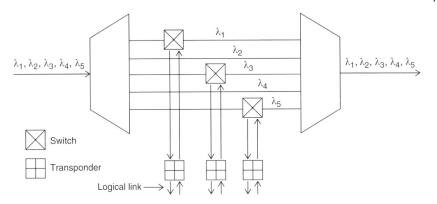

Figure 2.30 Partially flexible OADM in parallel architecture.

channels have to be dropped, otherwise cost can be quite significant because a number of devices have to be cascaded in that case.

The parallel design given in Figure 2.30 is a partially flexible one, as an arbitrary set of channels can be added or dropped and the rest can pass through. Hence, at the planning stage of the network, fewer design constraints are required.

There can also be a modular architecture in which the multiplexing and de-multiplexing are done in two stages. The first stage of de-multiplexing separates into wavebands and the second stage separates the bands into individual channels. Both serial and parallel architecture OADMs can have band dropping/adding instead of individual λ-channels.

Reconfigurable Optical Add-Drop Multiplexers   Reconfigurability in OADMs or any other network device (OXC, WSS, etc.) adds flexibility during the design and planning stages of the network. It allows dynamic connections to be set in the network so that any channel can be added or dropped when the network is in operation. The reconfigurability is added in the OADMs by the optical switches. Reconfigurable OADMs can be partly reconfigurable and also be fully reconfigurable. As the name indicates, in partly reconfigurable architectures part of the select channels can be added/dropped which are reconfigurable, i.e. any of them can be added/dropped or bypassed, but there is also a predetermined connectivity matrix between add/drop and through ports. These OADMs are not completely flexible [45] as the number and the particular channel to be added or dropped are fixed and planned at the design stage. This rigidity limits the ability of the network to adapt to changing traffic patterns. Figure 2.30 showed the partially flexible ROADM with the restriction of adding/dropping only certain wavelengths. The transponders are not tunable and are of fixed single wavelength. Fully reconfigurable OADMs provide

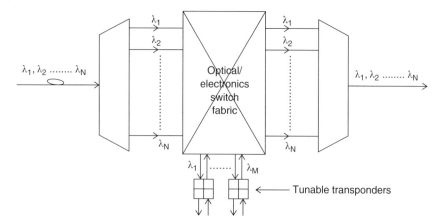

Figure 2.31 Fully reconfigurable OADM architectures.

complete flexibility, with all channels having the ability to be added/dropped or bypassed. They offer connectivity between add/drop and through ports, which enables flexible wavelength assignment with the use of tunable transmitters and receivers, as shown in Figure 2.31. The ROADMs can have parallel and serial architectures as in the case of fixed OADMs and can also have modular structure architectures with wavebands instead of wavelengths, as discussed earlier. Thus, the choice of add/drop wavelengths can be changed easily without affecting any of the other connections terminating at or transiting the nodes. Furthermore, these OADMs are normally remotely configurable through software.

2.11.2.2 Optical Cross-Connect

OXC is the basic module for routing optical signals in any large optical network with large port counts. In DWDM networks which serve multiple client nodes, have high traffic load, and require several connections, OXCs with large port count are used. The basic functions carried out by an OXC are bypassing traffic and adding/dropping channels, much like the ROADMs, with some ports connected to WDM equipment and others to terminating devices such as SONET/SDH ADMs, IP routers, or OTN switches. The other key functions carried out by the OXC are automatic service provisioning with remote reconfigurability, and performance monitoring, fault localization, and diagnostic can be done remotely. Besides multiplexing, OXCs have traffic grooming and wavelength conversion capability. They have sufficient reliability, protection, and fault tolerance capability added with enough intelligence in the switch. Thus, unlike OADMs, they perform many more complex functions besides the bypass and add-drop functions.

To support flexible path provisioning and network resilience, OXCs basically have two parts: one is the core switching fabric, discussed earlier, to enable routing of incoming wavelength channels from different input fibers to the appropriate output ports, and the second is a port housing to house port cards that are used to interface the core of the switch to communicate with the external line or the equipment. There are a number of OXCs available with different architectures based on different switching technology. The switching technology functions may be carried out either optically or electronically and accordingly we have the transparent, translucent, or opaque OXCs. These are illustrated in Figure 2.32a–c. An OXC based on O-E-O technology is shown in Figure 2.32a. The switch fabric or core is electronic, and each of the switch ports is equipped with a short-reach (O-E) interface (transponder) to convert the incoming 1310 nm optical signal to an electronic signal and vice-versa. These switching fabrics are high on cost and heat dissipation due to electronics used for switching. Each port has to operate at the speed of a λ-channel. Therefore, this type of OXC is power consuming, it is opaque, expensive, and more complex for high bit rate of the data channels of 40 Gbps or 100 Gbps. Nevertheless, the O-E-O architecture also offers some advantages. With electrical switching, sub-wavelength switching granularities can be supported by providing grooming capabilities for more efficient bandwidth utilization. Opaque OXCs also offer inherent regeneration, wavelength conversion, and bit-level monitoring.

Figure 2.32b illustrates the photonic or translucent switch in which the switch core is optical [46]. MEMS optical technology is often used to build such photonic switches. There are WDM transponders on both the input and output fibers side, which convert the incoming 1550 nm line signal first to electronics and then back to optics at 1310 nm for the core, as the switching is carried out at 1310 nm. A photonic switch configuration requires significantly less electronics. The other advantage of this opaque switch is that processing and wavelength conversion are available; it can, therefore, act as a wavelength interchange cross-connect. Here the optical switch fabric has ($MN \times MN$) ports but the transponder makes the OXC opaque and optical bypass is not supported.

Finally, in the case of (O-O-O) or all-optical OXC, optical bypass can be supported. The signals enter the switch fabric in optical form, have switching in optical form, and exit also in optical form [46, 47]. Therefore, all processing in the switch is in the optical domain. The number of ports on the switch can be large. In the case of (O-O-O), the switch fabric of the OXC has M ($N \times N$) switches, where M is the number of fibers each carring N wavelengths. The incoming signals are routed through an optical switch fabric without the requirement of opto-electronic conversions, thereby offering total transparency to bit rates and protocols. The switching granularity may vary and support switching at the fiber level, the wavelength band level, or the wavelength channel level.

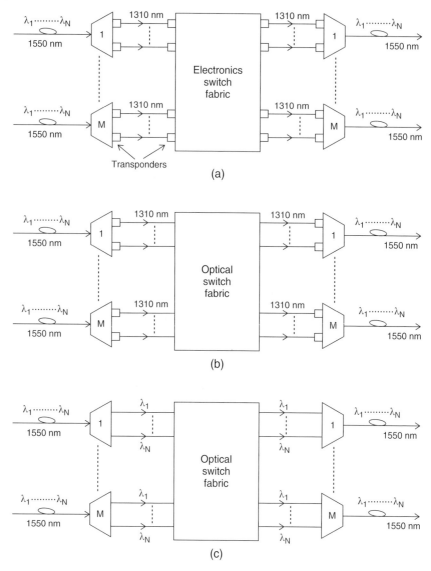

Figure 2.32 (a) OXC with OE-E-EO opaque configuration with electronic switching fabric. (b) OXC with OEO-O-OEO translucent configuration with optical switching fabric. (c) OXC with O-O-O configuration with transparent switching fabric.

Figure 2.33 gives an architecture of an OXC with 4 degree ports. Similar to Figure 2.32, where few ports allow signals to be switched in the optical domain and few ports can allow the signals to be routed down to the electrical layer which can provide the advantage of low data-rate grooming, O-E-O

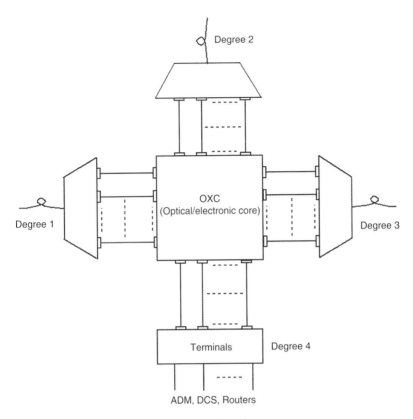

Figure 2.33 A typical OXC network node with 4° of port connections.

wavelength conversion, and 3R regeneration. The architecture minimizes the cost and maximizes the capacity of the network.

2.12   Summary

In this chapter, we have provided a brief overview of the enabling optical devices and components used in optical networks, as well as some insight into the technology used in their fabrication. We have studied many different optical components. With a very brief introduction to the fiber, commonly used lasers, and photodiodes in communications systems and optical amplifiers, we have concentrated more on the optical network devices. Both the static and reconfigurable devices used have been discussed. The couplers, optical filters, fiber gratings, thin-film multi-cavity filters, switches, and AWGs are the basic components used in commercial WDM systems. There is a variety of technologies available to build switches which are an integral part of any reconfigurable device, for example variable filters, routers, etc. MEMS-based optical switches are used for moderate or large port count switching fabrics. In the last section, details were given of the ONNs, including ROADMs, OXCs, and NATs, using the devices discussed above.

Problems

2.1 A 16×16 star coupler is made of 2×2 couplers with insertion loss of 0.5 dB for distribution of signals between the connected nodes. If each receiver at the node requires a minimum of 100 nW power, calculate the transmitter power at each input node.

2.2 In an optical communication system, if the transmitter has a tuning range from 1500 to 1600 nm and the receiver has a tuning range from 1550 to 1650 nm, how many 2.5 Gbps channels can be accommodated in the system if 10% guard-band is maintained between the channels?

2.3 (a) Determine the transfer matrix T, of a 2×2 non-reciprocal isolator with no power loss. The transfer matrix is governed by the equation $E_{out} = T E_{in}$. E is a column vector representing the input/output fields at the two ports.

(b) Design a 1×8 de-multiplexer using a 2×2 MZ interferometer that can hand handle a channel separation of 50 GHz. Let the shortest λ be 1550 nm; determine the differential length in each stage assuming n_{eff} to be 1.5.

2.4 (a) Prove that the finesse of a Fabry–Perot filter is given by $F = \pi \sqrt{R} / (1 - R)$, where R is the reflectivity of the two mirrors. Also derive the expression for the transmission spectrum of the filter considering multiple round trips inside the cavity filled with air.

(b) A FP filter is used to select 50 channels spaced apart by 0.4 nm. Determine the length of the filter cavity and the mirror reflectivity, assuming the refractive index to be 1.5 and the operating wavelength to be 1550 nm.

2.5 Show that when a $1 : N$ optical splitter is operated in the opposite direction, it is an $N : 1$ combiner with a combining loss of $1/N$.

2.6 There are three different 4×4 port devices: a 4×4 passive star with λ_1, λ_2, λ_3, and λ_4 wavelengths entering from input ports 1, 2, 3, and 4, respectively, a 4×4 passive router for 4 wavelengths, and a 4×4 wavelength space switch of four wavelength, all entering from each fiber. Tabulate the devices that can support the following simultaneous connections in the three cases. Explain your choices as well.

(a) Wavelength λ_1 from input fiber 1 to output fiber 2, Wavelength λ_2 from input fiber 2 to output fiber 1, Wavelength λ_3 from input fiber 3 to output fiber 1.

(b) Wavelength λ_1 from input fiber 1 to output fiber 1, Wavelength λ_2 from input fiber 2 to output fiber 1, Wavelength λ_3 from input fiber 1 to output fiber 3.

(c) Wavelength λ_1 from input fiber 1 to output fiber 1, Wavelength λ_1 from input fiber 1 to output fiber 2, Wavelength λ_2 from input fiber 2 to output fiber 1.

2.7 (a) Prove that an $N \times N$ Benes switch uses $N \log_2 N - N/2$ binary elements.

(b) In an 8×8 Benes switching fabric, the access request is for $\{1 \rightarrow 2, 2 \rightarrow 8, 3 \rightarrow 4, 4 \rightarrow 6, 5 \rightarrow 1, 6 \rightarrow 5, 7 \rightarrow 3, 8 \rightarrow 7\}$. The insertion loss for each switch is 10 db for the bar state and 20 db for the cross state. Show the matrix of the switches for making the above connections and determine the insertion loss for each connection. Compare the performance of the same connections in the case of the crossbar switching fabric.

(c) Design a 16×16 strict-sense non-blocking *Clos* switch with $P = K = 4$. Determine the parameters of the switch and the number of 2×2 switches used. Compare this switch fabric with a 16×16 Benes switch, in terms of cross points, number of switches used and their blocking characteristics. Substantiate with an example.

2.8 Design a three-node link with 10 wavelengths at 0 dBm each entering the fiber. At the first node 5 wavelengths are dropped, at the 2nd node 1 wavelength is dropped and another is added at 0 dBm, and at the 3rd node two wavelengths are dropped. You have OXC, series and parallel OADMs available to select from for the design. Discuss the choice of components selected in terms of the cost and link performance. What will the power be at the end of the link? Assume the fiber to have no attenuation and the devices to be lossless.

2.9 Consider an all-optical 128-port OXC with each WDM line system carrying 64 wavelengths; 75% of the light-paths pass through the node, while the remaining 25% are dropped and also added to the OXC. Each light path added and dropped onto a router takes up two ports of the OXC. Calculate the number of WDM line systems that the OXC can support.

References

1 Fujiwara, M., Dutta, N.K., and Dutta, A.K. (2002). *WDM Technologies: Active Optical Components*. Academic Press.
2 Dutta, A.K. and Dutta, N.K. (eds.) (2003). *WDM Technologies: Passive Optical Components*, vol. 2. Academic Press.
3 Ching-FuhLin (2004). *Optical Components for Communications: Principles and Applications*. Springer Science & Business Media.
4 Ramaswami, R., Sivarajan, K., and Sasaki, G. (2009). *Optical Networks: A Practical Perspective*, 3e. Morgan Kaufmann.
5 Stern, T.E. and Bala, K. (1999). *Multiwavelength Optical Networks*. Reading, MA: Addison Wesley.
6 Mukherjee, B. (2006). *Optical WDM Networks*. Springer.
7 Keiser, G. (2008). *Optical Fiber Communications*, 4e. Tata-McGraw-Hill.
8 I. Tomks, E. Palkopoulou, M. Angelou (2012). "A survey of recent developments on flexible/elastic optical networking," *14th International Conference on Transparent Optical Networks (ICTON), 2–5 July 2012, UK*, https://doi.org/10.1109/ICTON.2012.6254409
9 Napoli, A. and Lord, A. (2016). Elastic optical networks: introduction. *IEEE/ OSA J. Opt. Commun. Networking* 8 (7).
10 Agarwal, G.P. (2003). *Fiber-Optic Communication System*, 3e. Wiley-Interscience.
11 *ITU-T Recommendation-G.694.1 for DWDM frequency grid*
12 Agarwal, G.P. (2013). *Nonlinear Fiber Optics*, 5e. Elsevier, Academic Press.
13 Morthier, G. and Vankwikelberge, P. (2013). *Handbook of Distributed Feedback Laser Diodes*, 2e. Artech House.
14 Blumenthal, D.J., Buus, J., and Amann, M.-C. (2005). *Tunable Laser Diodes and Related Optical Sources*. Wiley-IEEE Press.
15 Wilmsen, C.W., Temkin, H., and Coldren, L.A. (2001). *Vertical-Cavity Surface-Emitting Lasers: Design, Fabrication, Characterization, and Applications*. Cambridge University Press,.
16 Seimetz, M. (2009). *High-Order Modulation for Optical Fiber Transmission*. Springer.
17 Personick, S.D. (2008). Optical detectors and receivers. *J. Lightwave Technol.* 26 (9): 1005–1020.
18 Kikuchi, K. (2016). Fundamentals of coherent optical fiber communications. *J. Lightwave Technol.* 34 (1): 157–179.
19 Latinovic, V. (ed.) (2015). *Essentials of Optical Amplifiers*. Clanrye International.
20 Desurvire, E. (1994). *Erbium-Doped Fiber Amplifiers: Principles and Applications*. New York: Wiley.
21 Connelly, M.J. (2002). *Semiconductor Optical Amplifiers*. Kluwer Academic Publishers.

22 S. Namiki and Y. Emori, "Recent advances in ultra-wideband Raman amplifiers", *Optical Fiber Communication Conference (OFC)*, https://doi.org/10.1109/OFC.2000.869427, 2000.

23 Maier, M. (2008). *Optical Switching Networks*. Cambridge University Press.

24 Ai-QunLiu (2009). *Photonic MEMS Devices: Design, Fabrication and Control*. CRC Press.

25 Renaud, M., Bachmann, M., and Erman, M. (1996). Semiconductor optical space switches. *IEEE/OSA J. Sel. Top. Quantum Electron.* 2 (2): 277–288.

26 E. J. Murphy, T. O. Murphy, R. W. Irvin, R. Grencavich, G. W. Davis, and G. W. Richards, "Enhanced performance switch arrays for optical switching networks", *Proceedings of the Eighth Euro. Conf. on Integrated Optics, Paper EFD5, Sweden*, 1997.

27 Tanaka, S., Jeong, S.-H., Yamazaki, S. et al. (2009). Monolithically integrated 8:1 SOA gate switch with large extinction ratio and wide input power dynamic range. *IEEE J. Quantum Electron.* 45: 9.

28 Lin, L.Y., Goldstein, E.L., and Tkach, R.W. (1998). Free-space micro-machined optical switches with sub-millisecond switching time for large-scale optical cross-connects. *IEEE Photon. Technol. Lett.* 10 (4): 525–528.

29 Vázquez Garcìa, C., Pérez Garcilópez, I., Contreras Lallana, P. et al. (2010). Liquid crystal optical switches. In: *Optical Switches – Materials and Design* (ed. S.J. Chua and B. Li). Woodhead Publishing.

30 B. B. Dingel, K.-I. Sato, W. Weiershausen, A. K. Dutta, *Optical Transmission Systems and Equipment for Networking*, 2006.

31 McGreer, K.A. (Dec. 1998). Arrayed waveguide gratings for wavelength routing. *IEEE Commun. Mag.* 36 (12): 62–68,.

32 Koonen, M.J., Smit, M.K., Herrmann, H., and Sohler, W. (2001, pp. 262–292.). Wavelength selective devices,. In: *Devices for Optical Communication Systems* (ed. H. Venghaus and N. Grote). Heidelberg: Springer-Verlag.

33 Cheung, K.-W. (1990). Acousto optic tunable filters in narrowband WDM networks: system issues and network applications. *IEEE J. Sel. Areas Commun.* 8 (6): 1015–1025.

34 Simmon, J.M. (2014). *Optical Network Design and Planning*. Springer.

35 El-Bawab, T.S. (2006). *Optical Switching*. Springer.

36 Tawfik, I. (2014). *Advanced Optical Packet Switches over WDM Networks*. LAP Lambert Academic Publishing.

37 Jue, P., Yang, W.-H., Kim, Y.-C., and Zhang, Q. (2009). Optical packet and burst switched networks: a review. *IET Commun.* 3 (3): 334–352.

38 Clos, C. (1953). A study of nonblocking switching networks. *Bell Syst. Tech. J.* 32: 406–424.

39 Spanke, R.A. and Benes, V.E. (1987). An n-stage planar optical permutation network. *Appl. Opt.* 26.

40 Strand, J., Doverspike, R., and Li, G. (2001). Importance of wavelength conversion in an optical network. *Optical Networks Magazine* 2 (3): 33–44.

41 Yoo, S.J.B. (1996). Wavelength conversion techniques for WDM network applications. *J. Lightwave Technol.* IEEE/OSA JLT/JSAC Special Issue on Multiwavelength Optical Technology and Networks 14 (6): 955–966.

42 Daikoku, M., Yoshikane, N., Otani, T., and Tanaka, H. (2006). Optical 40 Gbps 3R regenerator with a combination of the SPM and XAM effects for all-optical networks. *J. Lightwave Technol.* 24: 1142–1148.

43 Scavennec, A. and Leclerc, O. (2006). Towards high-speed 40Gbps transponders. *Proc. IEEE* 94: 986–996.

44 Lam, C.F., Frigo, N.F., and Feuer, M.D. (2001). A taxonomical consideration of optical add/drop multiplexer. *Photon. Network Commun.* 3 (4): 327–334.

45 Kataoka, N., Wada, N., Sone, K. et al. (2006). Field trial of data-granularity-flexible reconfigurable OADM with wavelength-packet-selective switch. *J. Lightwave Technol.* 24: 88–94.

46 Papadimitrion, G.I., Papazoglou, C., and Pomportsis, A.S. (2003). Optical switching: switch fabrics, techniques and architectures. *J. Lightwave Techol.* 21: 384–405.

47 Kabacinski, W. (2005). *Nonblocking Electronic and Photonic Switching Fabrics.* Springer.

3

Broadcast-and-Select Local Area Networks

3.1 Introduction

Broadcast-and-select (B&S) all-optical WDM networks are essentially transparent optical networks based on *static* optical network nodes. The passive splitting/combining optical nodes have no wavelength selectivity. Any signal entering the ONN distributes equally to all the links connected to the node. It floods the network without any optical spectrum selectivity. The advantage of B&S networks is in their simplicity and natural *multicasting* capability to transmit a message to multiple destinations. Also, the optical node being passive, it has higher reliability and costs much less compared with its active counterparts. But because there cannot be any *wavelength reuse* possible in the transparent network, they have the limitation of requiring a large number of wavelengths to increase the number of connections, typically as many as there are end nodes connected to the network. Thus, the networks are not scalable beyond the number of wavelengths supported by it. The other drawback is that they cannot span long distances since the transmitted power is split among all the connected end nodes and each station receives only a fraction of the transmitted power, which becomes smaller as the number of connected stations increase. For these reasons, the main application for broadcast-and-select is in high-speed local area networks (LANs) and access networks. These networks can be single-hop and multi-hop. Single-hop networks have no intermediate relaying nodes. The transmitting node can transmit on any one of the WDM channels and the signal can be received by any other node on the same wavelength. Multiple-hop requires intermediate relaying nodes and packets may have to travel through several nodes. Therefore, single-hop B&S networks have very low latency. We will be discussing single-hop B&S networks only in the present text.

In Section 3.2 we discuss the different physical topologies used in single-hop B&S networks. Section 3.3 gives the different multiplexing and access techniques followed in these networks. In Sections 3.4 and 3.5, we first

Optical WDM Networks: From Static to Elastic Networks, First Edition. Devi Chadha.
© 2019 John Wiley & Sons Ltd. Published 2019 by John Wiley & Sons Ltd.

introduce details of the general concepts of different types of traffic and connections in optical networks, followed by the resources and capacity sharing concepts in the underlying shared resources. With these general discussions of the traffic, capacity sharing, etc. in Section 3.6, specifically, these quantities for B&S single-hop networks are calculated. Section 3.7 gives details of packet switching in the optical layer, and finally the media access protocols used in B&S networks for their smooth and efficient working in Section 3.8 are summarized.

3.2 Physical Topologies of Single-Hop Networks

The general physical topologies of the B&S single-hop network are the *star*, *folded bus*, and *tree* or network of *connected trees*. The general representation of all of them can be considered to be a broadcast star, though each has its own advantages and shortcoming.

3.2.1 Star Topology

In the B&S star topology the information transmission from the sender to the recipient flows through the central passive optical node: the funneling point, to all the other connected nodes including the sender. An $N \times N$ star coupler is shown in Figure 3.1a. The signal entering from any input at a wavelength is equally divided to all the N connected nodes by the $N \times N$ passive star coupler. In the case of WDM systems, simultaneous multiple connections are possible with optical spectrum sharing by sending signals at different wavelengths from the input ports which are combined in the coupler and then broadcasted to all the ports. The destined receiver at the nodes can tune to the appropriate wavelength to receive the required signal. The passive star coupler, which is the ONN in this case, is all-optical and is static, therefore all functionalities or control required for network connection reside either in the *user network node* in the electronic layer or in the *network access terminal* (NAT). The multiple connections are made with wavelength multiplexing in the *optical layer* with respect to the hierarchical network layered structure and if time division multiplexing (TDM) is also used in order to increase the number of connections; in that case the multiplexing is carried out in the electronic end nodes in the *logical layer*. Therefore, the performance of the network, such as throughput, delay, and logical connectivity, will depend on the NAT and end nodes. The NAT has multiple optical transmitters and receivers for providing multiple WDM connections and electronic processors for providing the required processing. B&S single-hop networks used in LANs or in the access networks have gained much importance due to their use in the access network as *passive optical networks* (PONs).

The $N \times N$ star coupler can be made with multiple technologies as discussed in Chapter 2. The loss in dB between two connected stations in this case is the splitting and the excess loss as expressed below:

$$L_{star} = 10 \log N + 2L\alpha_p + 10 \log l_{ex} \tag{3.1}$$

where l_{ex} is the excess loss of the star coupler and α_p the attenuation coefficient in dB km^{-1} of the fiber connected to the each station L (km) away from the star coupler. The splitting loss in a star coupler thus increases logarithmically with the number of ports as the power gets equally divided in all the ports.

3.2.2 Folded Bus Topology

The folded bus topology is shown in Figure 3.1b with N nodes. Each node is connected by a T-coupler to the fiber line of the bus. The network of N nodes will therefore require $2N$ T-couplers. The loss on the bus with N number of nodes between the transmitting and the receiving stations will vary from the closest node to the farthest node. The worst case is when the first node communicates with the last node. The worst-case loss is, therefore:

$$L_{bus} = l_{ex}^{(2N-1)} . l_{cp}^2 . \left(1 - l_{cp}\right)^{2N-3} \tag{3.2}$$

The first factor in Eq. (3.2) is the excess loss between the first and the Nth node of the bus with l_{ex} as the excess loss of each T-coupler. The second factor is the coupling loss with l_{cp} as the coupling loss coefficient of a coupler, and the last factor is the through-power loss in the coupler. The optimum value of the coupling factor, l_{cp}, that minimizes L_{bus} is equal to $2/(2N-1)$ and the value of loss in dB for this optimum value for large N is:

$$L_{bus}^{opt} \approx 8.7 + 20 \log_{10} N + \left(2N - 1\right) 10 \log_{10} l_{ex} \tag{3.3}$$

From Eqs. (3.1) and (3.3) we observe that the excess loss in the star grows logarithmically with N while in the bus topology it increases linearly. Therefore, the nodes in the bus are restricted to a much smaller number compared with the star topology. But the bus topology has the capability to sense the activity of transmission from the upstream transmission and thus can avoid packet collision.

3.2.3 Tree Topology

The *tree topology*, as shown in Figure 3.1c, is essentially a connected star topology growing from the root to the different branches. Star is the generic topology for the single-hop B&S networks. Both bus and tree topologies can be derived from the star. Henceforth in the chapter we will be discussing the B&S single-hop network with respect to star topology only.

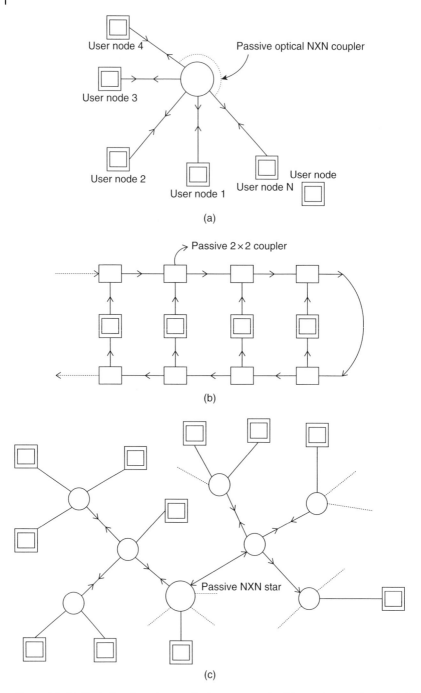

Figure 3.1 (a) Single-hop broadcast and select star coupler. (b) Broadcast-and-select folded bus topology. (c) Tree topology.

3.3 Multiplexing and Multiple Access in B&S Networks

Multiple transmissions of different logical connections and the access of these connections to/from the B&S shared-medium network is possible with various multiplexing schemes, the TDM/WDM/CDM and WDMA/TDMA/CDMA, respectively. Each user network node is connected by the NAT to the passive star coupler. The NAT is equipped with one or more fixed-tuned or tunable optical transmitters and one or more fixed-tuned or tunable optical receivers. The star coupler combines the received messages from the source nodes and broadcasts the combined messages to all the end nodes. The destination receiver receives the desired message by tuning it to the particular wavelength. In order to enhance the number of connections and utilize the spectrum capacity efficiently, instead of providing a full wavelength spectrum capacity of 2.5 GHz or more for a single connection, the channel capacity is time shared with TDM/TDMA for more connections. Similarly, one can use multiple code division multiplexing (CDM) and code division multiplexing access (CDMA) schemes as well in order to increase the number of connections in these networks.

In terms of the functional hierarchy of the layers, as the network node is passive in the case of static B&S networks, *optical path* cannot be selected by the ONN. Therefore, no control is possible for the connection by the *optical path layer*. The connection control is only by the *λ-channel* or *transmission channel* sub-layers, which can have signaling with the end node NATs. All the receivers can receive signals from all the transmitters. The receiver makes a point-to-point single-hop connection with any transmitter by tuning it to the transmitting wavelength and in the designated time-slot in the case of time division multiple access (TDMA), or accessing the required signal in the case of CDMA with the help of the receiver processor. This obviates the need for a separate optical network manager for making the connection.

Static B&S networks can be classified by the number of transmitters and receivers and whether the transmitters and receivers are tunable or not. The NAT, therefore, can have either a number of *fixed transmitters* and *fixed receivers* (FTx^i-FRx^m), or a number of *tunable transmitters* and *fixed receivers* (TTx^j-FRx^m), or a number of *fixed transmitters* and *tunable receivers* (FTx^i-TRx^n), or a number of *tunable transmitters* and *tunable receivers* (TTx^j-TRx^n).

3.4 Network Traffic

As discussed above, the B&S optical networks have common resources, such as the network nodes and channels, which are shared among the logical nodes and ONNs for providing connections to the upper client layers. The different traffic and connection types provided to networks are the same in the case of both B&S and wavelength routing networks (WRNs). We, therefore, step back here and first give the different models of the traffic and the connections provided to the source-destination pairs in the optical networks, in general, before we discuss specifically the B&S network traffic constraints.

There can be two types of traffic flows on the logical connection: the *synchronous stream* type and the *asynchronous packet* or *cell* traffic. Stream traffic typically supports voice, fixed bit rate video, and other real-time applications supported by Synchronous Optical NETworking (SONET) and synchronous digital hierarchy (SDH) services. In circuit switched services, the connection path is established before the traffic actually flows over it. In the case of asynchronous bursty traffic, the flow of information is carried by random length and rate of data packets or cells. Each packet or cell has its routing addressing information to its destination in the header. Therefore, the path is made while the traffic flows. The asynchronous bursty packet, viz., Internet Protocol (IP) type traffic, can also be supported by synchronous circuit-switch services of SONET or SDH like the stream traffic. In this case, the random sequences of packets separated by idle characters are encapsulated into synchronous data frames. For example, when a sequence of asynchronous transfer mode (ATM) cells or any asynchronous bursty data traffic has to be carried on the SONET connection, the data cells are carried within a synchronous transport signal (STS) frame at intervals with idle cells in between them. In this case, though the cells or the packets are asynchronous, the slotted frame stream carried by the SONET/SDH frame on the optical network is synchronous. The logical connection of this encapsulated bursty traffic is, therefore, a continuous synchronous stream and the *transmission channel* is once again modeled as a fixed capacity pipe, as in the case of synchronous stream traffic.

Another way of distinguishing the traffic to be synchronous or asynchronous is whether the optical nodes treat the traffic information in the data packets transparently or not. If the packets are processed transparently, without reading the header of each packet in the frame, it is considered to be stream type. If the individual packets headers are read for the purposes of scheduling, routing, etc., the traffic is considered to be asynchronous. The former is the circuit-switched synchronous, while the latter mode of operation corresponds to packet- or cell-switching within the optical layer.

In the case of WRNs, the packet or burst switching is carried out at the ONNs in the fiber path-layer level with the help of optical management and control system. In the case of B&S networks, as the optical node itself is static and transparent, this requires transmission and reception processors in the NAT to execute a media access control (MAC) protocol designed for sharing one or more λ-*channels* dynamically among many logical connections. Although synchronous traffic is always *connection oriented*, asynchronous traffic may be either connection oriented or *connectionless*.

3.4.1   Circuit-Switched Traffic

The network traffic demand is defined as the number of connections requests received by the network at any point in time. These demands can be in synchronous streams or asynchronous packet traffic as discussed above. We first discuss the circuit-switched traffic demands, both streamed and packet. The circuit-switched traffic can require either a *dedicated* connection or a *demand switched* connection. Dedicated connections are assumed to be held for a relatively long period of time and are pre-decided as the complete set of dedicated connections is known. Demand-assigned switched connections requests occur as *random sequence* and are established and released on demand, with demand *holding times* varying from a few minutes or less.

In the case of WRN, the requests for switched connections are made with the help of the network control and management system of the network as they occur. When a demand is received, the network management system (NMS) decides to accept or block the request. It also decides the routing and channel assignment for the request once the request is accepted. But in the case of B&S networks, as the optical nodes are passive and can have no wavelength selectivity, all the switched connections decisions are taken by the transmission/receiver processors at the NAT, which may be based on various considerations, including the current network load, pattern of currently active connections, etc.

In the following sections, we will discuss in brief all types of possible connections in optical networks in general.

3.4.1.1   Streamed Synchronous Traffic on Dedicated Connections

The traffic on the network with M sources and N destinations client-pairs can be represented by an $M \times N$ traffic matrix. Assuming the "fluid flow" model, the traffic matrix for the $M.N$ streamed synchronized traffic flow between the source-destination pairs is expressed as:

$$\left[\Gamma\right]_{MxN} = \left[\gamma_{ij}\right]_{MxN} \tag{3.4}$$

where γ_{ij} (bits/sec) is the synchronous traffic flow rate from source i to destination j on the logical connection. Therefore, the total *traffic* or *demand* on the network is the sum of all the flows, as:

$$\bar{\gamma} = \sum_{ij} \gamma_{ij} \tag{3.5}$$

In the case of static dedicated connection, the traffic will be represented by a static matrix with all entries to be constant, while in the case of dynamic traffic the entries in the matrix will change with time. We can represent the matrix as:

$$\bar{\gamma} = \begin{bmatrix} \gamma_{11} & \cdots & \gamma_{1N} \\ \cdot & & \cdot \\ \cdot & \cdot & \cdot \\ \cdot & & \cdot \\ \gamma_{M1} & \cdots & \gamma_{MN} \end{bmatrix} \tag{3.6}$$

The traffic matrix can also be normalized and expressed as:

$$[\Gamma] = R_0[T] \tag{3.7}$$

where T is the normalized traffic matrix and R_0 is the basic bit rate of the data. Thus, the total normalized traffic for fixed stream can now be expressed as:

$$\bar{T} = \sum_{ij} t_{ij} \tag{3.8}$$

where t_{ij} is the normalized traffic from source i to destination j.

The different traffic connections can be carried over the WDM network with TDM and WDM multiplexing on a wavelength or a *lightpath*. When all the demands are met or accepted by the network, it is said to have *zero blocking*. For the network to have zero blocking, the network capacity should match this total traffic in order to have no connection demand dropped. The network capacity or the total numbers of possible orthogonal flows which are carried by the network are built up by multiplexing the traffic either on WDM channels and/or in the TDM slots. Hence, for the total number of wavelengths (*λ-channels*) to be W in the network and with total S-time-slots in TDM fixed frame, the total connections capacity is the product of W and S. This capacity is fully utilized for the dedicated connections with synchronous traffic with no packet drop when it equals the total traffic demand. The synchronous traffic on dedicated connections is fixed and deterministic, therefore the network capacity required to carry it can be fixed corresponding to exactly the quantity of traffic being carried with some margin.

In the case of B&S single-hop networks with the passive optical node, this can be expressed as

$$WxS = R_0 \sum_{ij} t_{ij} \tag{3.9}$$

3.4.1.2 Packet Traffic with Fixed Frame on Dedicated Connections
In the case of stream traffic flow in a fixed frame, as discussed in the last section, all variables, such as the bit rate, packet length, etc., were deterministic and therefore network capacity could be exactly matched to the traffic demand. The network capacity thus could be fully utilized with no blocking. Packet

traffic, meanwhile, has a random flow and is bursty in nature with random burst size and rate. The packets possibly will have random lengths and arrive at random points in time. Therefore, here we need to handle the statistical parameters, the average packet length, in bits, and the average arrival rate, λ in packets per second. Therefore, if fixed frame type of synchronous traffic flow is to be used for random packet traffic, it will need different channel-capacity allocation scheme. This is an example of IP layer supported by the SONET/ SDH over the optical layer. In this section we consider that the packets to be transmitted on the optical layer are sorted and separated in the logical electronic switch node before being passed on to the transmitter processor of the NAT as a fixed frame which has to be transmitted synchronously. For example, the IP packets are encapsulated in a SONET synchronous frame by the SONET switch. The transmission processor in the NAT processes the packets transparently. Now the encapsulated random packet traffic is to be sent on the synchronous TDM *slotted fixed frame* for further transmission on the λ-channel. Modifications now need to be made in the allocations of the channel capacity (*WxS*) for the random packet traffic fluctuations. We cannot have the network capacity to support on the estimated *average* values; this may cause very poor performance in terms of packet blocking or drop whenever traffic is higher. Hence, one can either allocate excess capacity greater than the average estimated value which will improve performance but at the expense of lower capacity utilization, or have buffers for the packets to wait before they can be scheduled for transmission, which will add delay.

Figure 3.2 shows the model for random packet traffic transmission in a synchronous fixed frame over the network. Random packets, which will have been encapsulated by the logical switches, arrive at a random rate. These are the logical connections to be carried by the *transmission channel*. The single server, shown in Figure 3.2, is the *slotted fixed-frame transmission channel* model. If s_{ij} is the number of time slots allocated to the packets for $[i, j]$ connection in the frame, and assuming each time slot to carry only one packet. Also, as packets have random lengths, we assume the slot lengths to be equal to the *maximum*

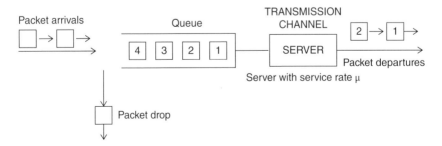

Figure 3.2 Transmission channel model of slotted-line fixed-frame structure.

packet length instead of the *average length*. To allow for random arrival rate of packets, the packets form a queue and wait in the electronic buffer for their turn for transmission in the allocated λ-*channel* frame time slot. With S as the number of slots in the constant frame length, and R_t the optical transmitter rate in bits/sec, the capacity allocated to the $[i, j]$ connection on a particular λ-*channel* is $(R_t s_{ij}/S)$ bits/sec. Using the time-slotted scheduling rule, the transmission processor in each NAT places packets belonging to a given logical connection in the corresponding slots allocated to that connection in each frame. If no packets are awaiting transmission at the beginning of an assigned frame, the slots in the frame are left empty.

The behavior of the *queue* in the buffer can be complex because of the random time each packet spends in the system and the possibility of buffer overflow. A queue is characterized by its *arrival process, queue discipline, service process*, and *buffer size*. Many queue models are available, with specific assumptions in order to obtain the performance of the network. We here make the following assumptions:

- The buffer size is assumed to be *infinite*.
- The arrival process is assumed to be a *Poisson process* with arrival rate λ packets s^{-1}.
- Queue discipline is assumed to be first-come-first served (FCFS).
- The *service process* is determined by time-slot assignment of the packets in the frame of length L in time. Therefore, the *service rate* μ, which is the number of packets served per second, is given as:

$$\mu = \frac{s_{ij} R_t}{Sl} \text{packets / sec.} \tag{3.10}$$

where l is the effective packet length.

With the arrival rate and the service rate defined, the behavior of the queue is determined by its *traffic intensity* ρ, which is the ratio of packet arrival rate and the service rate, (λ/μ). The traffic intensity ρ gives the probability of the server being busy. Therefore, the probability that the queue is empty is given by $(1 - \rho)$. When the queue is operating in the statistical steady state, the quantities of interest in describing queue behavior are the average *queue-length*, including both the packets in the buffer and in service, and the average *delay* experienced in the system. These parameters will take different values with particular types of queue.

A commonly used model for the packet traffic in optical networks is the Markovian queue model of M|M|1. In the M|M|1 queue model, the first letter is for the arrival process – M: *memory less*, with exponential inter-arrival time and Poisson model as the arrival process. The second letter stands for the service time distribution – M: *memory less*, with exponential distribution, and the third letter gives the number of servers to serve the queue, which is one in this case for considering one wavelength channel to serve. In the M|M|1 model, the packet length is assumed to be random and distributed exponentially with a mean value of l, giving an average queue length of [1]:

$$L_q = \frac{\rho}{1-\rho} \tag{3.11}$$

The service times of the packets are mutually independent of each other and are also independent of inter-arrival times. The transit delay, which includes the queuing delay and the service delay, is:

$$D_t = \frac{1}{\mu - \lambda} \tag{3.12}$$

Therefore, just the queue delay is:

$$D_q = \frac{1}{\mu - \lambda} - \frac{1}{\mu} = \frac{\rho}{\mu - \lambda} = \frac{\rho}{\mu(1-\rho)} \tag{3.13}$$

Equation (3.12) gives the mean response time using Little's Law [2]. We observe that the queue length and the queuing delay both become infinite as $\rho \to 1$. At this point the queue becomes unstable and a statistical steady state for the queue no longer exists. The queue is stable *only* if $\rho < <1$, i.e. when traffic density is low. This behavior is characteristic of virtually all types of queues with infinite buffers. Hence, for given packet traffic the NAT transmitter rate, R_t must be large enough to speed up the transmission of packets so that the traffic intensity ρ is *significantly* less than unity to maintain stable operation of the queue. If the channel capacity is less due to lower transmitter rate, ρ is very close to unity. This can cause a large number of packets to drop as a small deviation in average traffic from its prescribed value will overload the system. In addition to queuing delays, another important issue is buffer overflow in the case of its finite size. Because buffer capacity is always finite in any real system, there is always a nonzero probability that the buffer will be full when the packets arrive at a higher rate, in which case the packets will be dropped. Packet loss due to buffer overflow becomes larger as the traffic intensity ρ increases and also when the traffic is bursty. In the case of finite buffer with packet loss, we define *carried traffic* or *throughput* as a performance parameter. The carried traffic is defined as the *offered traffic* reduced by the packet loss. If P_L is the probability of packet loss, then throughput is expressed as:

$$Throughput = (1 - P_L)\bar{\gamma} \tag{3.14}$$

where $\bar{\gamma}$ is the offered traffic. The traffic intensity ρ should be much less than unity, of the order of 0.3–0.5 or so, to keep the packet loss probability to be low enough for obtaining acceptable throughput values. The performance deteriorates significantly whenever $\rho \to 1$ whether the buffer size is finite or not.

To reduce traffic intensity and obtain better performance, in terms of both delay and throughput, the network capacity in terms of number of wavelengths and time slots has to be increased. As the traffic is random, an excess capacity margin is needed to reduce packet drop. This channel capacity margin will depend on packet arrival rate, statistical arrival process, and burstiness of the traffic, which is the ratio of peak-to-average arrival rate. Sometimes the fluctuations of packet arrivals may be so high that it causes large variation in queue length, resulting in very high packet loss rates.

Example 3.1 Calculate the delay and throughput in a fixed capacity dedicated WDM optical network with bursty traffic for a link of 1000 km. The transmitter rate is 5 Gps, frame time with 50 time-slots of connection i,j is 100 μs, and the traffic intensity is 0.45. What are the different types of delays encountered by the packet in the network? Calculate the throughput with infinite and finite buffer and with packet loss probability of 10^{-5}.

Solution

D = 1000 km, R_t = 5 Gbps, L = 100 μsec, s_{ij} = 50, ρ = 0.45. Calculate delay and throughput for infinite and finite buffer and for packet loss = 10^{-5}.

$$\mu = s_{ij} / L = 50 / 100 \times 10^{-6} = 0.5 \times 10^6 \text{ packet / sec}$$

Delay

    (i)   Propagation delay = Dn/c = 1000 × 1000 × 1.5/(3 × 10^8) = 5 ms
    (ii) Transit delay = queue delay + service delay

Queue delay = $\rho/(\mu - \lambda)$ and $\lambda = \mu\rho = 0.45 \times 0.5 \times 10^6 = 0.225 \times 10^6$ packet s^{-1}

Therefore Dq = 0.45/(0.5 − 0.225) × 10^6 = 1.6 μs

And Ds = $1/\mu$ = 2μs and transit delay = 3.6 μs

For infinite buffer throughput = λ = 0.225 × 10^6 packets s^{-1}

For finite buffer = $(1-P_L)\lambda$ = $(1-10^{-5}) \times 0.225 \times 10^6$ packet s^{-1} or 0.225 × 10^6 packet s^{-1} approx.

3.4.1.3 Traffic with Demand-Assigned Circuit-Switched Connections

In this section, we treat the case of capacity allocation for demand-assigned circuit-switched connections. The traffic on the connections can be either stream or asynchronous packet traffic as discussed in the last section. When the demand for a connection is requested, it is granted on the availability of network resources. The requests are statistical in nature, having random connection establishment and termination times. A connection request may be accepted or blocked which will depend on a number of factors, such as the number and type of active connections and their channel assignments in the network, the network capacity, etc. This is the case when we deal typically with a large number of stations, each of which is connected only for a random short period of time. The key performance parameter in this type of system is *blocking probability* – a quantity that depends on the offered traffic demands, the network resources, and the connection control algorithms.

Next, we determine how the traffic for demand switched connections is handled. Assume that the source stations make connection requests at random and for random durations of time. If a request is accepted, the connection is held for a random length of time and is then released. To avoid complexity, we assume here that the traffic on each connection occupies one full λ-channel, or in other words TDM connections are not considered here. Also, the destination node is considered to be non-blocking, so the blocking considered is due only to the *busy channel* or the non-availability of any free wavelengths in the network. For the performance analysis of the network, the connection request model is selected to be Markovian.

As shown in Figure 3.3a, a source station is modeled as a two-state continuous time Markov chain. When the source station is in idle state, it generates connection requests at a rate of Λ requests per second. If the request is accepted by the network, the station moves to the active state. If a request is blocked, it remains in the idle state and continues to generate requests at the rate Λ, corresponding to the *lost calls cleared* (LCC) traffic model forgetting that any earlier request was made, much like the telephone traffic. In the case of non-blocking receivers as assumed earlier, i.e. a receiver is always available, a call request is accepted whenever there is at least one λ-*channel* available for the

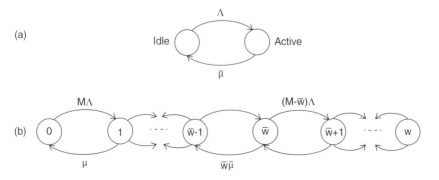

Figure 3.3 Markov chain model: (a) two-state single source model, (b) birth-death model with M > W.

route. An active station is assumed to release calls at a rate of $\bar{\mu}$ releases per second, which implies that the call holding time is distributed exponentially with a mean value of $1/\bar{\mu}$ seconds.

Similar to the telephone network, which is also a circuit-switched system, traffic flow is expressed in terms of a dimensionless unit, *Erlang*. Erlang is the product of two quantities, one the connection arrival rate and the other the average connection holding time. For M source stations and W total λ-*channels*, there is no blocking in the network when $M = W$, that is, each station has an available channel. In this case the two-state (idle and active) model generates $\{\bar{\rho}/(1+\bar{\rho})\}$ *Erlangs* of traffic [2, 3], where $\bar{\rho}$ is the connection traffic density given by $\Lambda/\bar{\mu}$. Thus, the total *offered traffic demands* with M source nodes in the network is:

$$T_{off} = M\frac{\bar{\rho}}{1+\bar{\rho}} \, Erlangs \tag{3.15}$$

Hence, T_{off} represents the average number of sources that would be active at any one time in the absence of blocking.

When we assume $M > W$, i.e. the number of sources is more than the number of channels in the network, there is the possibility of blocking. The complete Markov chain model for this system is the birth-death process shown in Figure 3.3b, in which the state $\bar{w}$ represents the number of calls in progress, i.e. the number of occupied channels. The birth rate, which is the connection arrival rate, is *state* dependent now and hence with state w_i, the birth rate is:

$$\Lambda_{w_i} = (M - \bar{w})\Lambda, \bar{w} = 0,1,\ldots\ldots(W-1) \tag{3.16}$$

and the death rate, which is the connection departure rate from state w_i, is:

$$\bar{\mu}_{\bar{w}} = \bar{w}\bar{\mu}, \bar{w} = 0,1,\ldots\ldots(W-1) \tag{3.17}$$

For low blocking probability in the network, the arrival rate Λ, of the connection request from any station must be small enough so that the aggregate offered traffic T_{off} is equal to or less than the network capacity, which is W in this case, assuming an *infinite population* (Erlang) model, in which if we let $M \to \infty$ and $\Lambda \to 0$, with $M\Lambda = \bar{\Lambda}$ held constant. The result is a simple birth-death Markovian model in which there is no longer any state dependence on the call arrival rate as given in Eq. 3.16. Without the state dependence of the arrival rate, the blocking probability equals the loss probability. In this case, the *carried traffic*, T_{rr} is equal to $(1 - P_L) T_{off}$, where P_L is the probability of connection loss or blocking probability.

The aggregate call arrival process is now a *Poisson process* with rate $\bar{\Lambda}$. Departures occur at a rate $\bar{\mu}_{\bar{w}}$, when the chain is in $\bar{w}$ state and the offered traffic is $T_{off} = (\bar{\Lambda}/\bar{\mu})$. This model leads to the well-known *Erlang-B* formula for blocking probability:

$$P_B = P_L \equiv B\left(T_{off}, W\right) = \frac{T_{off}^W}{W!} \frac{1}{\sum_{j=0}^{W} T_{off}^j / j!} \tag{3.18}$$

where P_L is loss probability and W the total number of wavelengths on which the connections are carried.

Example 3.2 For a network of 50 nodes and 5 wavelengths and blocking probability of 0.001, what will be the offered and carried traffic in Erlang? What are the assumptions made in the model?

Solution

Assumptions: W<< M so blocking will be there and the Erlang B formula with LCC model is used.

Erlang B formula: $\mathrm{P_B} = \dfrac{T_{off}^W}{W!} \dfrac{1}{\left[\displaystyle\sum_{j=0}^{W} \dfrac{T_{off}}{j!}\right]}$

$$0.001 = \frac{T_{off}^W}{5!} \frac{1}{\left[1 + \dfrac{T_{off}}{1} + \dfrac{T_{off}^2}{2} + \dots + \dfrac{T_{off}^5}{120}\right]}$$

Ignoring the higher-order terms in the denominator as $T_{off} <<$, $T_{of} = 0.77E$

$$T_{carr} = \left(1 - P_B\right) T_{off} = 0.73E$$

3.4.2 Optical Packet Switching

As discussed in the previous sections, in the case of asynchronous *packet* traffic if we provide fixed-frame synchronous circuit-switched services then sufficiently excess capacity of the network is to be provided in order to reduce packet loss. Therefore, it is less appropriate to provide a dedicated, guaranteed bandwidth for switched packet services except perhaps for the case of real-time services. When there are large numbers of low-throughput bursty user traffic with a high degree of logical connectivity, packet switching at the optical level is more appropriate for carrying the traffic instead of circuit switching. The packet switching in a purely optical network is then done in the optical layers. Instead of fixed capacity allocation in the switched case, now capacity allocation is done dynamically as and when the packets arrive, and then they switch the path for transmission over the network. In this case there is better capacity utilization and reduced network latency to a large extent, though this adds to large processing and control functions involving the optical layer.

The packet switching is more dynamic as the scheduling is done during the information transfer phase. Therefore, each node needs to have the instantaneous *global* state information of the traffic in all other nodes for efficiently scheduling the packets. This information is distributed geographically throughout the network.

In the case of B&S networks, a separate control channel is required which can give the information concerning packets awaiting transmission at all the nodes. This has to be communicated among the nodes to provide dynamic scheduling decisions. This significantly increases out-of-band communication overheads and real-time processing as the number of nodes increase in the network. We will be discussing in detail the MAC for dynamic capacity allocation and scheduling the packet traffic for the case of B&S networks in this chapter in Section 3.6.

3.5 Network Resource Sharing in Optical Networks

We understand that point-to-point dedicated connections are not possible to all the client-user nodes in any network. It is neither a scalable solution nor an efficient way of using the network resources as most of the connections require smaller capacity. Besides the shared spectrum of the fiber in terms of WDM channels, optical network hardware resources, i.e. number of optical transmitters and receivers, are also shared among all the users. The outcome of this problem is blocking of some of the connection requests in the network whenever the resources cannot be made available for the connection. This causes connection drops in the network if the resources available are insufficient to carry higher density traffic. With this shared medium and limited hardware capacity, we now discuss how we can increase the number of synchronous logical connections carried by circuit-switched networks for the given optical network resources and what the maximum achievable performance can be.

Capacity in any network including the B&S network can be enhanced by either increasing the number of optical channels, W for wavelength multiplexing or by increasing the number of slots in the frame for time multiplexing in the circuit-switched system. So, *why can't we keep increasing them, what limits them?* The number of channels, W and the number of time-slots, S in the frame are limited due to the reasons given below:

1) The number of wavelength channels, W cannot be increased after a certain extent due to the limited available optical spectrum of the fiber and the channel packing density. There is also the cost involved in increasing the number of optical transmitters and receivers in each NAT.

2) The increase in number of time-slots has the technical constraint of increasing the tuning capability of the transmitters and receivers. If the number of

slots is increased in a fixed frame, the tuning of the transmitters and receiver, especially in the case of single-hop transmission, has to be much higher.

3.5.1 Capacity Increase with Number of λ-Channels

For a given fixed optical spectrum of the fiber, the number of channels can be increased by increasing the *optical spectral efficiency* of the network and the *packing density* of the wavelength channels.

- *Optical spectral efficiency.* This represents the throughput of the shared channel system per unit of optical bandwidth. For the total optical bandwidth B_{op} available and assuming that all the W wavelength channels are fully utilized, the optical spectrum efficiency is expressed as:

$$\eta_{op} \equiv \frac{R_t W}{B_{op}} (bps / Hz) \qquad (3.19)$$

assuming all optical transmitters work at the transmission rate of R_t.
- *Packing density.* Besides increasing the speed of the transmitter (R_t), the capacity of the network ($R_t x \ W$) can be increased by high channel packing density, in other words, having more channels in the given optical spectrum of the fiber. This is limited by the following physical constraints:

- *Laser characteristics:* modulation bandwidth, line width, chirp, and wavelength stability.
- *Transmitter impairments:* nonlinearity, dispersion, and attenuation.
- *Receiver characteristics*: optical filter imperfections, noise, and the detection methods (direct/coherent).
- *Signal processing:* electronic speed limitation, modulation technique in transmitter, and receiver processor.

For the above reasons for time-wavelength multiplexed channels, as the optical bandwidth occupied by the channels increases, fewer number of channels can be accommodated. In order to have high optical spectral efficiency it is important to use a bandwidth efficient modulation scheme besides the other physical constraints mentioned above.

3.5.2 Capacity Increase with Number of Time Slots

The logical connection capacity can be increased with the number of time slots in the network, but there is a constraint. With increasing number of time-slots in the fixed time frame, the tuning speed of the optical transmitter has to be increased and also the receiver has to switch fast enough from one slot to the other. The limit on the optical transmitter to switch to different logical connections in fixed frame TDM connections will depend on the speed of electronics, modulation bandwidth of the laser transmitter, and capacity of the λ-*channel*. The switching can be done faster with the increase of electronic speed and higher capacity of the WDM channel. The transmitter tuning rate and, therefore, the bit rate, R_t, has to be less than the maximum tuning rate R_{max} of the transmitter or $R_t \leq R_{max}$.

3.6   Capacity of the B&S Network

In this section, we will first discuss the resources available in a B&S shared-medium network and then discuss the *constraints* in providing an increased number of logical connections with limited resources in the network with circuit-switched connections. The resources can be shared by the logical connections with different multiplexing schemes as discussed earlier.

Consider a passive star coupler as an ONN for the B&S single-hop network with N transmitting stations and N receiving stations connected to a $N \times N$ passive coupler, as shown in Figure 3.1a.We will be limiting the discussion for capacity calculation and performance measures of the networks when the connection requests are time multiplexed on a λ-*channel* of WDM system, i.e. the demands are time-wavelength division multiplexed (T-WDM) (other multiplexing schemes, such as CDM, subcarrier multiplexing, etc., can be treated on a similar basis). On the receiving end, the reverse process is applied, i.e. first tuning the receiver to the specific wavelength and then time de-multiplexing to access the specific slot in the frame, in other words there is a W-TDMA access. The logical connections are realized in these systems by aggregating the traffic by time-multiplexing different connections in the electronic domain in the *logical layer*, converting them to an optical signal to a particular λ-*channel* and then broadcasting on the shared transparent optical media after multiplexing the different λ-*channels*. On the receiving side, the optical receiver at the destination node NAT is tuned to the desired λ-*channel* to access the destined WDM channel and converting to an electrical signal. The receiver then selects the TDM slot and obtains the required logical connection.

For the network in Figure 3.1a, we consider a total of W wavelengths, all transmitters at the end nodes operating at a bit rate of R_t bits/sec and traffic time-multiplexed in a fixed frame of S number of slots. Thus, there is a total of (*WxS*) connections or slots in the frame to carry traffic flow. The maximum possible capacity of this network, therefore, is (*WxS*) for a single fiber. If s_{ij} is the number of slots allocated to a logical connection [*i,j*] periodically transmitted by the optical transmitter, then the capacity required for this connection on a λ-*channel* or the effective bit rate of the connection [*i,j*] is:

$$\frac{R_t s_{ij}}{S} \, bits \, / \sec \qquad\qquad (3.20)$$

In the case of dedicated connection with synchronous traffic and no blocking, the capacity required for the connection has to match the total traffic, or:

$$\frac{R_t s_{ij}}{S} = R_o t_{ij} \tag{3.21}$$

When there is no packet drop, or for s_{ij} (number of channel slots) equal to t_{ij}
(normalized traffic), the transmitter rate is obtained from Eq. (3.21) as:

$$R_t = SR_0 \tag{3.22}$$

From (3.22) it is seen that the frame length of S slots is equal to R_t/R_0, which
is the *speed-up* factor of the transmitter, assuming that each slot is one bit
long. The speed of the transmitter decides the frame size. If the frame length
is small, the transmitter speed required can be reduced. The constraint for
frame length is decided by R_t which is limited by R_{max}, where R_{max} is the maximum tuning speed of the transmitter or receiver, otherwise the traffic cannot
be supported.

3.6.1 Scheduling Efficiency

As discussed above, there will not be any connection drop if the network
resources in a B&S network satisfy traffic requirements. These network
resources are the transmitter rate R_t, which has to be less than or equal to R_{max},
and the total number of available channels, W. The other factor which needs
consideration is the scheduling of the traffic in the frame (WxS). These scheduling options within a frame are limited by the number of transmitters/receiv-
ers and their tunability. The scheduling efficiency, η_s, is defined as the ratio of
total traffic demands to the total capacity available for traffic, which can be
expressed as:

$$\eta_s = \frac{\sum \gamma_{ij}}{R_t W} \tag{3.23}$$

Using Eqs. (3.7) and (3.22) in Eq. (3.23), the scheduling efficiency can also be
written as:

$$\eta_s = \frac{\sum t_{ij}}{SxW} \tag{3.24}$$

The scheduling η_s can attain the maximum value of unity when there are no
idle slots in the (WxS) frame. For the given total traffic this can be achieved by
a balanced traffic matrix (all links are equally loaded), by high transmitter and
receiver tunability, or by increasing the number of channels. The required R_t
depends on how we schedule the traffic in the wavelength-time frame.

3.7 Packet Switching in the Optical Layer in B&S Networks

In Section 3.4 we discussed the traffic flow of stream and asynchronous data packets for connection-oriented services with *circuit switching*. This was done with fixed-capacity allocation for the traffic in a periodic frame structure. This approach for asynchronous packet traffic will have very poor performance because of the randomness of the traffic. The B&S networks which are in the access part of the network are closest to the end users; there are a large number of logical connections each carrying a very small fraction of the total traffic, which is dynamic. In other words, in the access segment the packet traffic requires high dynamic logical connectivity but each with low traffic intensity. This makes the aggregate traffic on the network high. Hence, if we have to provide fixed capacity, it has to be high enough to reduce blocking or packet drop, as discussed in earlier sections. Therefore, the network capacity is highly underutilized.

A more efficient way of utilizing capacity is to have dynamic capacity allocation of the packets in the *optical layer* by packet switching and improving transmission efficiency. Instead of providing a fixed connection-oriented service, we should provide a connectionless service with the transmission switched only for the duration when the packet is being transmitted after its arrival. The dynamic packet switching in the optical layer of the transparent B&S network provides higher logical connectivity, which increases the throughput and reduces the delay entailed in the circuit switching with fixed frame capacity allocation. In the single-hop B&S network, as the ONN is a transparent passive star, packet switching has to be carried out electronically in the *transmission channel* sublayer in the end node access station. For packet switching in the optical layer, addressing, scheduling, and capacity allocation are implemented on a packet-by-packet basis dynamically. Thus, there will be an additional communication and processing load on the optical layer. In the case of B&S networks, significant improvements in throughput and latency can be made with dynamic switching at the packet level using good MAC protocols modified specifically for the WDM shared-channel media. The MAC protocol addresses the problems of contention resolution, packet loss, and retransmission, if required.

There are several MAC protocols used for optical B&S networks, each with various tradeoffs among performance, hardware cost, and processing complexity. Many require large numbers of separate control channels for dynamically scheduling packets and there is always some probability of packet loss in most of these protocols. But the important point to note is that due to the broadcasting nature of the star node, there is full connectivity in the fiber sublayer to all the connected nodes. This comes as an advantage for packet switching, which requires global state information of the nodes before transmission of packets. This information now is simultaneously obtained because of the fully connected client nodes/NATs in the transparent star network.

3.8 Medium Access Protocols

In the B&S single-hop network we have a transparent shared transmission medium with no wavelength reuse or conversion possible. Therefore the number of logical connections is restricted to the number of available orthogonal WDM channels and time-slots. In order to increase throughput with limited resources, the main networking challenge in these networks is to have coordination of transmissions between various nodes to avoid collisions; therefore, it is important for these networks to have efficient MAC protocol to get access to the network with minimum blocking.

In the single-hop, B&S networks, connectivity between the source-destination nodes is made in a single hop through a shared transparent passive optical medium. The destination receiver therefore needs to get tuned to the same wavelength for the designated slot during the duration of the transmission. The tuning time of the receivers and transmitters has to be very small to increase the capacity of the network, and the MAC protocols have to take the tuning time delay into consideration during the design. The MAC protocol also has to consider two more situations when access to the network can be denied. One is the *channel collision*, i.e. when two sources at the same wavelength send the packet in the same time-slot. This is due to the *distinction wavelength constraint* in optical node. The second reason when the connection can be dropped is due to *destination contention* when there is a single tunable receiver at a node and more than one transmitter of different wavelengths in the same time-slot contend for the same destination.

In general, the MAC protocols used in the WDM optical B&S networks are classified in two categories [4]:

(i) *No pre-transmission coordination*. When there is no transmission coordination among user nodes, the scheduling of a logical connection is *pre-assigned* to avoid contention; hence, there is a *deterministic* data transmission schedule. In the case of circuit switching, one assigns a predetermined wavelength and time-slot in the synchronous W-TDM data frame. The deterministic scheduling MAC protocol cannot be used for larger network as the frame length will then become very large to accommodate all the stations and thus the protocol has the limitation of scalability. With the increase in the number of nodes, the latency increases and also will not be able to accommodate the higher-speed connections.

(ii) *Pre-transmission coordination.* Pre-transmission coordination protocols have one common out-of-band shared *control channel* through which all nodes arbitrate their transmission requirements. There are few shared or otherwise *data channels* over which the data is transferred by the source nodes after coordination with the end nodes. All the nodes monitor the control channel and once a source node gets right to transmit, it transmits data on one of the data channels. This is a dynamic capacity allocation MAC protocol which is used for packet switching in the optical sub-layer.

3.8.1 Non-pre-transmission Coordination

In the non-pre-transmission coordination protocol, as there is no coordination before transmission of the signals, for no collision and higher throughput therefore there has to be fixed scheduling of traffic for the case of switched services. The circuit-switching protocols normally have slot synchronization among all the network nodes. That means that all nodes have suitable time references so that signals transmitted in different slots do not collide anywhere in the network. The connections are allocated time-slots on the orthogonal channels in the WDM system. There are basically three types of these protocols: *fixed allocation, partial fixed allocation,* and *allocation free.* In fixed allotment, each possible connection in the network is allocated a slot on a particular wavelength channel. Each node has a tuned transmitter and a tuned receiver. The tuning range is over W available channels. Each station will transmit to a particular destination in a fixed time-slot on a fixed wavelength channel. As an example, the schedule matrix, for a 2 channel, 3 nodes, and 3 slot network, is given in Table 3.1. This allocation will have no channel collision and no receiver contention. Hence, its throughput will be one, but the delay will be large as the number of nodes increases. This also cannot accommodate non-uniform and time-varying traffic loads as it is insensitive to the dynamic bandwidth requirement.

To remove the excessive restrictive nature of the fixed allocation protocol and in the case of lower traffic density, partial fixed protocols are used. As the name indicates, multiple connections can be allocated one time-slot on the same wavelength, hence there can be a possibility of channel collision or destination collision. Destination allocation can happen when more than one source

Table 3.1 Fixed assignment with W (number of wavelength channels) = 2, N (nodes) = 3, and *S* (slots) = 3.

Channel no.	t	t + 1	t + 2
1	(1,2)	(1,3)	(2,1)
2	(2,3)	(3,1)	(3,2)

transmits to the same destination station in the same slot and wavelength. This is *channel collision*. In another case, there can be a destination collision but no channel collision. This happens when several source nodes are allowed to transmit on different channels to the same destination node in the same time-slot so that there will not be a channel collision. This is called the *source allocation*.

In allocation-free protocol, all source destination pairs have full rights to transmit on any channel over any time-slot, as shown in Table 3.2. So, there will be a possibility of channel as well as destination contention. The throughput will be less than one but the delay can be reduced considerably. This protocol is for larger networks but with lower density traffic.

3.8.2 Pre-transmission Coordinated MAC Protocol

When there is no pre-assignment of transmission schedule then in the case of a shared optical medium, such as the transparent star topology in the single-hop B&S network, a random access MAC protocol is required to avoid contentions. In the case of fixed traffic, capacity allocation and scheduling are done once for all at connection setup time, but in the dynamic case, allocation and scheduling have to be done during information transfer. With pre-assignment transmission, the end nodes knew "*when to broadcast*" and "*what to select*," but in the case of no pre-assignment of traffic there is a need for coordination.

In this section, we describe the *pre-transmission coordinated*, MAC protocols which are used in the case of random asynchronous packet-switched transmission when there is random arrival of data with random packet lengths. In such a case, an individual packet with its source-destination address in the header, switch the connection in the optical layer. The transmission protocol coordinates among the stations to avoid contention before scheduling the transmission of data.

The packet-switched B&S optical star network uses an out-of-data band control channel for the coordination of transmission from different stations. It has a N × N broadcast star and all connected user stations have one or more transceivers for data and an additional transceiver for control signal transmission and receiving. The common *control channel* controls the transmission of packets on all the data channels. The instantaneous global state of the traffic is distributed geographically through the control channel and the traffic is scheduled dynamically. The systems operate in a slotted mode, with stations picking up the slot synchronization time as well as other control information by virtue of the broadcast nature of the star node. All nodes have a suitable time reference so that signals transmitted in different slots do not collide anywhere in the network. There are several proposed MAC protocols for these networks, with different throughput and delay achievable. They are designed assuming tunable

Table 3.2 Allocation free with W = 2, N = 2, and S = 2.

Channel no.	t	t + 1
0	(1,2); (3,2)	(1,3), (2,1)
1	(2,3)	(3,1), (1,2)

transmitters and/or receivers with minimum tuning time. The protocol has to take into consideration the delay in tuning while designing.

There is a variety of MAC protocols in the packet-switched network which are classified according to the scheduling control of the packets on the shared channel. These can be categorized as:

- random-access protocols with uncontrollable scheduling
- controlled traffic scheduling with packet loss
- controlled traffic scheduling without packet loss
- perfect scheduling.

The common protocols used in the above categories in the optical WDM networks are the SA (Slotted Aloha)/SA, DT-WDMA (Dynamic Time-WDMA), CSMA/CD (carrier sense multiple access with collision detection), and several others with reservations and perfect scheduling. In the following sections, we discuss in brief a few of the more commonly used protocols. In deriving performance parameters, we assume in general that the tuning time of the transmitters and receivers is negligible compared with slot duration.

3.8.2.1 Slotted Aloha/Slotted Aloha Protocols

Slotted Aloha/Slotted Aloha (SA/SA) [5] is a random-access protocol. This protocol is the variation of the popular random-access Aloha protocol with changes required to WDM networks and subject to control channel inclusion. The protocol in brief is explained here.

The SA/SA MAC protocol finds application in a packet-switched optical WDM network with a large number of end nodes (N), each having very low traffic density. There are a small number of wavelength channels ($\lambda_1, \lambda_2, ...\lambda_W$) for transmitting data with number of data channels, W much less than the number of nodes or $W << N$. There is one separate common control channel (λ_C), which is out of the data band. Each node has a fixed transmitter and fixed receiver for the control channel, λ_C and one tunable transmitter and tunable receiver which is tunable over all the W wavelengths.

Both the data channels and the control channel are time-slotted. The control and data slots from all sources have to be perfectly synchronized to avoid any collision of slots due to misalignment of the slots reaching from different nodes. The data channels have slots of, say, L times longer than the control slot of, say, 1 unit. The control slot has the information of source and destination address and the data channel wavelength. Each data channel can overlap in time with $(L - 1)$ slots that precede it and $(L - 1)$ slots that follow it from other data channels (λ_w). Figure 3.4 illustrates the SA/SA protocol with L = 4 for the data channels and their transmission from transmitters at different instants for successful/unsuccessful transmissions to their destinations.

Algorithm. When a user node has a packet to send, it transmits a control packet in the first control slot. Immediately after transmission of the control packet, it transmits the corresponding data packet over one of the available data channels chosen at random, as shown in Figure 3.4 for the transmitters 1–4. The control packet has the destination node identity and the identity of the data channel wavelength. All the nodes receive the control slot broadcasted by the passive star ONN, and hence do not send any of their control packet if they have data to transmit. Therefore, unless two nodes transmit control packets simultaneously in the same slot, the destination node will receive the control packet, say in slot $[1 + d, 1 + d + 1)$, sent by the source node in slot 1 by transmitter 1 (Figure 3.4), where d is propagation delay of the packet from one node to the other, assuming all nodes are at equal distance from the ONN. The destination node tunes its receiver to the transmitted data wavelength in the time when it will receive the data packet in the interval $[1 + d + 1, 1 + d + 1 + L)$, and will successfully receive the packet if there is no control packet collision and no data packet collision on the particular data channel. If a collision occurs on either the control channel or the data channel (as in the case of transmitter 3 in Figure 3.4), the node waits for some random number of control slots before retransmitting the control packet and the original data packet as described above. Therefore the transmitted traffic has both new and retransmitted packets. This is a *tell-and-go* algorithm when a transmitter just tells on the control channel regarding its data transmission in the next slot.

The upside of a *tell-and-go* algorithm [4] is the low access delay and hence the transmission speed can be increased, which also helps in reducing buffer size at the transmitting node. But the downside is that it reduces the throughput if there is a collision of the control packet. This is due to channel collision of the control signal in the event when another station also sends a packet in the same control slot simultaneously even if it is to some other destination on another data channel. There will be a control packet collision and none of the destinations will receive the control packets. So, for the successful transmission of the data packets, both the control and the corresponding data packet should not have any collision.

The throughput of the SA/SA can be improved by using a "*wait and see*" protocol. The source station waits for $(1 + d)$ time to see the successful reception of the control packet sent by it and then transmits the data packet in the next slot. This will avoid the wasteful transmission of the data packet in the "*tell and go*" protocol at the expense of added access delay of the network. Thus, in *wait and see* or delayed *SA/SA* protocol [4], the node transmits a new control packet in

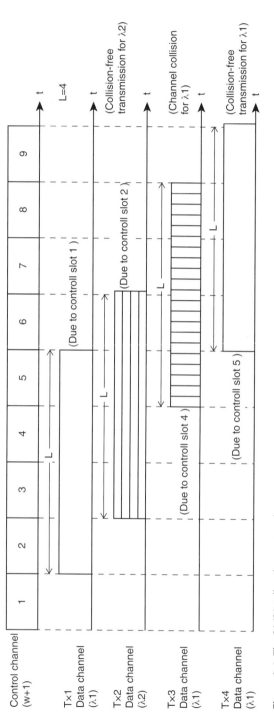

Figure 3.4 The SA/SA *tell and go* protocol.

the first control slot after the data packet is generated. If the control packet is transmitted successfully, immediately after that the node transmits the corresponding data packet over one of the data channels chosen randomly. If the collision occurs on either the control channel or the data channel, the node waits some random control slot times before retransmitting the earlier control packet and the corresponding data packet as described above.

Throughput calculation. The throughput of a data channel is defined as the number of successful data packets in the given channel. Therefore, it can be expressed as:

$$Throughput = G_d * \Pr(s) \tag{3.25}$$

where G_d is the offered traffic per channel in terms of data packets per data slots and Pr(s) is defined as the probability of successful transmission of a given data packet on a given data channel.

The following assumptions are taken for both "wait and see" and "tell and go" protocols:

1) The transmission probability of a packet in any slot to any station is independent of other slots and other stations.

   The number of nodes in the network is very large compared with the number of data channels available, i.e. $N >> W$ and the traffic from each node is very small.

   Considering W to be fixed, with $N \to$ infinity and the transmission probability, p_t at each node to be very small or $p_t \to 0$, then the product $(N.p_t)$ is finite. If G is the total load on the network, then $N.p_t \to G$. The expected number of control packets to be transmitted will be G as well. Therefore, assuming G to be a random variable with a *Poisson distribution*, we can express the probability of the traffic to be:

$$\Pr(G = k) = e^{-G} \frac{G^k}{k!}, k \geq 0 \tag{3.26}$$

2) Because $N \to$ infinity, therefore, the probability of a particular receiver being free is large and hence there will be no receiver contention.

3) Each data packet is assumed to be equally likely to be transmitted on any of the W data channels or the traffic can be said to be *uniformly distributed* over W.

To summarize, for the successful receipt of data packets, the following events have to occur:

i) Successful transmission of the control packet. Only one packet is transmitted on λ_c in a slot, say [0,1), or the probability of successful transmission of the control packet, $Pr(\lambda_c)$ in the slot [0,1) can be expressed as:

$$\Pr(\lambda_c) = \Pr(G = k = 1) = e^{-G}.G \tag{3.27}$$

ii) Probability of selecting one free data channel (W), $Pr(\lambda_w)$ can be expressed as:

$$Pr(\lambda_w) = 1 / W \tag{3.28}$$

Each control slot triggers a data slot of length L, hence the offered data traffic per channel, G_d, can be expressed as:

$$G_d = \frac{L}{W} G e^{-G} \tag{3.29}$$

iii) As there can be an overlap over (L – 1) slots preceding and following the particular data slot over other data wavelengths as discussed earlier, therefore there should not be any data packet transmission in these 2(L – 1) slots on the same data channel, to avoid channel contention. The total 2(L – 1) events are independent, therefore the probability of these is:

$$Pr(s) = Pr(k = 0) = \left[e^{-G/W} \right]^{2(L-1)} \tag{3.30}$$

iv) The *throughput* of the data packets on a data channel as defined in Eq. (3.25), for the *tell and go* protocol from Eqs. (3.29 and 3.30) can be expressed as:

$$Throughput = G_d * P(s) = \frac{L}{W} G e^{-G} * \left[e^{-G/W} \right]^{2(L-1)} \tag{3.31}$$

The throughput of the network will depend on the resources, i.e. W and L, and the traffic load and it is quite low – for example, for L = 10, W = 16, and G = 0.5, the maximum throughput is only 0.11.

In the calculation of the throughput for the case of delayed SA/SA or the *wait and see* protocol, all factors given for SA/SA above remain unchanged except factor (iii) and therefore the throughput calculation in (iv). For the case of *wait and see*, the throughput is obtained as:

$$Throughput = G_d * P(s) = \frac{L}{W} G e^{-G} * \left[1 - \frac{G}{W} e^{-G} \right]^{2(L-1)} \tag{3.32}$$

The maximum throughput now increases to 0.15 for the same values of L and W, as in the case of *tell and go* but for a traffic load somewhat higher. The latency increases with the added delay in the access of the network.

3.8.2.2 DT-WDMA Scheduling Protocol

The throughput obtained in the case of SA/SA protocol is quite small even for reasonably low loads as well. This is because the number of nodes is much greater which try to access the network with very limited resources, i.e. the number of data channels. We can improve the throughput value by adding more data channels and thus reduce packet loss. DT-WDMA [6] MAC protocol has much higher throughput and is commonly used in the optical B&S networks with the many variants. DT-WDMA is an example of a scheduling protocol with reduced packet loss.

In DT-WDMA each node is equipped with a tunable receiver and a fixed-tuned data transmitter, each tuned to a unique wavelength λ_k, therefore the total number of wavelengths in a N × N network will be N. In addition, each node has a fixed-tuned control transmitter and fixed-tuned control receiver operating at the control wavelength λ_c, i.e. each node has a fixed control channel transmitter and receiver and a data channel with a fixed transmitter and a receiver tunable to all data channel wavelengths. The control channel as well as the data channels are slotted. The data channel slot size is N (or W) times of a control slot length. The data slots do not overlap because they all have independent own wavelength. The numbers of control slots in the control wavelength frame are equal to the number of nodes N, so that each node sends the request only in its own assigned control slot.

DT-WDMA algorithm. When a source node has data to send, it sends a control packet in its designated control slot, then sends the data packets immediately on its own data channel (Refer Figure 3.5). The control channel carries the destination address of the packet. The destination receiver on receiving the particular control slot from the source identifies the wavelength on which the data slot will be sent and tunes its receiver to receive the message coming from *source* in the next slot. Each node always monitors the control channel. The data packets sent simultaneously on the data channels can never collide as they are on different wavelengths. Also, the control packets do not collide as the transmitters transmit a control packet in their assigned slots, hence no channel collision can occur. The receiver contention can occur when two transmitters contend for the same receiver at the same time. This is shown in Figure 3.5 for receiver 5 receiving signals from two sources at the same time. The algorithm used for contention resolution is known to all the nodes and then the contending transmitting stations can retransmit their packets next time accordingly.

Throughput calculations. The assumptions in this protocol are of uniform traffic distribution. Each data packet is equally likely to be sent in any slot independent of all other slots and is destined to any of the nodes except of the transmitter node with a probability p.

To calculate the throughput, let us consider, say, tunable receiver Rx_D at the destination node to receive data packet in a data slot [1, N + 1). This data packet will be successfully received if receiver Rx_D does not receive the data from more than one transmitter in the data slot [1, N + 1). The probability of any transmitter at any time having data to transmit to any receiver is:

$$\frac{p}{(N-1)} \tag{3.33}$$

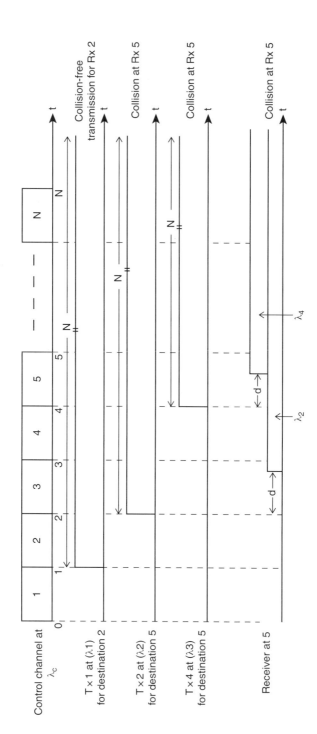

Figure 3.5 DT-WDMA algorithm.

Therefore, the probability of no other station transmitting in $[1, N+1)$ slot is:

$$\left(1 - \frac{p}{(N-1)}\right)^{(N-1)} \tag{3.34}$$

Hence, the throughput of the protocol is:

$$Throughput = 1 - \left(1 - \frac{p}{(N-1)}\right)^{(N-1)} \to \left(1 - e^{-p}\right) \dots N \to \infty \tag{3.35}$$

Now if $p = 1$, i.e. when each node has a packet to be transmitted in every slot, then the throughput is:

$$\left(1 - e^{-1}\right) \approx 0.632 \tag{3.36}$$

Thus, in DT-WDMA, as the number of available channels and slots in the TDM frame has been increased, the channel conflicts have been eliminated but receiver conflicts remain. These conflicts are reduced by proper scheduling and maximum throughput achieved can be substantially improved as compared with the SA/SA protocol.

3.8.2.3   MAC Scheduling Protocols

To further achieve higher throughput in the above MAC protocols, more stringent scheduling of control and data packet features has to be added in order to avoid all contentions. Hence, once a control slot announces the availability of transmission of a data packet from a source to the destination, all the nodes become aware, but the transmitter station "waits" for a small finite time to become aware whether any other station has a packet to be transmitted in the same slot and same destination. Then, with the known scheduling protocol, all the transmitting stations schedule their packets in cooperation with each other on different time-slots. There are many protocols to this effect. Using these protocols, one can ideally reach a throughput of 1 but each packet suffers access delay. In the case of DT-WDMA *with scheduling*, because of scheduling, the delay is much higher than DT-WDMA when the load is less, but once the load is above 60%, the delay with scheduling becomes lower than DT-WDMA [7].

Protocols with perfect scheduling can be designed with very high throughput but they are always coupled with increased latency. Besides throughput and delay in the case of a packet-switched network, one also needs to take into account packet loss as this is always there in packet-switched networks. These dropped packets need to be rescheduled for transmission to improve throughput. Thus, both the original and the rescheduled packet traffic has to be scheduled in this case.

3.9 Summary

This chapter has provided an overall picture of the static single-hop B&S network topologies, constraints, advantages, and limitations. The broadcast star topology as an example has been used throughout as a shared communication medium. A generic description was given in the beginning for the different traffic classes and services provided to the networks, which was later applied specifically to B&S networks to obtain mathematical models for their performance parameters. It was seen that the best performance in terms of capacity utilization and throughput is achieved when dedicated connections are implemented for synchronous traffic. With randomness, be it in traffic type or service connection of the network, performance generally reduces. In packet-switched systems, vulnerability to traffic fluctuations is reduced by dynamic capacity allocation. MAC protocols have been discussed for both circuit-switched and packet-switched services.

The most important current application of static networks is in the access network. The broadcast-and-select networks which are used in LANs and access networks have the advantage of simplicity, natural multicasting capability, and low cost.

Problems

3.1 (a) Power is distributed to the connected stations in the star coupler and the folded bus assuming negligible excess losses in the devices and no attenuation in the fibers. With the help of a neat diagram, show that in the 16×16 star coupler using 2×2 couplers each output port receives the sum of the powers entering at all input ports attenuated by a factor of 16.

(b) In an optical bus of 8 nodes, if the 1st station transmits $1\,mW$ of power over the bus, calculate the power received by the 3rd, the 6th, and the 8th node. Assume each optical tap couples 10% of the power to the node and has 1-dB insertion loss.

3.2 What are reasons for contention in a single hop B&S optical network? Suggest a few techniques for contention resolution.

3.3 (a) A single-hop partial fixed destination allocation protocol is used in a passive star network. There are 6 nodes and 2 channels available. What can be the shortest time slot scheme used? Show it in a tabular form.

(b) A single-hop TTx-TRx network has 4 nodes connected via a star coupler with three channels and 4 time slots in each frame.

Assume that the data corresponding to every s-d pair arrives with probability p every four time slots. Calculate the throughput with *fixed assignment* and partial *source allocation* protocols that can have more than one connection in each slot.

(c) The normalized traffic matrix of a 4×4 B&S network with synchronous dedicated stream traffic is:

$$
\begin{matrix}
0 & 2 & 1 & 0 \\
2 & 0 & 2 & 1 \\
2 & 0 & 0 & 2 \\
0 & 2 & 2 & 0
\end{matrix}
$$

Assume that all transmitters have the same speed and there is no blocking.

(i) For the case of TDM/TDMA, show a suitable slot allocation and determine the frame length for the same. At what bit rate will each transmitter operate for the same if $R_0 = 1$ Mbps.

(ii) For the case of W-TDM/T-WDMA and TT-TR system with two wavelengths what will be the frame length and bit rate of each transmitter? What is the scheduling efficiency? What is the throughput and delay in this case?

(d) For the given normalized traffic matrix draw the channel allocation schedule with 2 wavelengths and for (i) TT-FR schedule; (ii) FT-TR; (iii) TT-TR. Explain the reason for different frame lengths in the three cases.

$$
T = \begin{bmatrix} 0 & 1 & 3 \\ 2 & 0 & 2 \\ 2 & 2 & 0 \end{bmatrix}
$$

In case (iii) if the bit rate of the 1st transmitter is twice that of the 2nd and 3rd transmitters, what will be the required bit rates for the system as well as the channel allocation schedule.

3.4 For a network of 50 nodes and 5 wavelengths and blocking probability of 0.001, what will be the offered and carried traffic in Erlang and the assumptions made in the model?

3.5 (a) Calculate the blocking probability for a network with 20 nodes and 5 wavelengths and an offered traffic of 1.5 Erlang. Also calculate the arrival rate of the traffic when the holding time of the connection is 50 ms.

(b) What is the difference between the Engset and Erlang model of the blocking probability?

3.6 Calculate the delay in a fixed capacity dedicated WDM optical network with bursty traffic on a fiber link of 1000 km long. Consider the average transmission rate of the packets to be 5 Gbps, frame time to be 100 μs, and number of time slots in the frame to be 50. Name the different types of delays encountered by the packet in this network link. Calculate the throughput with infinite buffer and also with finite buffer with packet loss probability of 10^{-5}. Assume that the traffic intensity is 0.5, only one packet can be accommodated in one slot, and the service time includes the time for serving the packets and the transit time of the packet in the server.

3.8 Define throughput, and prove that for *modified SA/SA* ("wait and see") protocol it is given by $\dfrac{LGe^{-G}}{W}\left(1-\dfrac{Ge^{-G}}{W}\right)^{2(L-1)}$ for uniform, independent packet transmission, where G is the arrival rate of control packets in each control slot, W is the number of data wavelengths and the data slot is L times the control slot size.

3.9 A shared broadcast network has 21 stations; each station has its own specific fixed data transmitter and a tunable receiver of 21 wavelengths. Each station also has a common control channel transceiver. Which protocol should the stations use to access the network if the desired throughput is 0.35? Explain in brief the protocol. What will be the probability of the node having data for transmission?

3.10 In DT-WDMA protocol, with number of wavelengths W equal to the number of stations N, there is no control packet nor channel collision. The only collision that can take place is the destination collision, when more than one source attempts to reach the same destination, limiting the throughput to 63%. Give a suggestion with justification for an optimal algorithm to maximize the throughput with controlled transmission.

3.11 We have a shared broadcast medium. A large number of users are connected to the network, and have very little traffic to transmit, which is bursty in nature. Suggest the type of protocol you would like to use for providing the connections and give the reason for selecting this.

References

1 Leon-Garcia, A. (2008). *Probability, Statistics, and Random Processes for Electrical Engineering*, 3e. Prentice Hall.
2 Ibe, O.C. (2011). *Fundamentals of Stochastic Networks*. Wiley.
3 Flood, J.E. (2016). *Telecommunication Switching, Traffic and Networks*. Pearson Education India.

4 Mehravari, N. (1990). Performance and protocol improvements for very high speed optical fiber local area networks using a passive star topology. *J. Lightwave Technol.* 8 (4,): 520–530.

5 Habbab, I., Kavehrad, M., and Sundberg, C. (1987). Protocols for very high-speed optical fiber local area networks using a passive star topology. *IEEE J. Light. Technol.* 5 (12): 1782–1794.

6 Chen, M.S., Dono, N.R., and Ramaswami, R. (Aug. 1990). A media access protocol for packet-switched wavelength division multiaccess metropolitan area networks. *IEEE J. Sel. Areas Commun.* 8: 1048–1057.

7 R. Chipalkatti, Z. Zhang, and A. S. Acampora, "High- speed communication protocols for optical star networks using WDM," Proceedings, IEEE INFOCOM '9.2, Florence, Italy, pp. 2124–2133, May 1992.

4

Optical Access Networks

4.1 Introduction

The access network is the first leg of a communication network in terms of reach which connects the *central office* (CO) of the service provider to the clients. These clients are the individual residential subscribers, the business establishments, or institutions. Therefore, this is known as the *first-mile* network from the user end or also the *last-mile* network as seen from the core end. The individual home subscriber, who not very long back was satisfied with a digital subscriber loop modem providing 1.5 Mbps of downstream and 128 Kbps of upstream bandwidth, now needs Gbps of speed to access the multitude of high-bandwidth multi-media services, and also at a competitive low price. Meanwhile, corporate business subscribers demand broadband infrastructure, such as leased lines at various speeds ranging from STS-N to OC-192 (10 Gbps), and Ethernet-based data technologies to provide high-speed access through which they can connect to their local-area networks (LANs) and then to the Internet backbone and other high-speed services. They need bandwidth for applications such as video-on-demand (VoD), high definition television (HDTV), interactive gaming, two-way video conferencing, cloud-computing etc., delivered via access networks. All the different types of services have to be supported by an access network, therefore faster access-network technologies are clearly desired for next-generation broadband applications.

There is a predominance of Internet Protocol (IP) traffic these days, which is inherently bursty, asymmetric, and random in nature. The current nature of traffic presents many challenges, especially with new real-time applications. At the same time, these networks are required to continue to support legacy traffic and protocols. At present, although broadband copper-based access networks provide much higher data rates than 64 Kbps dial-up lines with DSL and cable TV (CATV) technologies [1], they are unable to provide enough bandwidth for the tremendous growth of Internet traffic and the emerging

Optical WDM Networks: From Static to Elastic Networks, First Edition. Devi Chadha.
© 2019 John Wiley & Sons Ltd. Published 2019 by John Wiley & Sons Ltd.

services. The physical reach and channel capacity of these technologies have almost reached the limits of the available copper cabling.

Optical fibers have been widely used over decades in backbone and in metro networks because of their large bandwidth and very low loss. However, optical fibers in access networks are far behind, and until not very long time ago, access networks were based on DSL and CATV technologies only. But now, with so much of traffic demand required till the last mile, the access network is, therefore, truly the bottleneck for providing different broadband services to end users. The fiber-based access networks, FTTx (fiber to the *x*), promise to bring fiber closer to the home and offer the potential for high access bandwidth to end users. These technologies aim at providing fiber directly to the subscriber home, or very near the home, from the backbone network.

In Section 4.2, we start with the generic architecture of an access network and then take stock of the present available copper-based wired and RF wireless access technologies. In Section 4.3, we start with the details of fiber access networks followed by the passive optical networks (PONs) in Section 4.4. In Section 4.5, the Ethernet passive optical network (EPON) architecture and its protocols are given in detail. Section 4.6 discusses the wavelength division multiplexing (WDM) PONs, Section 4.7 gives the details of the next generation PONs, and in Section 4.8 another class of upcoming optical access networks, free space optics (FSO) and hybrid PON with RF wireless-optical broadband access network (WOBAN), are discussed briefly. This is followed by the conclusion of the chapter in Section 4.9.

4.2   Available Access Technologies

Figure 4.1 shows the basic architecture of an access network. It has three blocks: *hub*, *remote nodes* (RNs), and *network interface units* (NIUs).The part of the network between the hub and the RN is called the *feeder* network, and the part between the RN and the NIUs is called the *distribution* network. The hub, as the name indicates, sits at a central location inside the *central office/local exchange/ head end* of the internet service provider (ISP) facility. One end of the hub is connected to the metro or backbone network and the other end is connected to several RNs deployed in the field, which in turn are connected to several NIUs in a tree or bus topology. An NIU either may be located at the subscriber location or may be the nodal point serving several subscribers. It is through the RN and NIUs that the data to all subscribers reaches from the hub, and similarly through the NIUs the traffic from the subscriber reaches the hub, or in other words, the subscriber gets connected to the metropolitan area network/wide area network (MAN/WAN) through the hub via the RN and NIU.

The *feeder* network could either assign each NIU its own dedicated bandwidth at different frequency bands or have the total bandwidth shared by all the NIUs in the time domain. When the traffic is bursty, it is more cost effective

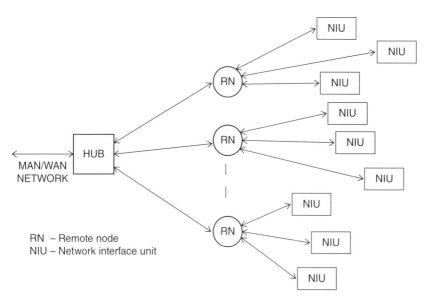

Figure 4.1 Architecture of access network.

to share the total bandwidth among many NIUs rather than assign each NIU its own dedicated bandwidth. Each NIU then could potentially access the entire bandwidth for short periods of time. Frequently the traffic from the hub is broadcasted downstream to all the NIUs, but for the upstream transmission from the NIUs to the hub it is point-to-point unicast. As the upstream bandwidth is time shared, we need some form of media access control (MAC) protocol to coordinate the access of the NIUs to the hub.

The *distribution* network could be either broadcast or unicastor switched topology. For the broadcast distribution network, the RN simply broadcasts the data it receives from the feeder network to all its NIUs, while in the switched case the data received from the feeder for different NIUs are sent as separate data streams to the desired different NIUs. As an example, cable television network has a broadcast distribution network, while the telephone network has the distributed switched network. In the case of switched networks, the NIUs can be made simpler as the *intelligence*, i.e. the control and other administrative functions, lies in the RNs. Also, switched systems have more security compared with the broadcast system.

4.2.1   Access Network Classification

Access networks are broadly of two types: *wired* and *wireless*. We discuss briefly first the wired networks followed by prevailing wireless broadband access technologies.

4.2.1.1 Digital Subscriber Line

DSL [1–2] is one of the predominant broadband access solutions deployed today. DSL technology runs over the telephone network, which was originally built for carrying voice signals. The telephone network which carries the digital signals provides very little bandwidth per home but has advanced switching equipment and operations and management systems. The DSL uses advanced data compression, modulation, and coding techniques to get a capacity of a few megabits per second over the twisted pair. This requires each line from CO and the home to have a DSL modem. The DSL is a switched distribution network, with each subscriber getting its own dedicated bandwidth. This can provide 1.5 Mbps of downstream bandwidth and 128 Kbps of upstream asymmetric bandwidth with the limitation of reach distance of less than 6 km because of signal distortions. The 1.5 Mbps bandwidth with DSL technology is sufficient to receive compressed videos.

4.2.1.2 Cable Television

CATV networks [3] provide Internet services by dedicating some of its radio frequency channels in co-axial cable for data services. The cable network is a broadcast network with all subscribers sharing the total cable bandwidth. The head end (hub) of the cable company is connected to the RNs with fibers and the subscribers are connected in the distribution network in bus topology. Different channels are frequency multiplexed to form a composite signal. The composite RF then modulates the laser before broadcasting to the RNs from the head end. From the RN after the optical to electrical conversion, the RF is carried on the coaxial cables to each home. The downstream cable bandwidth used is between 50 MHz and 550 MHz, with up to 78 AM-VSB (amplitude-modulated vestigial sideband) television signals. A window between 5 MHz and 40 MHz is kept for upstream data. One RN serves 500–2000 homes. The cable companies also carry the video channels in digital format. The cable network provides large bandwidth to each home with no switching and with simple management control. As the cable networks are mainly built for delivering broadcast services, they are limited for distributing access bandwidth.

4.2.1.3 Hybrid Fiber Coaxial Network

Hybrid fiber coaxial (HFC) network is basically an upgraded version of CATV architecture. The network architecture is essentially the same as that of CATV. The feeder network is with optical fiber. Like the existing cable network, the downstream data is broadcast from the head end to RNs by using a passive optical star coupler. High-power 1.55 μm transmitters are used along with booster amplifiers to achieve a high split ratio. To provide additional bandwidth for selected groups of users, 1.3 μm signals are multiplexed on the same set of fibers to be used in *narrow casting* mode. From an RN, several coax trees branch out to the NIUs. The function of the NIU is to separate the signals into telephone signals and broadcast video signals, and to send the telephone signal on twisted pair and the video signal on coax to each home that it serves. Each coax leg is a broadcast bus architecture serving about 50–500 homes. Downstream broadcast video to the home would be sent on analog subcarrier channels. Video signals could be sent as analog AM-VSB streams, compatible with existing equipment inside homes. Downstream video is between 52 MHz and 550 MHz ,with 550 MHz to 1 GHz also made available for down-streaming. Digital video, as well as telephony and data services, can be carried over the same infrastructure. In addition, upstream channels can be provided in the 5–40 MHz band, which is not used for downstream traffic.

In order to provide increased bandwidth per user, the transmitted frequency range can be increased up to 1 GHz. Each subcarrier channel uses a 256 QAM (quadrature amplitude modulation) digital modulation technique providing spectral efficiency of 8 bits/Hz. By driving the fiber deeper into the network, the number of homes served by an RN can be reduced to about 50 homes instead of 500 and providing larger bandwidth to each. The overall capacity can also be increased by using multiple fibers and having a WDM system.

HFC maintains compatibility with existing analog equipment and is a cost-efficient approach to deliver broadcast services, but it has the disadvantages of limited upstream bandwidth and reliability.

4.2.1.4 Fixed Wireless Access Networks

The fixed wireless access networks [4], though they have limited bandwidth and reach, have the advantage of being deployed quickly and at lower cost without digging cables. The common wireless access services are the *multichannel multipoint distribution system* (MMDS), *local multipoint distribution system* (LMDS), WiMax, and WiFi. MMDS are also known as *broadband radio service* (BRS) and as *wireless cable,* as it was commonly used as an alternative method of cable television reception. MMDS works in the range of 2.5–2.7 GHz, with a reach of 15–55 km, depending on the transmitted power. LMDS commonly operates at much higher frequencies in the range from 26 GHz to 30 GHz bands. It was conceived for fixed wireless, point-to-multipoint technology for utilization in the last mile and for digital television transmission. LMDS has shorter reach coverage of 3–5 km in dense urban areas with a bandwidth of 1.3 GHz.

IEEE802.16, commonly known as WiMAX, has different bands available for applications in different parts of the world. The frequencies commonly used are 3.5 GHz and 5.8 GHz for 802.16d and 2.3 GHz, 2.5 GHz, and 3.5 GHz for 802.16e standards. These standards can provide up to 70 Mbps of symmetric bandwidth for distances of about 50 km. They have a variety of applications, including point-to-point links and portable Internet access. Another common wireless access technology to the Internet by laptop computers and other personal computing devices is the IEEE 802.11 wireless LAN technology: WiFi. It operates in the 2.5 GHz UHF (ultra-high frequency) and 5 GHz SHF (super-high frequency) public spectrum and can provide data rates of about 50 Mbps. They are limited by a very short range of tens of meters to an access point.

4.2.1.5 Satellite Wireless Access Networks

Finally, satellite systems have their own space in broadband wireless access networks. The direct broadcasting geostationary satellites with a large footprint on the ground can be a very efficient downlink broadband broadcasting technology. But it is not easy to provide support for upstream traffic. Also, due to the large footprint of each satellite, frequency reuse is limited. For distance communication and providing high-speed Internet access, satellites are used for downstreaming the signal, but for the upstream direction signals are carried over a regular telephone line or under-oceanic fiber cables.

4.3   Optical Fiber Access Networks

With the requirements of the access bandwidth increasing from hundreds of MHz to GHz, optical fiber reaching the access networks was inevitable. Fiber access systems can be point-to-point or point-to-multipoint [5–9]. Figure 4.2a–d depicts different possible architectures of fiber access networks, assuming N end nodes. First, a direct point-to-point link between the hub and each subscriber is shown in Figure 4.2a. This gives dedicated fiber bandwidth to each subscriber but is expensive as it requires N fibers and $2N$ transceivers to support N subscribers with one fiber reaching each home. Figure 4.2b is the architecture of switched services with a switch at the RN. This requires one transceiver at the hub and $(N+1)$ at the RN switch along with one each at the customer side: a total of $2(N+1)$ transceivers. The third case is where the RN and end users are connected in a point-to-multipoint configuration with WDM links. In comparison with the second architecture, now we have a wavelength de-multiplexer/multiplexer at the RN. Each user has its own wavelength transceiver and the corresponding WDM transmitter/receiver at the hub end, hence a total of $(N+1)$ transceivers will be needed. The last case in Figure 4.2d is most economical, sharing the same topology as the previous one but with a passive optical splitter/coupler replacing the multiplexer/de-multiplexer at the RN. A single transceiver at the hub side is shared among all the end users and therefore only $(N+1)$ transceivers are needed in total, reducing the cost considerably. There will be time sharing of the terminal resources at the hub, therefore the speed of the hub equipment equals the sum of data rate for all end users.

Fiber access systems as given in Figure 4.2c,d, in general, are referred to as a *fiber-to-the-x* (FTTx) passive system, where x can be *home, building, curb*, etc.,

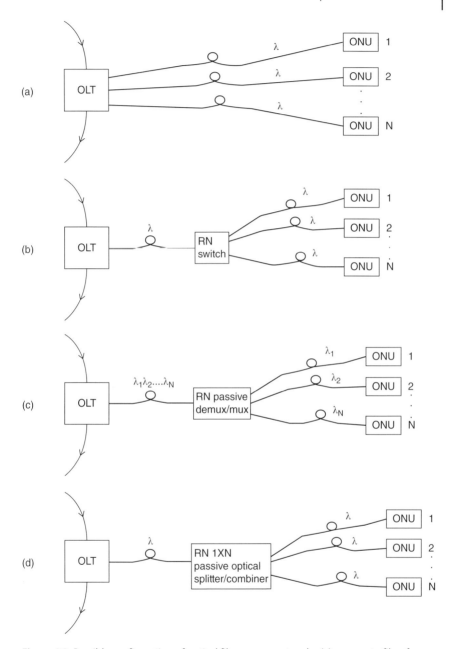

Figure 4.2 Possible configuration of optical fiber access networks: (a) a separate fiber for each ONU, (b) switched connections from the RN to each ONU, (c) WDM PON with a passive MUX/DEMUX at the RN, (d) PON with passive splitter/combiner at the RN.

depending on how deep in the field fiber is deployed or how close it is to the user. In a fiber-to-the-home (FTTH) system, fiber is connected down to the home from the service provider; in case of building the fiber reaches to a building and from there to individual apartments or offices over the copper cable. In an FTTC system, fiber is connected to the curb of a community where the optical signal is converted into the electrical domain and distributed to end users through twisted pairs, cable, or even with RF wireless. FTTx which brings high-capacity optical fiber networks closer to end users brings larger bandwidth and hence appears to be the best candidate for the present multimedia access network. FTTx is considered an ideal solution for broadband access networks because of the inherent advantages of optical fiber in terms of low cost, huge capacity, small size and weight, and its immunity to electromagnetic interference and cross-talk.

4.3.1 Passive Optical Network Topology

The FTTx is commonly used to describe an architecture in which the signals are broadcasted from the hub to the RN, and the RN shares the total bandwidth in time or spectrum with all the end users. The feeder network is the part of the network between the hub, called the optical line terminal (OLT), and the RN. The distribution network is the part between the RN and the NIU, which is the optical network unit (ONU).The RN deployed in the field is a simple passive device, such as an optical star coupler or a static wavelength router. The downstream traffic is broadcasted by OLT to all ONUs and the upstream traffic uses time division multiple access (TDMA) in the case of passive coupler or a dedicated wavelength in the case of passive DEMUX/MUX at the RN. These FTTx are referred to as a PONs [10] as only passive optical devices are used: splitters/combiners, MUX/DEMUX, or a static router. With all passive components in the field, the PON access network is more reliable, less expensive, and easy to operate and service by the service provider. As there are no active switching components in the RN, it needs no control for operation, and power is required only at the CO or at the customer premises. High-powered lasers and other optical components within the OLT are expensive but these are shared among several subscribers. Meanwhile, the ONU is kept simple and cost-effective. Thus, the PON services are made cost-effective to be available for the subscriber requiring limited services.

The optical path from the RN to the ONUs is called the optical distribution network (ODN), shown in Figure 4.2c,d, depicting a common PON architecture for the FTTx solutions. As explained earlier, the optical line terminal is located at the CO connecting the optical access network to the backbone network for IP or Synchronous Optical NETworking (SONET) services. The ONU, also known as the optical network terminal (ONT), is located at the curb or at the end user location. FTTC may be the most economical deployment among the FTTx solutions.

There is no direct communication possible between the ONUs as they are not connected directly with each other – all transmission over the ODN occurs from or toward the OLT. In the downstream there can be point-to-point or point-to-multipoint transmission from the OLT to ONUs, and for the upstream point-to-point transmission from ONUs to the OLT takes place through the RN. In the feeder section, commonly separate fibers for up- and down-link traffic are used. The wavelengths of uplink and downlink signals can be the same, but usually separate wavelength bands are used as this reduces optical reflections in the network, and also reduces the losses associated with up- and downlink combination and splitting.

4.4 PON Architectures in Access Networks

4.4.1 Broadcast-and-Select Passive Optical Networks

There are three basic architectures of PONs: the TPON, the WPON, and the WRPON. The TPON or the PON with telephony architecture, also known as broadcast-and-select PON, is most popular when traffic requirements from the different subscribers are not very high. WPON is the WDM PON which is also in the B&S category, and WRPON is the wavelength routing PON, used for subscribers requiring larger bandwidth of the capacity of a wavelength [11–14].

The signal transmission in the TDM/TDMA PON of TPON category is illustrated in Figure 4.3. Though the number of ONUs connected to the RN can be much higher, the figure shows only four ONUs connected to the RN. The node at the CO connected to the core caters to both low data rate T1/E1 and the broadband STS-N signals. The TDM/TDMA PON have reach of over 20 km without amplification. At the OLT end, a high-power transceiver is needed and

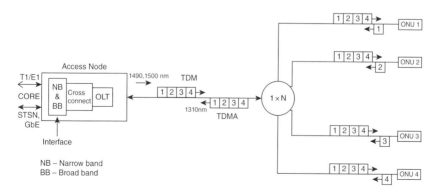

Figure 4.3 Architecture of TDM/TDMA PON.

other high-speed optical and electrical equipment is also required. As large numbers of subscribers share the fiber and these more expensive equipment at the CO, the cost for individual users reduces. Thus, TDM PON provides an economical solution for the last mile problem and is among the most efficient broadband access technology used today.

The TDM multiplexed data is sent downstream by the transmitter at the OLT and from the OLT to the ONUs. This is broadcasted as point-to-multipoint from the OLT to the ONUs on a single wavelength. Before broadcasting the data downstream, the OLT labels each packet with the intended recipient ONU's ID for identification. In the upstream direction, the traffic is transmitted point-to-point from ONU to OLT but the bandwidth is time shared in the feeder section of the PON. Hence, to avoid collision between signals transmitted by the different ONUs, a suitable TDMA MAC protocol is used. The OLT allocates time slots to each ONU for upstream communication. The ONUs can transmit data within their pre-allocated fixed or variable size slots time and thus use the capacity of the single wavelength more efficiently.

For determining the optical power budget of the PON, total losses including fiber losses due to the farthest ONU and the power reduction in the broadband splitter or combiner are taken into consideration along with the optical power launched in the fiber in the downstream and the receiver sensitivity. Most of the commercial PON systems have a splitting ratio of 1:16 or 1:32. The splitting ratio directly affects the system power budget and transmission loss. In order to support large splitting ratio, high-power transmitters and high-sensitivity receivers are required.

Example 4.1 Consider a PON which has 64 ONUs. Let the transmission power of the OLT transceiver be 0.1 mW and the receiver is able to detect signals of at least 0.0001 mW for a reasonable bit-error rate. Assume that the splitter is 10% lossy. The optical fiber has an attenuation of 0.2 dB/km. What is the maximum possible distance of an ONU from the OLT ignoring the loss in the feeder line.

Solution

Splitter loss = 10% and $\alpha = 0.2\,\text{dB/km}$

$$\text{Power at receiver} = \frac{(P_{tx} - \alpha L) \times 0.9}{64}; \text{or} \frac{0.0001 \times 64}{0.9} = P_{tx} - \alpha L = 0.1\,\text{mW} - (0.2\,\text{dB/km})L$$

or L = 58.9 km.

In the case of TDM-PONs, all the ONUs need to be synchronized to a common clock for the proper functioning of the system. This is done by a process called *ranging*, where each ONU measures its delay from the OLT and adjusts its clock such that all the ONUs are synchronized relative to the OLT. As the downstream data is broadcasted, the signal received by each ONU is continuous, with no rapid change in the amplitude and phase of the signal. In the upstream the data sent by the ONUs in assigned slots and therefore works in *burst mode*. Each ONU turns its laser on, sends the data, and shuts it off as it has to transmit the data in its granted time slot. Also, each ONU is at a different location, therefore the received signal at the OLT has different optical loss. For these reasons the amplitude and phase of the upstream signals vary from burst to burst at the OLT on their arrival. The power level received at the OLT may be different for different ONUs because the received signal has the *near-far* problem. Therefore, if the receiver at the OLT is adjusted to receive a high-power signal from a nearer ONU, it may read as zero when receiving a weak signal from a distant ONU, and vice versa in the opposite case. Therefore, to properly detect the incoming bit stream, either the OLT receiver should adaptively adjust its zero-one threshold quickly while receiving from different ONUs or it should operate in burst *mode*. Therefore, in a TDMA-PON a burst-mode receiver is used to perform on-the-fly in amplitude, clock-phase, and data recovery. Burst-mode receivers are expensive and more complex. But there is a need for a burst-mode receiver at the OLT. As ONU receives constant broadcasted signal in downstream, a burst mode receiver will not be required.

There are a number of TPON standards. There is the *ATM passive optical network* (APON). It is based on the asynchronous transfer mode (ATM) protocol established by the *full-service access network* (FSAN) working group in 2001. APON specifies a downstream bit rate of up to 622 Mbps and an upstream bit rate of up to 155 Mbps. It uses lasers at the OLT and ONUs and a power splitter allows a 16- to 32-way split with total fiber attenuation in the 10–30 dB range. The targeted distance with this attenuation is 20 km.

Another PON, the *broadband PON* (BPON), is basically an APON with some improvements and was developed by the International Telecommunications Union (ITU). These improvements include supporting survivability and dynamic allocation of upstream bandwidth. This standard supports more broadband services, including high-speed Ethernet and video distribution. The TPON standards at the gigabit rates are the ITU's *Gigabit PON* (GPON, Rec.G.984, ITU-T, 2003), which is an upgrade of APON for full service support, including voice (TDM over SONET/SDH), *Ethernet* (10/100 BaseT), ATM, leased lines, wireless extension, etc., by using a convergence protocol layer designated GFP (*generic framing procedure*). It gives support to several data rate options: a symmetrical link at 622 Mbps or 1.25 Gbps, or an asymmetrical link at 2.5 Gbps downstream with 1.25 Gbps upstream. GPON uses WDM so a single fiber can be used for both downstream and upstream data. A laser on a wavelength of 1490 nm transmits downstream data. Upstream data is transmitted on a wavelength of 1310 nm.

The IEEE's *Ethernet PON* (IEEE802.3 standard, EPON) or G-EPON is the *Gigabit Ethernet* over PONs. For G-EPON, the downstream bandwidth can be either 1.2 Gbps or 2.5 Gbps and the upstream bandwidth can be 155 Mbps, 622 Mbps, 1.2 Gbps, or 2.5 Gbps, with the restriction that it cannot exceed the downstream bandwidth. A 10G-EPON is also available with symmetrical up and downstream bit rate of 10 Gbps and an asymmetrical architecture with 10 Gbps down and 1 Gbps upstream bit rate. The 10 Gbps versions use different optical wavelengths on the fiber, 1575–1580 nm downstream and 1260–1280 nm upstream, so the 10 Gbps system can be wavelength multiplexed on the same fiber as a standard 1 Gbps system. EPON [15–17] is fully compatible with other Ethernet standards, so no conversion or encapsulation is necessary when connecting to Ethernet-based networks on either end. As Ethernet is the primary networking technology used in LANs and also in MANs sometimes, there will be no requirement for protocol conversion.

4.5   TDM/TDMA EPON Operation

GPON and EPON are the most common TDM/TDMA PONs. The primary differences between EPON and GPON lie in the protocols used for downstream and upstream communications [16–18]. Here in this section we will be basically discussing the operation of EPON.

4.5.1   Upstream Communication in PON

As mentioned above, the downstream traffic is broadcasted by the OLT to all the ONUs; it passes through a passive splitter and reach the ONUs. But the upstream data frames from any ONU can only reach the OLT and not to any other ONU due to the passive optical combiner which has directional properties. As there is a common feeder network between the power splitter and the OLT, the data frames from different ONUs can collide if transmitted simultaneously. Therefore, the ONUs need to employ an arbitration mechanism to avoid data collisions in the upstream direction. The arbitration scheme used is based on a TDM/TDMA MAC protocol, where each ONU sends its upstream signal in the allocated time slot provided by the OLT to avoid any collision.

The OLT is responsible for allocating the upstream time slot to the ONUs. The transmission delay from each ONU to the OLT may be different, as discussed above. In order to take into account the delay, the OLT measures the delay and sets a register by the signaling messages for each ONU according to its delay with respect to all the other ONUs. Once the delay of all ONUs has been set, the OLT transmits the grants messages to the individual ONUs through which it grants permission to the ONUs to use a defined time interval for their upstream transmission. The grant to all the ONUs is dynamically recalculated every few milliseconds. This calculated grant then allocates time-slot bandwidth to all ONUs as per the allocation scheme used, such that each ONU receives required bandwidth for its service needs in time. The time-slot allocation schemes can be either static or dynamic statistical multiplexing schemes.

For the signaling scheme to work precisely, first all ONUs are synchronized to a common time reference and each ONU is allocated a time-slot to transmit. Each time-slot has a length in which several data frames can be transmitted. When the subscriber sends data, it is buffered at the ONU and is transmitted only when the allotted time-slot arrives. On the arrival of the time-slot, all the stored frames are sent as a burst at full channel speed by the ONU. If there are no frames in the buffer to fill the entire time-slot, an idle pattern is transmitted. In the dynamic adaptable schemes, the OLT gets the information of the collected queue sizes from the ONUs and then allots time-slots to each; this leads to more efficient use of bandwidth. We will be discussing the dynamic bandwidth allocation (DBA) protocol in detail for EPON, which is similar to the GPON DBA. In GPON there are two forms of DBA: status-reporting (SR) and non-status reporting (NSR). In NSR DBA, the OLT continuously allocates a small amount of extra bandwidth to each ONU. If the ONU has no traffic to send, it transmits idle frames during its excess allocation. When the OLT observes that a specific ONU is not sending idle frames, it increases the bandwidth allocation to that ONU. Once the ONU's burst has been transferred, the OLT observes a large number of idle frames from the given ONU and reduces its allocation accordingly. In SR DBA, the OLT polls ONUs for their backlogs. A given ONU may have several transmission queues of priority or traffic class. The ONU reports each queue size separately to the OLT. The report message contains a measure of the backlog in the queue.

The DBA mechanism in EPON is equivalent to GPON's SR DBA solution. The OLT polls ONUs for their queue status and grants bandwidth using the GATE message, while ONUs report their status using the REPORT message. There can be several other variations of DBA. Before discussing the DBA we first explain the supporting signaling protocol for the same.

4.5.2 Multi-Point Control Protocol

In the case of EPON, to increase upstream bandwidth utilization, the OLT dynamically allocates a variable time-slot size to each ONU based on its instantaneous demands using a *polling* scheme. This arbitrating mechanism of upstream transmission by the multiple ONUs to the OLT is the Multi-Point Control Protocol (MPCP) of IEEE standard 802.3ah. The main functions of MPCP [19] are the allocation of time-slots to the ONUs to avoid packet collisions and to allow efficient transmission of data in the upstream direction. MPCP also provides the signaling infrastructure for timing reference to synchronize ONUs, auto discovery, registration, and Round Trip Time (RTT) computation or ranging operation for newly added ONUs. We discuss them in the following.

4.5.2.1 Auto-Discovery and Registration

Discovery and registration is the process in which any newly connected or offline ONU registers in the network. In this step, an ONU is discovered and registered in the network while compensating for the RTT. The steps involved in the discovery process are shown in Figure 4.4.

1) First, the OLT periodically makes available a discovery time window during which the offline ONUs are given the opportunity to register themselves with the OLT. A *discovery GATE* message is broadcasted to all ONUs in the downstream. This message includes the starting and ending time of the discovery window.
2) Next, any offline ONU that wishes to register waits for a random amount of time within the discovery window and then transmits a *REGISTER_REQ*

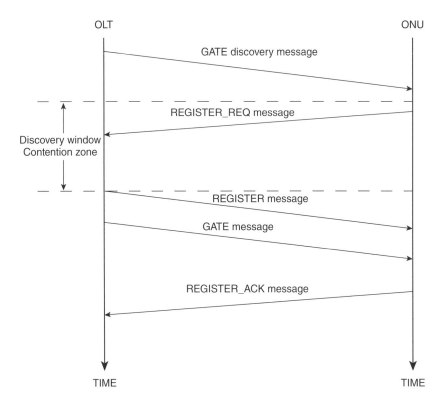

Figure 4.4 Discovery-phase message exchanges.

message. The *REGISTER_REQ* message contains the ONU's MAC address. The random wait is required to reduce the probability of *REGISTER_REQ* messages transmitted by multiple ONUs colliding.

3) The OLT, after receiving a valid *REGISTER_REQ* message, registers the ONU and allocates a *logical link identifier* (LLID) to it. The OLT now transmits a *REGISTER* message to the newly discovered ONU which gives *LLID* to the ONUs.

4) Next, the OLT transmits a standard *GATE* message, indicating a time-slot to the ONU to transmit data.

5) Upon receiving the *GATE* message, the ONU responds with a *REGISTER_ ACK* message in the assigned timeslot. Upon receipt of the *REGISTER_ACK*, the discovery process is complete and normal operation may start.

At each step, timeouts are maintained at the ONU and the OLT. If an expected message is not received before a timeout, the OLT issues a *DEREGISTER* message, which makes the ONU register again.

4.5.2.2 Ranging and Clock Synchronization

As discussed above, synchronization of the clock at the OLT and at all the ONUs is essential in the TDM PONs for the proper functioning of upstream traffic slot allocation and data transmission signaling protocol between the OLT and ONUs. For the many signaling events between the OLT and all the ONUs, the local clocks have time stamping done. Each ONU clock needs to be synchronized with the OLT local clock, as the operation of MCPC depends on the correct determination of the RTT between the OLT and the ONU. The RTT is expected to be different for each ONU as they may be located at different distances from the OLT. With clock synchronization, the RTT is compensated and then the OLT does not have to keep track of the different RTTs of different ONUs when it issues time-slots in GATE messages to the ONUs.

How is clock synchronization achieved? Figure 4.5 illustrates how this happens and the RTT of an ONU is calculated from the OLT. Whenever the ONU receives a GATE message from the OLT, it resets its local clock time from the time-stamp of the sent OLT message, i.e. T_1. Next, when the ONU sends its REPORT message, the time-stamp on it is relative to the time-stamp corresponding to the OLT clock, which is T_2 as in Figure 4.5. Now when the OLT receives the REPORT message it calculates the RTT as the difference between its local time and the time-stamp of the REPORT message; i.e. $(T_3 - T_2)$. Any significant change in RTT implies that the OLT and the ONU clocks are no longer synchronous and the OLT now issues a *DE-REGISTER* message for that particular ONU. The ONU will then attempt to register in the network again through the discovery process.

Example 4.2 Assume an EPON to have equal downlink and uplink time. When a REPORT message arrives from an ONU_i at the OLT, the OLT's local time is 10. The time-stamp on the REPORT message sent by the ONU_i is 5. What would have been the current local time at ONU_i if it was not reset with the OLT clock? How is it related to RTT?

Solution

With reference to Figure 4.5:

- Timestamp on the REPORT message sent by the ONU_i: $T_2 = 5$.
- OLT's local time when REPORT arrives: $T_3 = 10$.

RTT which is equal to the sum of downlink and uplink time delay

- From Figure 4.5: RTT $= (T_3 - T_1) - (T_2 - T_1) = T_3 - T_2 = 5$

Therefore, downlink time = 2.5 and the local time of the ONU_i at the time of sending the REPORT message is 7.5.

4.5.2.3 Signaling Messages Used for Arbitration

There are two messages used during arbitration of transmission between the ONU and OLT: REPORT message by the ONU and the Gate message sent by the OLT.

- *Report handling*: ONU generates REPORT messages through which bandwidth requirements are transmitted to the OLT. The OLT needs to process the REPORT messages so that it can allot bandwidths to the individual ONUs accordingly. The REPORT messages are sent by ONUs in their assigned transmission windows along with the data. A REPORT message is

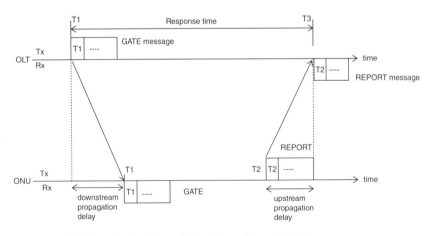

Figure 4.5 Clock synchronization and calculation of round-trip time.

generated in the MAC client control layer of the ONU and is time-stamped. Typically, REPORT would contain the desired size of the next timeslot based on the ONU's queue size. REPORT messages are sent periodically, even when no request for bandwidth is being made. This prevents the OLT from deregistering the ONU. At the OLT, the REPORT is processed and the data is used for the next round of bandwidth assignments.

- *Gate handling*: GATE messages are used by the OLT to grant a time-slot at which the ONU can start transmitting data. Time-slots are computed at the OLT while making bandwidth allocation. The transmitting window for an ONU is indicated in the GATE message from the OLT, with the transmission start and transmission length time specified. Upon receiving a GATE message, the ONU transmits the data queue by registering the transmission start and length times.

Figure 4.6 shows how the DBA agent in the OLT and an ONU interact with the MPCP protocol for data transmission arbitration for the upstream data queue classes in the ONU using the REPORT and GATE messages. This is an intra-ONU transmission arbitration and a similar type of arbitration can be used for inter-ONU case. In Figure 4.6 the REPORT message is used by the ONU to report bandwidth requirements in the form of queue occupancy to the OLT. Upon receiving the REPORT message, the OLT sends it to the DBA module. The module calculates the upstream transmission schedule of all the queues in this ONU and also of other ONUs and grants through the GATE message the time-slot for a queue to avoid any collision with other upstream transmissions.

4.5.3 Dynamic Bandwidth Allocation Algorithms

The time-slot or the bandwidth allocation in the upstream TDMA MAC schemes could range from a static allocation to a dynamically adaptive

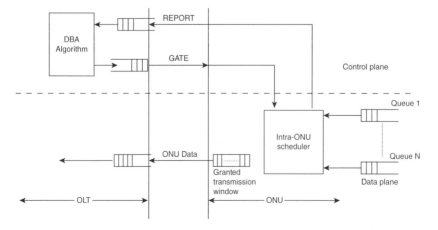

Figure 4.6 Signaling with MPCP protocol for intra-ONU upstream transmission arbitration.

allocation scheme. The static time-slots allocation assigns to each ONU a fixed value. This results in a simple solution as the OLT no longer has to poll the ONUs and schedule the timeslots. This avoids the need for *REPORT* messages in the MPCP protocol altogether. However, this approach is not bandwidth efficient. The network packet traffic being random in nature results in a situation where some time-slots overflow while others are very light loaded, resulting in packets being delayed for several time-slot periods, while a large number of other slots remain underutilized. Hence, there is need for DBA algorithms to achieve statistical multiplexing, so that the OLT schedules the time-slots according to the packet queue lengths at the ONU or other service-level agreement (SLA) decisions. DBA in PONs provides bandwidth efficiency and fair utilization of the PON upstream bandwidth while supporting the quality of service (QoS) requirements of different traffic classes. There are certain differences in the DBA for the two standards, EPON and GPON [20]. The EPON DBA is based on a simple standard with not so stringent hardware requirements, while GPON is based on a relatively complex standard with tighter hardware requirements and more focus on QoS assurance. Here we describe the DBA with reference to EPON; for GPON DBA details, the reader is referred to [20, 21]. One of the first DBA protocols proposed was *interleaved polling with adaptive cycle time* (IPACT) [22]. We describe IPACT based on MPCP described earlier.

4.5.3.1 Service Disciplines

In a DBA, the OLT has to authorize each ONU a time-slot after processing the request from other ONUs on the basis of a certain scheme. We discuss a few service disciplines for IPACT. The size of the buffer or the requested time-slot size at the ONU is conveyed to the OLT using a REPORT message. The DBA agent at the OLT decides the size of the next time-slot to be allocated to the ONUs based on the following service disciplines:

1) *Fixed service* always grants a fixed time-slot to each ONU. This corresponds to synchronous TDMA.
2) *Limited service* grants the requested time-slot size but not more than a maximum scheduling time-slot: max *Length*. It is the most conservative scheme.
3) *Gated service* does not impose any limit; rather, the DBA agent allocates as much time-slot size as is requested by the ONU.
4) *Constant-credit service* adds a constant credit to the requested time-slot size and gives the credit of a few more bytes over and above requested time-slots in the preceding REPORT message sent by the ONU.
5) *Linear-credit service* uses a similar approach as the *constant-credit service* scheme. However, the size of the credit is proportional to the requested window. This is with predictive reasoning. The network traffic has a certain

degree of predictability; specifically, if we observe a long burst of data, the burst is likely to continue for some time into the future.

6) *SLA*-based allocation for each ONU. These service disciplines could be fixed, gated, constant credit, linear credit, or elastic [21].

4.5.3.2 Interleaved Polling with Adaptive Cycle Time DBA Protocol

Next, we explain the IPACT protocol [23] taking the case of a limited-service scheme using maximum scheduling time-slot:

1) Whenever a *REPORT* message with the requested time-slot from the ONU arrives at the OLT, the DBA agent at the OLT has to calculate the start time for the transmission of the time-slots from the ONU.

   For maintaining high utilization of the upstream channel, the DBA agent allocates the next time-slot immediately adjacent to an already allocated time-slot to the ONU with only a small guard time interval separation. Therefore, the start time is computed as:

$$T_{start} = T_{schedule} + T_{guard} \tag{4.1}$$

$T_{schedule}$ is the previous scheduling time, which is a variable and changes after each allocated time-slot.

Sufficient start time is also to be provided so that the ONU has enough time to process the received *GATE* message from the OLT before the granted time-slot is to start. Hence, if T_{local} is the local time and $T_{processing}$ denotes the processing time of the *GATE* message, the start time T_{start} is therefore updated as:

$$T_{start} = T_{local} + T_{processing} \tag{4.2a}$$

if,

$$T_{start} < T_{local} + T_{processing} \tag{4.2b}$$

2) Now, in the case of a maximum scheduling time-slot DBA using a *limited-service scheme*, the length of the allocated time-slot is calculated as follows:

- Let the constant time-slot to transmit a *REPORT* message be $T_{REPORT,}$
- *REPORT.length* be the queue length at the ONU and
- *max Length* denote the maximum scheduling time-slot by the DBA.

The length of the time-slot is computed as:

$$T_{slot} = REPORT.length + T_{REPORT} \tag{4.3a}$$

and if

$$T_{slot} > \max Length, \tag{4.3b}$$

then,

$$T_{slot} = \max Length \qquad (4.3c)$$

3) Once the length is computed, the corresponding *GATE* message with $T_{schedule}$ is transmitted by the OLT with:

$$T_{schedule} = T_{start} + T_{slot} \qquad (4.5)$$

First, the OLT issues individual GATE messages to all the ONUs. The ONUs send their REPORT messages in their respective time-slots regarding the queue size of the packets. When the REPORT from an ONU arrives, the OLT schedules timeslots as in Eq. (4.3(a) and (b)). Whenever any ONU requests a time-slot that is larger than the maximum length, only the maximum scheduling time-slot is granted and the remaining bytes in the queue of the ONU are scheduled later for its next scheduled slot in the TDMA frame.

4.6 WDM PON Network Architecture

The TDM/TDMA PON has limited bandwidth and reach; first due to a single optical channel being shared by all the connected ONUs to the passive splitter, and second, the attenuation due to splitting limits power available to each ONU. With the increasing demand of bandwidth in the access, the PON can be upgraded by employing WDM so that multiple wavelengths may be supported in the upstream and downstream directions. Instead of point-to-multipoint, the WDM-PON can be a point-to-point access network in which there can exists a separate wavelength for each ONU. The different types of WDM PONs are discussed in the following sub-sections.

4.6.1 WDM PON with TPON Architecture

WDM technology can be applied to TPONs to increase the bandwidth of the TDM/TDMA discussed earlier. WDM can increase the capacity and flexibility of TPONs but still keeping the costs low. Architecture to implement WDM into a TPON was shown in Figure 4.2c, where the single transceiver at the OLT is replaced with a WDM array of transmitters or a single tunable transmitter to make a WDM PON (WPON) [23, 24]. This approach allows each ONU to have electronics running only at the rate of its own data, and not at the aggregate bit rate of all ONUs. It has a power splitter at the RN, limiting the power to each user by splitting. A B&S WDM PON is an upgraded version of the basic PON architecture. In this case, the OLT broadcasts multiple wavelengths to all the ONUs and each ONU selects a particular wavelength. As in a conventional TPON, the ONUs time-share an upstream channel at a wavelength different from the downstream wavelengths. One of the earliest WPON architecture proposals employed WDM in the 1550 nm band in downstream and a single upstream wavelength in the 1300 nm band shared through TDMA.

4.6.2 Wavelength Routing WDM PON

The WPON with a power splitter discussed above increases the downstream bandwidth availability to the PON as each has one full wavelength capacity available. However, it is still limited by the power splitting at the star coupler. We can further upgrade the PON by replacing the splitter with a passive arrayed waveguide grating (AWG) at the RN and make a wavelength routed PON (WRPON) [25]. Now each wavelength is routed by a passive AWG. In a *WRPON*, different ONUs can be supported at different bit rates, if necessary. Each ONU can operate at a rate up to the full bit rate of a wavelength channel, therefore it does not have to share the available bandwidth with any other ONU in the network. Moreover, unlike the WPON, the WRPON does not suffer power-splitting losses. Use of individual wavelengths for each ONU also facilitates privacy and reduces security concerns as it is now a point-to-point network in the downstream as well. Finally, because of the periodic routing pattern of AWG, the WRPON is easily scalable. Keeping in view such advantages, WRPON has been recommended as an upgrade to the PON in the ITU-T G.983.

However, the various architectures proposed in the literature differ in the amount of resources used in the upstream direction from ONU to OLT. Upstream communication differs from downstream communication due to two main reasons: ONU transmitters must be inexpensive, and it is preferable to not have wavelength-specific equipment at the ONU because it is difficult to manage and maintain different kinds of inventory. Several types of WRPONs have been proposed and demonstrated. They all use a wavelength router, typically an AWG for the downstream traffic, but vary in the type of equipment used for the OLT and ONUs, and in how the upstream traffic is supported.

4.7 Next-Generation PONs

For higher scalability and capacity increase in GPON, FSAN and ITU-T have proposed the next-generation PONs (NG-PONs), or XGPONs. They envisage NG-PONs growing in two phases, NG-PON1 and NG-PON2, respectively, in order to use the laid-down infrastructure with minimal changes and cost. ITU-T G.987 is the standard for XG-PONs. Mid-term upgrades in PON networks are defined as NG-PON1, while NG-PON2 is a long-term solution in PON evolution. Major requirements of NG-PON1 are the coexistence with the already deployed GPON systems and the reuse of outside plant, which accounts for 70%

of the total investments in deploying PONs. Therefore, it is crucial for the NGPON evolution to be compatible with the deployed networks. Though NG-PON1 has clear goals for its development, that is not the case with NG-PON2, where there are many candidate technologies. Nevertheless, the NG-PON2 system must outperform NG-PON1 in terms of ODN compatibility, bandwidth, capacity, and cost-efficiency [26].

4.7.1 NG-PON1

The first phase of growth of GPON, i.e. the NG-PON1, was in the direction of low cost, increased capacity, wider coverage, and interoperability with existing technology of GPON. In the roadmap to the growth of GPON, NG-PON1 is a mid-term upgrade, which is compatible with legacy GPON ODN.

NG-PON1 [27, 28] uses the existing GPON distribution network part in order to reduce cost. NG-PON1 ODNs have single mode optical fiber in the outside plant with optical splitters and have duplex operation so that both upstream and downstream share the same fiber on separate wavelengths. Also, NG-PON has similar reach as in previous standards but supports a higher split ratio of 128 users per PON or more using reach extenders or amplifiers. This means any NG-PON1 network should be upgradable by changing the ONU and OLT terminals at each end with no change to the fiber itself. After a 10G interface board is added to the OLT, smooth evolution from GPON to XGPON1 can be achieved.

NG-PON1 is an asymmetric 10G system with rates of 10G downstream and 2.5G upstream in order to meet the higher downstream bandwidth demands as compared with upstream bandwidth demands. The selected NG-PON1 system is essentially an enhanced TDM PON from GPON. Therefore, the downstream bandwidth of XG-PON1 is four times that of GPON, while the upstream bandwidth of XG-PON1 is twice that of GPON. In order to extend the reach with support for 128 splits, XG-PON1 supports a range of optical budgets from 33 dB to 35 dB. A PON with a 35 dB optical budget could span 25 km or more and can be shared among 128 subscribers. Some ONUs can receive a broad range of optical spectrum from 1480 nm to 1580 nm, so that the NG-PON1 downstream signal is visible to G-PON receivers. In this case ONUs need to block the unwanted downstream signals with a wavelength blocking filter, which is a small passive optical device.

As an enhancement to GPON, XG-PON1 inherits the framing and management from GPON. It provides full-service operations via higher rate and larger split to support a flattened PON network structure. It inherits the point-to-multipoint downstream architecture of GPON and is able to support FTTH, FTTB, or FTTCurb etc. as in GPON. The upstream/downstream wavelength of XG-PON1 is different from that of GPON (upstream: 1260–1280 nm, downstream: 1575–1580 nm), as in Figure 4.7.

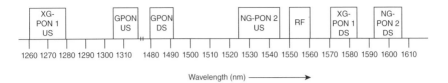

Figure 4.7 Optical spectrum including the GPON, NG-PON1, NG-PON2 wavelengths.

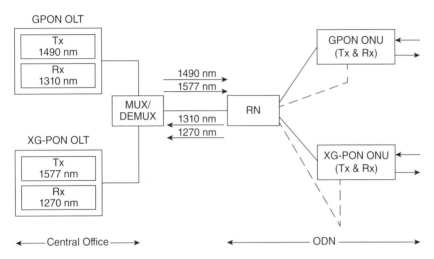

Figure 4.8 XG-PON1 architecture with GPON and XG-PON1 coexistence using WDM stacking.

Figure 4.8 illustrates the NGPON1 architecture with two OLTs, one supporting the GPON and the other the NGPON1. The framing is like GPON but it uses different wavelengths from GPON, using a MUX/DEMUX to separate them, so that GPON subscribers can be upgraded to NG-PON1 while GPON users continue on the original OLT. Compatibility between XG-PON1 and GPON is achieved by implementing WDM in the downstream and WDMA in the upstream by deploying a multiplexer at the CO and a WDM band filter (WBF) at the user side to multiplex or de-multiplex wavelengths on multiple signals in downstream and upstream directions. The 10 Gigabit PON wavelengths (1577 nm down/1270 nm up) differ from GPON and EPON (1490 nm down/1310 nm up), allowing it to coexist on the same fiber. The DBA mechanism in XG-PON1 is basically upgraded by offering better flexibility. XGPON is not yet widely implemented but provides an excellent upgrade path for service providers and customers.

The ITU-T standard of G.987 for the NG-PONs, which defines a management and control interface for administering ONUs, is referred by the G.988 recommendations. The G.988 is the standard for the ONU management and control interface (OMCI) specification. The ONU receives the downstream data and also uses time-slots allocated by the OLT to send the upstream traffic in burst mode. The OLT connects the PON to aggregated backhaul uplinks, allocates time-slots for ONUs, and ONUs then transmit upstream data. The OLT transmits shared downstream data in broadcast mode over the PON to users, much the same as in GPON.

4.7.2 Long-Term Evolution – NG-PON2

In NG-PON2, unlike NG-PON1, several types of prospective technologies were proposed [25, 29, 30]. Here the objective is to improve the rate to 40G from 10G by following the TDM technology, or to use a second method with WDM PON to achieve 40G access. The possible multiplexing schemes could be CWDM or DWDM.

Finally, according to 2015 ITU standard 989.2 [29, 30], NG-PON2 has an architecture capable of total network throughput of 40 Gbps, corresponding to up to 10 Gbps symmetric upstream/downstream speeds available at each subscriber. NG-PON2 is compatible with existing GPON by replacing OLT at the CO and the ONU near each end user. Unique to this standard is the use of both active filters and tunable lasers in the ONU. TWDM-PON coexists with the commercially deployed G-PON and the XG-PON1 systems.

The proposed architecture of NG-PON2 is shown in Figure 4.9. The architecture has both time and wavelength division multiplexing (T-WDM) in the upstream and downstream directions. Wavelength division multiplexing is

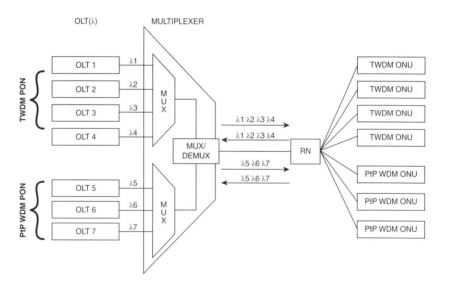

Figure 4.9 Proposed architecture of NG-PON2.

provided in the downstream direction by combining light from four fixed wavelength OLT lasers with a wavelength multiplexer. The light is then filtered at each ONU with an active tunable filter that passes only the desired downstream wavelength to its receiver. In the upstream direction, tunable lasers at each ONU are dynamically assigned a wavelength. Fibers from all ONUs are combined with a passive multiplexer/splitter. Each upstream/ downstream wavelength is capable of providing up to 10 Gbps symmetric bandwidth to each subscriber if the channel is not time-division multiplexed between several ONUs. With wavelength-division multiplexing on four available wavelengths, NG-PON2 can provide up to 40 Gbps throughput to the entire optical network. Wavelength allocations include 1524–1544 nm in the upstream direction and 1596–1602 nm in the downstream direction. A full prototype system offers 38 dB power budget and supports 20 km distance with a 1 : 512 split ratio.

Deployments for several downstream/upstream subscriber rates are described within the standard, including 10/10 Gbps at each subscriber, 10/2.5 Gbps, and 2.5/2.5 Gbps. Additionally, some wavelengths are reserved for potential use in point-to-point applications, for example three wavelengths as shown in Figure 4.9. NG-PON2 is designed to include backwards compatibility, or coexistence, with previous architectures to ease deployment into existing ODNs. Wavelengths were specifically chosen to avoid interference with GPON, NG-PON1, and RF video.

At the time of writing, there have been successful field trials reported of NG-PON1 and NG-PON2. To name a few, Verizon in the USA had a pre-standard XG-PON2 in 2010 [31], Portugal Telecom had s field trial of NG-PON1, again in 2010 [32], and BT in the UK is providing a trial using XGPON.

4.8 Free Space Optical Access and WOBAN

We have discussed the optical fiber-based technologies, e.g. the different types of PON structures. They are well suited to support high bandwidth services and thus can alleviate bandwidth bottlenecks in the access networks. However, laying fiber infrastructure incurs significant cost, requires many permissions, and therefore requires a long time to set up the network. The other limitation is untethered access, which has become so important, especially if one is mobile. Wireless technologies can support mobility and untethered access. But wireless access is constrained due to limited bandwidth. Therefore, combining the complementary features of optical and wireless technologies can potentially provide ubiquitous broadband access.

FSO (Free Space Optics) or *optical wireless* systems using lasers transmitting over free space into the home are being developed as an alternative approach. Another upcoming technology in this direction is the hybrid WOBAN (Wireless-Optical Broadband Access Network).

4.8.1 Optical Wireless Access System

FSO [33, 34] technology uses laser beams with a line of sight optical atmospheric channel for connection to transfer data, video, or voice communications. FSO can provide a transmission capacity of about 100 Mbps to 2.5 Gbps at 1550 nm, with a reach from hundreds of meters to a few kilometers in line of sight. Using FSO wireless networks eliminates the need to secure licensing found with RF signal solutions and also the expensive costs of laying fiber optic cable.

Figure 4.10 shows the block diagram of a simple FSO communication link with a transmitter and focusing optics at one end and a receiver with collection optics on the other end. In practice, the links are bidirectional, with each terminal unit housing an optical transmitter and receiver with focusing/collecting optics, respectively, for data transmission. The unit at one location transmits a beam of focused light carrying the information directly to the unit at the receiving location where the light beam is received by a high sensitivity receiver. As the reach of these links is limited, they can be used for the last mile access which can be set up fast and at much lower cost. FSO has several advantages over other wireless systems: it does not suffer from radio frequency interference, it has a license free operation, and inside installation is possible which is unaffected by operation through glass. Though FSO wireless networks can operate only as point-to-point links between two transceiver units, when combined with LAN or WLAN networks they can provide effective solutions as a complementary technology. It provides speeds comparable to those of optical fiber connections, with the flexibility and practicality of being part of a wireless network providing bandwidth speeds, with possible speeds of up to 10 Gbps becoming likely in the future with the use of WDM technology. Certain drawbacks associated with FSO are the atmospheric channel losses due to fog and rain which restrict their reach to a few kilometers. The technology also suffers from fading due to atmospheric turbulence.

There has been limited commercialization of FSO systems by a few companies, such as Light Pointe and MOSTCOM.

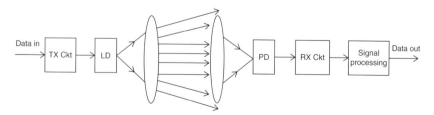

Figure 4.10 Free-space optical link.

4.8.2   Hybrid Wireless-Optical Broadband Access Network

WOBAN [35] is a high-capacity fiber optical backhaul with a RF wireless access. This hybrid network architecture consists of PON and wireless access of WiFi and/or WiMAX in the front-end. In WOBAN, a PON segment starts from the CO with an OLT at its head end. Like the PONs described earlier, each OLT can drive several ONUs, and each ONU in turn can support several wireless routers of the wireless front end in WOBAN. The wireless routers directly connected to the ONUs are called wireless gateways. The wireless front end in turn can consist of other wireless routers to provide end-user connectivity, making the front end of a WOBAN effectively a multi-hop wireless mesh network. In the reverse architecture of radio-over-fiber (RoF), where radio signals are carried over an existing optical fiber infrastructure using hybrid fiber radio (HFR) technology, the proposed WOBAN architecture focuses on the networking aspects of the wireless-optical converged architecture. Figure 4.11 shows the architecture of a WOBAN prototype.

As shown in Figure 4.11, the OLT is located at the CO and the tail end of the PON has a number of ONUs, which typically serve end users as in a standard PON architecture. In the proposed hybrid WOBAN, the ONUs will connect to wireless base stations (BSs) for the wireless portion of the WOBAN. The wireless portion of the WOBAN may employ standard technologies such as WiFi or WiMax.

In the downstream from the optical gateway ONU in the direction of the wireless front to a wireless user, WOBAN is a unicast network, i.e. a gateway will send a packet to its specific destination or user only. But in the upstream direction from a wireless user to the ONU, the WOBAN is any cast network; an end user can try to deliver its packets to any one of the gateways in the wireless mesh network, from where the packet with multiple hops reaches the ONU, is converted to the optical domain, and is finally sent through the optical part of the WOBAN to the OLT.

4.9   Summary

Access networks are advancing toward high data rates and flexibility with scope for scalability and reach. After taking a brief look at the existing access networks, the chapter discussed in detail the broadband optical access networks, the PONs. With increasing broadband traffic demand upto the last mile, and development of components especially optimized for PON applications, very fast expansion and development of PONs are being witnessed in their features, speed, and reach. Both EPON and GPON standards are discussed. EPON systems are more prevalent in Asia and Europe while GPON are

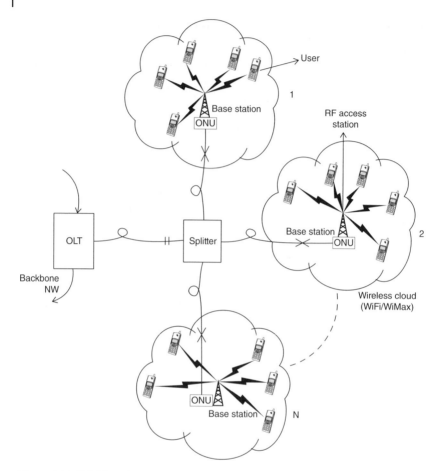

Figure 4.11 Hybrid wireless-optical broadband access network (WOBAN).

used in America. The next-generation GPONs, the twins NGPON1 and NGPON2, are discussed in detail. Also introduced are the FSO and WOBAN technologies for last mile connectivity.

Problems

4.1 (a) Why may it be more important to encrypt information in the downstream direction in a PON and not necessarily in the upstream path?

(b) The upstream wavelength used is 1310 nm while the downstream is 1550 nm in the EPONs. Is there any reason for this, or it can be interchanged?

(c) An access network is to provide services to 100 customers with 200 digital video channels and a maximum of 1 Gbps data rate for the internet access. Select out of (i) DSL, (ii) HFC, (iii) STS-N, (iv) TDM-PON, and design a system using one of the appropriate networks. Give your justification for the selection.

(d) What are the advantages/disadvantages of WRPON over WDM TDM PON? Draw a neat sketch of WRPON showing the upstream and downstream route.

4.2 Assume an EPON to have equal downlink and uplink delay. When a REPORT message arrives from an ONU, the OLT's local time is 12. The timestamp on the REPORT message is 5. The wait time at the ONU is 2. What would have been the current local time at the ONU if it was not reset with the OLT clock, which had a timestamp of 1.5? How is it related to RTT?

4.3 Consider a PON with a transmitter power of 0 dBm by the OLT transmitter. The connected ONUs have receivers that are able to detect signals of at least −35 dB for a reasonable bit-error rate. Assume the splitter has an insertion loss of 2 dB and optical fiber has an attenuation of 0.2 dB/km. How many ONUs can be connected in the PON to have a reach of 50 km assuming all the ONUs are at the same distance from the remote node.

4.4 An IPACT, DBA protocol is used for an EPON with a maximum allowed window size of 2500 bytes. Draw a timing diagram along with the corresponding table at the OLT at each time instant for the EPON with the given report sequence from the ONUs. Assume three ONUs in the system and the entire ONUs to be already registered with the OLT. The polling starts from ONU1. The report sequence from the ONUs is given in the table below.

ONU	Report 1	Report 2	RTT
1	1800	1000	100
2	1500	4500	200
3	3000	500	300

References

1 Abdallah, S., Maier, M., and Chadi, A. (2009). *Broadband Access Networks: Technologies and Deployments*. Springer.

2 V. K. Bhagavath, "Emerging high-speed xDSL access services: Architectures, issues, insights, and implications," *IEEE Communications Magazine*, 37(11): 106–114, Nov. 1999.

3 Ovadia, S. (2001). *Broadband Cable TV Access Networks: From Technologies to Applications*. Prentice Hall.

4 J. F. Mollenauer, "Functional Requirements for Broadband Wireless Access Network", *IEEE 802 Broadband Wireless*, Access Study Group, March 5, 1999.

5 Stern, J.R., Ballance, J.W., Faulkner, D.W. et al. (1987). Passive optical local networks for telephony applications and beyond. *Electron. Lett.* 23: 1255–1257.

6 D. E. A. Clarke and T. Kanada, "Broadband: The last mile," *IEEE Communications Magazine*, 31, 94–100, 1993.

7 Kazovsky, L.G., Cheng, N., Shaw, W.-T. et al. (2011). *broadband optical access networks*. Wiley.

8 Frigo, N.J. (1997). A survey of fiber optics in local access architectures. In: *Optical Fiber Telecommunications, III A* (ed. I. Kaminow and T.L. Koch), 461–522. Academic Press.

9 Harstead, E. and van Heyningen, P.H. (2002). Optical access networks. In: *Optical Fiber Telecommunications, IV B* (ed. I. Kaminow and T. Li), 438–513. Academic Press.

10 Wagner, S.S., Kobrinski, H., Robe, T.J. et al. (1988). Experimental demonstration of a passive optical subscriber loop architecture. *Electron. Lett.* 24: 344–346.

11 ITU-T Recommendation G983.4, "A broadband optical access system with increased service capability using dynamic bandwidth assignment," Nov. 2001.

12 ITU-T Recommendation G983.1, "Broadband optical access system based on passive optical networks," Nov. 2001.

13 ITU-T Recommendation G983.3, "A broadband optical access system with increased service capability by wavelength allocation," 2001.

14 ITU-T Recommendation G.984, Gigabit-capable Passive Optical Networks (GPON), 2003.

15 Payne, D.B. and Davey, R.P. (2002). The future of fiber access systems. *BT Technol. J.* 20: 104–114.

16 Lam, C.F. (2007). *Passive Optical Networks: Principles and Practice*. San Diego, California: Elsevier.

17 Kramer, G. (2005). *Ethernet Passive Optical Networks*. McGraw-Hill Communications Engineering.

18 McGarry, M., Reisslein, M., Maier M. "WDM Ethernet Passive Optical Networks" ,*IEEE Optical Communications*, S18-S25, Feb. 2006.

19 Ryosuke Nishino, Yoshihiro Ashi, "Development of Dynamic Bandwidth Assignment Technique for Broadband Passive Optical Networks", *FTTH Conference*, 2004.

20 Hood, D. and Trojer, E. (2012). *Gigabit-capable Passive Optical Network*. Wiley.

21 Björn Skubic, Jiajia Chen, Jawwad Ahmed, Lena Wosinska and Biswanath Mukherjee, "A Comparison of Dynamic Bandwidth Allocation for EPON,

GPON, and Next-Generation TDM PON", *IEEE Communications Magazine*, S40-48, March 2009.

22 G. Kramer, B. Mukherjee, and G. Pesavento, "IPACT: A dynamic protocol for an Ethernet PON," *IEEE Communications Magazine*, vol. 40, no. 2, pp. 74-80, Feb. 2002.

23 Banerjee, A., Park, Y., Clarke, F. et al. (2005). Wavelength-division-multiplexed passive optical network (WDM-PON) technologies for broadband access: a review. *J. Opt. Networking* 4 (11): 737–758.

24 K. Grobe and J.-P.Elbers, "PON in adolescence: from TDMA to WDM-PON", *IEEE Communications Magazine*, 46(1):26–34, Jan. 2008.

25 F. J. Effenberger, H. Mukai, S. Park, T. Pfeiffer, "Next-generation PON-part II: Candidate systems for next-generation PON", IEEE Communications Magazine, vol. 47, 11, Nov. 2009.

26 https://www.broadband-forum.org/standards-and-software/major-projects/ng-pon2viewed: May 2018

27 Nesset, D. (2017). PON roadmap [invited]. *J. Opt. Commun. Networking* 9 (1): A71–A76.

28 Abbas, H.S. and AGregory, M. (2016). The next generation of passive optical networks: a review. *J. Networking Comput. Appl.* 67: 53–74.

29 Wey, J.S., Nesset, D., Valvo, M. et al. (2016). Physical layer aspects of NG-PON2 standards—Part 1: optical link design [invited]. *J. Opt. Commun. Networking* 8 (1): 33–42.

30 Luo, Y., Roberts, H., Grobe, K. et al. (2016). Physical layer aspects of NG-PON2 standards—Part 2: system design and technology feasibility [invited]. *J. Opt. Commun. Networking* 8 (1): 43–52.

31 https://www.lightwaveonline.com/articles/2010/06/verizons-second-field-trial-of-10-gbps-xg-pon-fttp-affirms-fios-network-design--97045684.html viewed: june 2018.

32 http://www.lightreading.com/broadband/next-gen-pon/portugal-telecom-trials-10g-gpon-with-huawei/d/d-id/681323 viewed: June 2018.

33 Khaligh, M.A. and Uysal, M. (2014). Free space optical communication: a communication theory perspective. *IEEE Commun. Surv. Tutorials* 16 (4).

34 Chadha, D. (2013). *Terrestrial Wireless Optical Communication*. McGraw-Hill Education.

35 Sarkar, S., Dixit, S., and Mukherjee, B. (2007). Hybrid wireless-optical broadband access network (WOBAN): a review of relevant challenges. *IEEE/OSA J. Lightwave Technol.* 25 (11): 3329–3340.

5

Optical Metropolitan Area Networks

5.1 Introduction

In the last chapter we discussed the access segment of a large optical network which had diverse network technologies and protocols – digital subscriber line (DSL), hybrid fiber coaxial (HFC), passive optical networks (PONs), etc. – with large variation of traffic types and rates from the subscriber end, to be finally carried by the backbone network. Access networks are the closest to the end users and the backbone long-haul networks are at the core of the global network. The application of these long-haul networks is to transport large capacity pipes. Between these two different networking domains lies the metropolitan area network (MAN) with a reach in the order of tens to hundreds of kilometers (Figure 5.1). Therefore, the MAN on one end has to match the characteristics of the access networks, such as diverse networking protocols, channel speeds, and traffic dynamism. On the other end, this type of traffic has to be aggregated, fed, and matched on to the large capacity and more or less static optical channels of the large area core network. Also, it has to provide intelligence, dynamic service provisioning, and survivability to the network. With the fast growth of Internet data traffic, there is a significant increase in the highly dynamic and diverse mixture of client services and protocols, such as synchronous and asynchronous data, IP, Ethernet – 10/100 Mbps and 1G/10G, multiplexed time division multiplexing (TDM) voice, and other data protocols and multimedia services from the access network. All these signals have to be transported by the long-haul networks having tera-bit capacities with WDM. Metro architectures which were designed initially for legacy *voice centric* synchronous traffic now have more asynchronous data traffic. The new metro platforms have to offer high bandwidth and scalability and carry multiple protocols over a common infrastructure which is also compatible with the existing legacy network [1, 2].

Metro networks have been traditionally based on Synchronous Optical NETworking (SONET)/synchronous digital hierarchy (SDH) rings and point-

Optical WDM Networks: From Static to Elastic Networks, First Edition. Devi Chadha.
© 2019 John Wiley & Sons Ltd. Published 2019 by John Wiley & Sons Ltd.

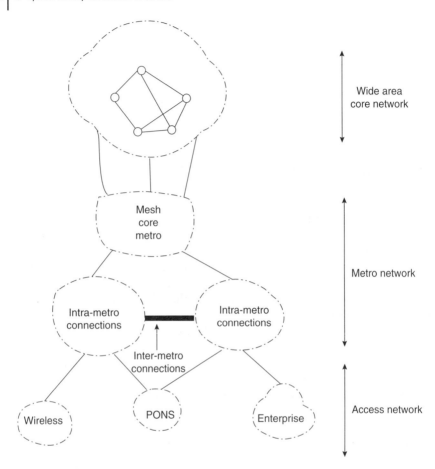

Figure 5.1 Connectivity of a metro network with the core wide area network and access network.

to-point topologies. They were designed for voice traffic and have been very successful in delivering synchronous traffic. Towards the edge, there are smaller rings with lower rates OC-3 (155 Mbps) or OC-12 (622 Mbps). This lower rate traffic is aggregated onto larger core rings that interconnect central office (CO) locations of higher bit rates traffic: OC-48 (2.5 Gbps), OC-192 (10 Gbps), or OC-768 (100 Gbps). The goal of the SONET standard was to specify interfaces that allow interconnection of voice-centric transmission systems of different manufactures and carriers, and to provide multiplexing and de-multiplexing of different rate signals which is cost effective and not complex.

ATM (asynchronous transfer mode), optical transport network (OTN), Gigabit and 10-Gigabit Ethernet, fiber channel and RPR (resilient packet ring) [3–6] are some other predominant metropolitan network protocols in the client layers of the optical network. With the exception of WDM rings and OTN, all these *client layers* process data in the electrical domain. They can have either fixed TDM or statistical TDM for packet switching. Standards exist today to map services as IP, Ethernet, and ATM in the SONET frames or OTN wrappers. With more data traffic and DWDM systems now, SONET/SDH networks are being slowly superseded by the OTN standard [4–7]. The interface with the OTN is better suited with the optical layer of the WDM optical networks carrying data traffic. The OTN has a complete and flexible set of operation, management, and monitoring features for very high transmission rates. But there is still a great deal of deployed legacy SONET/SDH-based equipment. We will be discussing these two protocols in more detail in this chapter.

Ethernet traffic is carried over all communication physical media, including twisted pair, wireless, and fiber cables. It can have a wide range of data rates: 10 Mbps, 100 Mbps, 1 Gbps, and 10 Gbps. IEEE 802.17-RPR media access control (MAC) protocol has the architecture and technology designed to meet the requirements of a packet-based MAN. It gives carrier-class service with 50 ms recovery for any single network fault and facilitates services over a ring network [5].

We start the chapter with Section 5.2 giving details of SONET/SDH as they still exist in most of the legacy networks and more so because many of their protocols, specifically with fault tolerance and survivability, are used in the current networks with some variations. Section 5.3 gives details of the OTN which are the currently deployed technology, followed by the conclusion of the chapter in Section 5.4.

5.2 Synchronous Optical Network/Synchronous Digital Hierarchy

SONET and SDH are the transmission and multiplexing standards for high-speed signals for the metropolitan networks in North America, and Europe and Japan, respectively [3, 4, 7]. In the following sub-sections, we will be giving details of these standards. SONET and SDH have some minor differences between them. Some key differences are the terms used to describe the layers, minor differences between the overhead bytes, lines rates, etc. Therefore we will be giving the details of the SONET standard only while mentioning the SDH standard at appropriate places.

5.2.1 SONET Networks

SONET standards were initially designed for voice and constant bit-rate (CBR) connections up to 51 Mbps only, which now carry traffic from different client layers and packet traffic with transmission rates in the tens of gigabits per second [2]. With the data link layer protocols and the GFP (generic framing procedure) adaptation method, all of data networks – IP, Ethernet, fiber channel, etc. – can be served by the SONET protocol. An important feature of SONET/SDH is that it provides carrier-grade service of high availability.

The first SONET standard was completed in 1988 and was defined by Telcordia. The SONET standard was designed to time multiplex digital signals (DS-N) that specify how to multiplex several voice calls (voice signals sampled at 64 Kbps rate produces one byte every 125 μs) onto a single link and transmit them optically. ITU-T adopted the SDH as the international standard. The electrical side of the SONET signal is known as the *synchronous transport signal* (STS) and that of the SDH signal is known as the *synchronous transport module* (STM). The optical side of both the SONET and SDH signals is known as the *optical carrier* (OC) signal. The TDM multiplexed digital signals are supported by corresponding optical carriers, OC-1 to OC-768, constructed by scrambling the STS-N signal and converting it to optical form. The basic STS-1 is carried in a 125-μs frame containing elaborate transport overheads plus the information payload. Higher rate STS-N signals of 125-μs frame are formed by byte-interleaving STS-1s. The frame structure is fairly complex, incorporating overhead bytes for communications, administration, and maintenance (O&M) functions with payload data.

In Figure 5.2a, typical metro network architecture with the traditional SONET/ SDH rings are shown. The SONET/SDH architecture has three levels of a ring hierarchy: first, the edge rings which have span distances from a few kilometers to a few tens of kilometers, and are composed of the ADMs (add/drop multiplexers) that electronically aggregate the access traffic onto the fiber. The traffic from these edge rings is transported to the second level through a hub to a CO location which is a combination of ADMs and DCSs (digital cross-connects switch).

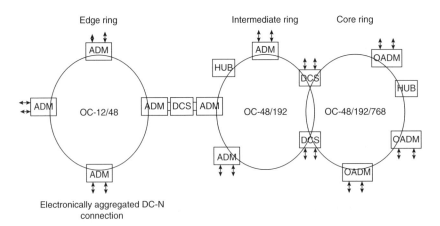

Figure 5.2 SONET metro ring network architecture.

At the second level the DCSs' rings have a span of possibly hundreds of kilometers. They aggregate the traffic from the connected edge rings and finally aggregate the traffic for transmission at the third-level interoffice rings which are mostly optical DWDM rings with optical cross-connects (OXCs), or are also interconnected to mesh topology to the core network, carrying OC-48 and higher rates.

SONET/SDH is globally deployed by a large number of network operators. The many benefits provided by SONET/SDH are listed below:

- *Simplification of multiplexing.* In an asynchronous system (PDH – *plesiochronous digital hierarchy*) the bit rates of different multiplexed signals are not exact integral multiples of the basic data rate but at somewhat higher data rates. Therefore, bit stuffing has to be done. This makes it difficult to draw out a low bit rate stream from the higher rate stream for de-multiplexing. On the other hand, in the synchronous SONET/SDH, the multiplexing and de-multiplexing of different data rate signals are very convenient and cost-effective as all the clocks are perfectly synchronized to the master clock. The lower bit rate streams can be extracted from the higher streams in one step, which makes the multiplexer very simple in design and hence cost-effective.
- *Management.* There are extensive management overheads in the SONET and SDH standards, which makes network management, performance monitoring, and failure reporting very effective.
- *Network availability.* The network protection techniques and protocols are well developed, which makes the availability of the network very high, with the restoration time after any failure less than 60 ms in SONET and SDH networks.
- *Inter-operability.* SONET and SDH have well-defined standards for different line and optical interfaces, which makes the interoperability of different vendor products possible on the link.

In SONET technology, multiple signals of different rates are aggregated by means of a *byte-interleaved* multiplexing scheme. Byte interleaving simplifies multiplexing and offers end-to-end network management. The complete SONET/SDH signal rates are shown in Table 5.1.

5.2.2 SONET Multiplexing

How is multiplexing of signals done in SONET? Frames of lower rate can be synchronously time-division multiplexed into a higher-rate frame by byte interleaving. In SONET, all clocks in the network are locked to a master clock. First, the lowest level or the base signal is generated. In SONET, this base signal is the STS-1 at 51.84 Mbps. The higher-level signals (STS-N) are integer multiples of STS-1, as given in Table 5.1. Table 5.1 also includes the optical

Table 5.1 SONET/SDH signal rates.

Optical signal	SONET (electrical)	SDH (electrical)	Data rate (Mbps)	Overhead rate (Mbps)	Payload rate (Mbps)
OC-1	STS-1		51.840	1.728	50.112
OC-3	STS-3	STM-1	155.520	5.184	150.336
OC-9	STS-9	STM-3	466.560	15.552	451.008
OC-12	STS-12	STM-4	622.080	20.736	601.344
OC-18	STS-18	STM-6	933.120	31.104	902.016
OC-24	STS-24	STM-8	1244.160	41.472	1202.688
OC-36	STS-36	STM-12	1866.240	62.208	1804.932
OC-48	STS-48	STM-16	2488.320	82.944	2405.376
OC-96	STS-96	STM-32	4976.640	165.888	4810.752
OC-192	STS-192	STM-64	9953.280	331.776	9621.504
OC-768	STS-768	STM-256	39813.120	1327.104	38486.016
OC-N	STS-N	STM-N/3	Nx51.840	Nx1.728	Nx50.112

counterpart for each STS-N signal, i.e. the OC-N. Each STS-1/STM-1 consists of a number of DS-1/E1 signals. We also have concatenated SONET/SDH links which are commonly used to interconnect ATM switches and IP routers which are packets over SONET. In the concatenated *structures* (OC-3c, OC-12c, etc.), the frame of the STS-3 or OC-12 payload is filled with ATM cells or IP packets packed in PPP (point-to-point protocol) or HDLC (high-level data link control) protocol frames.

5.2.2.1 Virtual Tributaries

SONET is designed to carry broadband payloads. Many present digital hierarchy data rates, however, are at a lower rate than STS-1. To make SONET backward compatible with the current hierarchy, its frame design includes a system of virtual tributaries (VTs). A virtual tributary is a partial payload that can be inserted into an STS-1. The different types are:

Type	Frame rate[a] (frames/ seconds)	No. of columns[a]	No. of rows[a]	No. of bits/ byte	Signal rate (Mbps)
VT1.5	8000	3	9	8	1.728
VT2	8000	4	9	8	2.304
VT3	8000	6	9	8	3.456
VT6	8000	12	9	8	6.912

a) Discussed in the following section.

5.2.3 SONET Frame

Each STS-N is composed of 8000 time frames. Each frame is a two-dimensional matrix of bytes with 9 rows and 90 × N columns. A standard STS-1 frame has 9 × 90 (bytes). The frame consists of two parts: the *transport overhead* and the *synchronous payload envelope* (SPE). The first three bytes (8-bit byte) of each row represent the *section* and *line* overheads (discussed later). These overhead bits comprise framing bits and pointers to different parts of the SONET frame and many other bytes for OAM (operation, administration, maintenance) procedures. The combination of the section and line overheads comprises the transport overhead. The remainder 87 columns of the frame, the SPE, consist of user data and additional overheads referred to as the *path overhead*. The STS-1 frame is shown in Figure 5.3. The frame is presented in matrix form and it is transmitted row by row from left to right and from top to bottom. Each cell in the matrix corresponds to a byte. Each byte in a SONET frame can carry a digitized voice channel. For STS-1, a single SONET frame is transmitted in 125 μs, or 8000 frames per second, regardless of the bit rate of the SONET signal. At the data rate of 51.84 Mbps (8000 × 810 bytes), of which the payload is roughly 49.5 Mbps, the STS-1 frame has enough capacity to encapsulate 28 DS-1s, a full DS-3, or 21 E-1 carriers.

The STS-1 SPE may begin anywhere in the STS-1 envelope. Typically, it begins in one STS-1 frame and ends in the next. There is one column of bytes in the SPE that represents the STS path overhead. This column frequently *floats* throughout the frame. Its location in the frame is determined by a pointer in the Section and Line overhead of STS-1 pointer in the transport overhead. If there are any frequency or phase variations between the STS-1 frame and its SPE, the pointer value will be increased or decreased accordingly to maintain synchronization.

The STS-N frame is of (N × 810) bytes. The first 3N columns contain the transport overhead sections, three columns for each STS-1, and the rest is for SPE. For an STS-N frame, the SPE contains *N* separate payloads and *N* separate path overhead fields. In essence, it is the SPE of N separate STS-1s packed together, one after another, hence it becomes easy to de-multiplex them. The

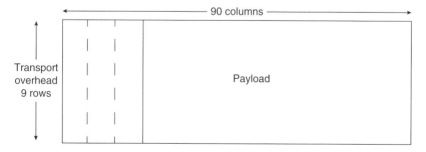

Figure 5.3 STS-1 frame.

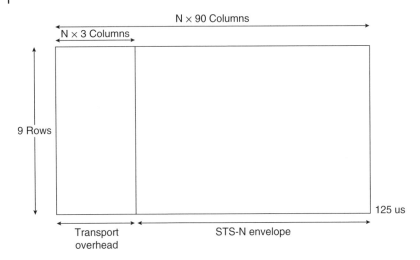

Figure 5.4 Frame for STS-N.

concatenated STC-Nc has the transport overheads same as that of STS-N. But unlike STS-N, there is only *one path overhead* field for the entire SPE. The SPE for an STS-Nc is a much larger version of a single STS-1 SPE. Figure 5.4 shows how the frame of an STS-N signal of 125 μs is multiplexed. The duration of an STS-1, STS-3, or STS-N frame is same and equal to 1/8000 seconds, or 125 μs. A SONET STS-N signal is also transmitted at 8000 frames per second.

STM-1 is the SDH equivalent of a SONET's STS-3. For STM-1, a single SDH frame is also transmitted in 125 μs, but the frame is 9 × 270 bytes long, or 155.52 Mbps, with a nine-byte header for each row. The nine-byte header contains the multiplexer and regenerator overhead. This is nearly identical to the STS-3 line and section overheads. In fact, this is where the SDH and SONET standards differ. SDH and SONET are not directly compatible, but they differ in only a few overhead bytes.

5.2.4 SONET/SDH Devices

There are various kinds of equipment used in different network topologies of SONET. Although the network elements (NEs) are compatible at the OC-N level in SONET, they may differ in features from vendor to vendor. The most commonly used pieces of equipment are the *terminal multiplexer (TM)*, the *regenerator*, ADM, and DCS.

5.2.4.1 Terminal Multiplexer/De-multiplexer

The terminal multiplexer multiplexes or packs a number of *DS-N/E1* or STS signals into a single *OC-N* signal and de-multiplexes the *OC-N* at the end of the connection. It consists of a software-driven controller, a low-speed interface for *DS-N* or *E1* signals in the input side, an *OC-N* interface in the output side, and a time-slot interchanger which feeds signals into higher-speed interfaces. It works also as a de-multiplexer.

5.2.4.2 Regenerator

The regenerator converts the optical signal at the input to the electronic domain, does the amplification and all processing of the signal in the electronic domain, and converts the regenerated electronic signal back in the optical domain. After amplifying the received signal, it replaces the *section overhead* bytes only before retransmitting the signal. The *line OHs*, *path* OHs, and the *payload* are not altered.

5.2.4.3 Add/Drop Multiplexer

The ADM does de-multiplexing, cross-connecting, adding, and dropping of chan-nels, and then re-multiplexes them. It receives an OC-N signal from which it can de-multiplex and drop any number of DS-N or OC-M signals, where M < N, while at the same time it can add new DS-N and OC-M signals into the OC-N signal. It can be used in intermediate ADM sites or in the hub configurations. Besides add-ing and dropping signals, the rest of the traffic is bypassed. An ADM performs operations similar to a DCS except that the ADM has only two inter-nodal ports.

5.2.4.4 Digital Cross-Connect

The function of a DCS is to de-multiplex, route, and re-multiplex the signals with which it interfaces. DCS accepts various optical carrier rates, accesses the STS-1 signals, and switches at this level. DCS can be used for grooming of STS-1 signals and can support hub network architectures. It is also used to interconnect multiple SONET rings. The major difference between a DCS and an ADM is that a DCS may be used to interconnect much larger numbers of STS-1 signals and have more ports.

5.2.5 SONET Protocol Hierarchy

Similar to open systems interconnection (OSI) layered structure, the functions of the SONET can be modeled by the layered structure. The SONET interface layers have a hierarchical relationship. Each layer builds on the services pro-vided by the adjoining lower layer. Each layer communicates to peer equip-ment in the same layer and processes information and passes it up or down to the next layer, much the same as the OSI standards. The four-layered hierarchy with *section*, *line*, *path* forming the data link and the *photonic* layer is the physi-cal layer of the SONET, shown in Figure 5.5. Each layer in the hierarchy termi-nates its corresponding header fields in the SONET payload. The functionalities of each layer are described further in detail.

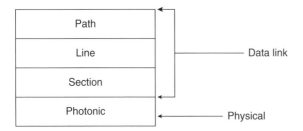

Figure 5.5 The protocol hierarchy.

Section layer. A section is a single fiber link terminated at a section terminating equipment (STE), which can be any network element (NE). As the name indicates, STE can access, modify, or terminate the section header overhead (OH) field. The main functions of the section layer are formatting the SONET frames, scrambling, error monitoring, and converting the electrical signals to optical signals, or vice versa. The section OH in the SONET frame is associated with the transport of STS-1 frames over a section. The first nine bytes in the transport header correspond to the section overheads. The section overhead is recalculated for each NE.

Line layer. The line OHs are associated with the transport of SPEs over a *line*. The overhead bytes from 10 to 27 are the line overhead bytes. Line-terminating equipment (LTE) originates or terminates one or more sections of a line signal. The LTE does the synchronization and multiplexing of information on SONET frames. Multiple lower-level SONET signals can be mixed together to form higher-level SONET signals. An ADM is an example of LTE.

Path layer. The *path* layer provides end-to-end connections between the point where the SPE originates and the point where it terminates on the nodes. Path-terminating equipment (PTE) interfaces non-SONET equipment to the SONET network. At this layer, the payload is mapped and de-mapped into the SONET frame. This layer is concerned with end-to-end transport of data. The first column of the SPE is the path overhead.

Example 5.1 Consider the SONET network operating over the photonic layer shown in Figure 5.6a. Trace the path of the connection through the network indicating the termination of different layers at each network element.

Solution
The trace of the path is shown in Figure 5.6b. At the source node, PTE maps different low data signals and the path overhead to form an STS-N SPE and hands this to its line layer. In the end terminal LTE MUX the three input low data rate SPE signals are multiplexed and line overhead are added. This

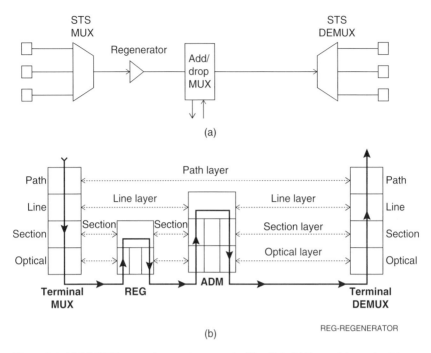

Figure 5.6 (a) A SONET network operating over the fiber link. (b) Trace of the path with different layers shown in the NEs.

combined signal is then passed to the section layer where framing and scrambling are performed and section overhead are added to form an STS-N signal. Finally, the electrical STS signal is converted to an optical signal for the photonic layer and transmitted over the fiber to the distant node, the regenerator. The signal is regenerated, passed through an ADM, which is an LTE-level device, and eventually terminated at a node at the PTE. At the distant node, the process is reversed from the photonic layer to the path layer where the lower rate signals terminate.

5.2.6 SONET Network Configurations

SONET is commonly used in ring topology. Towards the access network side it can be used as point-to-point, point-to-multipoint with ADM in hub for aggregation of traffic, and toward the core network side it is used in mesh configuration as well.

5.2.6.1 SONET Ring Architecture

Ring architecture is most commonly used in SONET/SDH networks, as rings have inherently more survivability because of the two-point connectivity as compared with the point-to-point linear topology, while still keeping the network topology relatively simple. The SONET/SDH ring architectures are *self-healing*, that is, they can automatically recover from link or node failures. The increasing demand for survivable services, fast restoration services, and flexibility have made rings a popular SONET topology. Most SONET/SDH rings can provide availability of five nines or 99.999% of the time. The time to restore the services has to be less than 50 ms.

Fiber links in the SONET rings are connected with ADMs. Multiple ADMs can be put into a ring configuration for either bidirectional or unidirectional traffic. In the case of any fiber cut or any node failure, the ADM sends the affected traffic through an alternate path in the ring without interruption with the help of network control signaling. This is achieved with the switch at each ADMs operated with the control signal using APS (automatic protection switching) protocol. In the case of a WDM SONET ring network, with wavelength multiplexing, multiple logical rings are formed on a single fiber. Each wavelength in such a WDM SONET network is operated at a particular OC-N line rate which can be TDM multiplexed to carry high data rate signals.

Different ring architectures have been developed based on the three features: number of fibers, direction of transmission (unidirectional or bidirectional), and *line* or *path* switching. In the case of *line switching* all the traffic that passes through a failed link is restored, while in the case of *path switching* the traffic on the specific source-destination path affected by a link failure is restored end-to-end.

Figures 5.7a,b show the connection for the two- and four-fiber ring configurations. In Figure 5.7a, fibers 1, 2, 3, and 4 are used to form the *working ring* (clockwise) and fibers 5, 6, 7, and 8 are used to form the *protection ring* (counter-clockwise). In another variation of the two-fiber ring, the capacity of each fiber is divided into two equal parts, one for working traffic and the other for protection traffic. Thus, each fiber can be both a working and a protection ring. In a four-fiber SONET/SDH ring there are two working rings and two protection rings.

Based on the above mentioned features, we have the following two-fiber or four-fiber possible ring architectures: the unidirectional *line* switched ring (ULSR), the bidirectional *line* switched ring (BLSR), the unidirectional *path* switched ring (UPSR), and the bidirectional *path* switched ring (BPSR). In path switching, a complete alternative route is provided for the source to destination end nodes connection in case of failure of any link along the path. In the case of line switching, total traffic on the particular link is given an alternate route only around that link. We will be discussing this in more detail in Chapter 7.

Out of the four types of ring topologies, the following three are used the most:

- two-fiber unidirectional path switched ring (2F-UPSR)
- two-fiber bidirectional line switched ring (2F-BLSR)
- four-fiber bidirectional line switched ring (4F-BLSR).

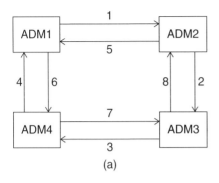

(a)

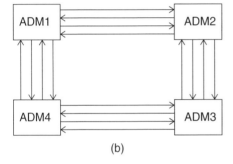

(b)

Figure 5.7 Self-healing SONET rings; (a) two fibers and (b) four fibers.

The 2F-UPSR, as shown in Figure 5.8a, has the working ring consisting of fibers 1, 2, 3, and 4, and the protection ring of fibers 5, 6, 7, and 8 have unidirectional transmission of signal, i.e. clockwise for the working and anticlockwise for the protection ring. In Figure 5.8a, *A* end node transmits to *B* over fiber 1 of the working ring in the clockwise direction, and *B* end node transmits over fibers 2, 3, and 4 of the working ring also in the clockwise direction to *A*. For the *path*-level protection, which is a $(1+1)$ protection scheme, the signal transmitted by *A* is split into two parts. One copy is transmitted over the fiber 1 to *B*, and the other copy is transmitted over the fibers 6, 7, and 8 in the anticlockwise direction to *B*. During normal operation, *B* receives two identical signals from *A* and selects the one with the best quality. If fiber 1 fails, *B* will continue to receive *A*'s signal over the protection ring path. The same applies if there is a node failure. These types of rings are used in the edge rings of the metropolitan network where traffic is not too high.

2F-BLSR which is used in metro core WDM rings is shown in Figure 5.8b with six nodes. On ring 1 made of fibers 1, 2, 3, 4, 5, and 6, the transmission is clockwise, and on ring 2 made of fibers 7, 8, 9, 10, 11, and 12 the transmission is counter-clockwise. Both the rings 1 and 2 carry working and protection traffic by dividing their capacity in two parts in terms of number of wavelengths in

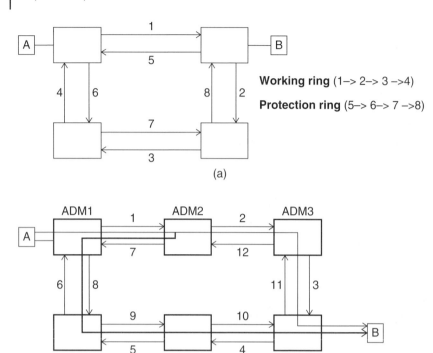

(a)

(b)

Figure 5.8 (a) Two-fiber unidirectional path switched ring. (b) Two-fiber bidirectional *line switched ring* (2F-BLSR).

the case of WDM rings for each fiber. Half of the capacity is used to carry working traffic and the other half protection traffic, therefore the nodes can choose the shortest route over any of the rings. For example, if node *A* is to transmit to *B*, it can select either a clockwise or an anti-clockwise path to reach the destination over any of the rings. In the case of a cut in fiber 2, signal loops back from ADM2 automatically with APS to take the path over fibers 7, 8, 9, and 10 to reach node *B*, as shown in Figure 5.8b. This ring architecture provides *line* switching. For 4F-BLSR, there is a redundancy of two fibers, otherwise the routing is the same as for 2F-BLSR.

5.2.7 Traffic Grooming in SONET/SDH Networks

Traffic *grooming* in general refers to aggregating traffic with similar characteristics, such as destinations, quality of service (QoS), or traffic type, so that they can be handled by the network as a group instead of individually, saving network infrastructure and processing. Aggregation of low-speed SONET connections onto a high-speed OC-N connection is said to be the traffic *grooming* problem [8, 9]. This helps in reducing the number of ADMs required in the network.

The WDM metro core rings are interfaced with high bit rates of 10, 40, and 100 Gbps long-haul networks (as shown in Figure 5.2), which carry large granularity wavelength channels in the Gbps range. Therefore, the metro edge rings which are interfaced with the access network's lower speed dynamic, multi-protocol, and multi-media traffic require intelligent opto-electronic solutions to perform the multi-protocol aggregation/grooming of this access traffic onto higher bit-rate WDM channels. Also, normally an ADM is required for each wavelength at every node in the SONET ring to perform traffic add/drop on that particular wavelength. But when there are large numbers of wavelengths supported by a WDM-SONET ring and not all of them have to be dropped or added at every node, it is not cost-effective to have large number of ADMs for every wavelength at each network node in a WDM-SONET ring. Most of the traffic on the ring can be bypassed by the intermediate nodes as it does not have to be dropped. Therefore, intelligent algorithms for traffic grooming are required in order to reduce the cost of the WDM-SONET ring design as it is possible for a node to bypass most of the wavelength channels optically with optical add-drop multiplexer (OADM) and only drop the wavelengths carrying the traffic destined to the node [9].

5.2.8 Scalability of SONET/SDH Networks

Capacity expansion or scalability of the laid-down network is an important issue for increasing higher bandwidth applications and services. Capacity expansion of SONET/SDH metro networks can be accomplished in several ways: first, with upgrade of equipment from lower rate to a higher OC-N rate, i.e. the TDM-SONET; second, with installation of another new wavelength ring on the existing fiber or the WDM-SONET; and lastly, adding a new fiber ring itself as WRN-SONET.

After TDM-SONET electronic processing in the ADMs, the next level of scalability is with the DWDM technology but with electronic processing. By applying WDM technology, several logical rings, each on a different wavelength, can be formed on a physical ring network but with ADMs at the nodes. Installing more fibers with wavelength routing in OXC nodes and overcoming the O-E-O conversions at the ADM or DCS by bypassing the traffic is the third level of scalability. WDM rings with OXC offer excellent bandwidth scalability, data transparency, and multiple data rates. The other advantage one gets with WDM rings is the enhanced survivability at the optical layer in metro networks. Much like TDM-SONET rings, survivability is achieved by reserving bandwidth of wavelengths and fiber for protection purposes. In a multi-fiber ring network, OXCs are used in place of OADMs to provide flexibility in switching a wavelength from one fiber to another. WDM optical ring networks have several attractive features over mesh networks, such as ease of planning and management, simple control, faster failure restoration, and lower cost. The *wavelength routing and assignment* problem is much simpler in a WDM ring network when compared with the routing problem in a mesh. In a unidirectional ring network, there is only one route between any two nodes. In a bidirectional ring network, there are only two routes between any two nodes and usually the shorter route is preferred as it consumes lesser bandwidth. All in all, the SONET ring architecture with its advanced survivability has excellent scalability features to extend from 2.5 Gbps capacity to hundreds of Gbps with WDM routing SONET rings.

To summarize, SONET/SDH has several advantages with respect to multiplexing standards, availability, capacity scalability, and grooming, and that is the reason for their survival for that long a time. Nevertheless, there are certain limitations. There is bandwidth limitation due to electrical processing, cost, power, and non-transparency. The multiplexing hierarchy is rigid, therefore there can be enough stranded bandwidth when more capacity is needed with the next multiple rate, resulting in an outlay for more capacity than is required. Also, since the hierarchy is optimized for voice traffic, there are inherent inefficiencies when carrying data traffic with SONET frames, and special encapsulation is required to transport for each data protocol.

5.3 Optical Transport Network

The SONET/SDH protocol, discussed in the last section, had many good features of fault tolerance, availability, etc., but it lacked many key features required in WDM large area optical networks, such as end-to-end network performance monitoring and fault detection in large networks and standard communication channels for OAM of WDM networks to name a few. In large area OTNs WDM architecture is the norm. These networks are operated by multi-carriers and their many interconnected domains are operated by multiple operators across the network boundaries. In this scenario, the lack of OAM standardization for WDM large networks becomes a big limitation in running them. Any fault in one portion of the network cannot be communicated to other portions of the network, especially across carrier boundaries. Hence, a new standard was required. With an OTN layer [10–13] between the IP and the DWDM layer we provide an efficient and cost-effective multi-service provisioning platform which aggregates traffic with efficient convergence of multiple protocols and services, including SONET/SDH and data services, while also providing scope for future client protocols. There is grooming of the optical signals at sub-wavelength level in OTN, which makes efficient use of available DWDM bandwidth and can carry larger volume of traffic. The OTN layer effectively maps different client protocols and rates into the same 10G, 40G, or 100G pipes, providing high bandwidth capabilities at comparatively low cost. Transit traffic can bypass the network's IP routers through the intermediate nodes with OTN switching applied at the OXC on these nodes. Thus, the router capacity is used efficiently along with reducing power consumption. There is a standard frame structure, the ITU-T G.709 OTN standard *digital wrapper*, used in OTN to encapsulate the payload and overheads, including the OAM bytes required for WDM networks. Unlike the SONET/SDH, OTN is an asynchronous system and can allow different traffic types, including Ethernet, digital video, and the legacy SONET/SDH to be carried over the digital wrapper frame more efficiently.

The OTN infrastructure has a set of network elements – transponders, muxponders, switches, etc. – connected by optical fiber links. These optical elements provide the functionality of transport, multiplexing, routing, management, supervision, and survivability of optical channels according to ITU G.872 recommendation. OTN is now an important networking standard for metro networks because of its many distinguishing characteristics: it provides transport for any digital signal independent of client-specific aspects; it can adapt to ever-changing customer requirements and provide more effective optical network management. Illuminating the other several advantages, OTN has an intelligent control plane enabling automated mesh connectivity and 50 ms mesh restoration for all client services, providing better performance with strong FEC (forward error correction) and more levels of tandem connection monitoring (TCM) across different network domains. It also improves network efficiency by conducting multiplexing of data channels for optimum capacity utilization, reducing network operational cost with effective OAM and protection capability which can provide remote manageability and all monitoring needed for reliable service provisioning. OTN being asynchronous, there is relatively less complexity and cost associated with timing hierarchy rates of SONET/SDH.

5.3.1 Layered Hierarchy of OTN

ITU-T standard G.709 [10] defines a number of layers in the OTN hierarchy. The layered hierarchy for OTN describes the functionality that is needed to make OTN work. It is based on the network architecture defined in ITU G.872 [14]. OTN layered hierarchy is composed of optical and electronic layers, as shown in Figure 5.9a. The OTN optical layer functions are wavelength multiplexing, switching and routing, and monitoring network performance at various levels in the network. The optical layer has three sub-layers: optical channel (OCh) layer, optical multiplex section (OMS), and optical transmission section (OTS). The different

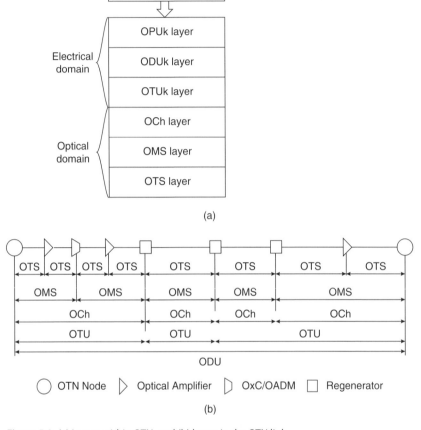

Figure 5.9 (a) Layers within OTN, and (b) layers in the OTN link.

sub-layers are there to delineate the various management functions. The services they provided are as follows:

- *OTS layer*. The lowermost optical sub-layer manages fiber link segments between any two optical devices in the span, such as optical amplifiers and WDM multiplexers. OTS provides transmission of optical signal based on the type of fiber, such as a dispersion shifted fiber (DSF). The services provided are the correct signal generation and reception at the section level.
- *OMS layer*. This layer manages fiber links between WDM multiplexers and switches. It provides multiplexing/de-multiplexing, and manages multiplex section backup and recovery. It gives support to WDM signals and manages each signal as an optical channel or a *lightpath*.

- *OCh layer.* This layer manages optical connections between 3R regenerators. It provides end-to-end optical channels s between two optical nodes and support client payload. It provides services of routing, monitoring, provisioning, and fault tolerance features in optical domain. Through the OCh layer the optical layer provides optical links to the electronic layers, and vice versa.

The electronic layers also have three digital sub-layers: the digital layer above the OCh is the OTU (optical transport unit), followed by the ODU (optical data unit), and the OPU (optical path unit).

- *OTU layer.* All client signals are mapped into an optical channel via the OTU layer. This layer has the same role as the SONET's *section layer.* The OTU contains the FEC. After the FEC are added, the signal is then converted to the optical domain.
- *ODU layer.* The electronic sub-layer above the OTU layer is the ODU layer, which is for connections composed of multiple links. SONET's *line layer* has a similar role to OTN's ODU layer. Different client services are converted into ODU packets and these are directed to their destination ports and converted into OTU for optical transmission.
- *OPU layer.* The OPU encapsulates the different client signals and carries out the required rate justification if needed. It is analogous to the *path layer* in SONET/SDH in that it is mapped at the source, de-mapped at the sink, and not modified by the network.

In Figure 5.9b, the different sub-layers are shown for a network link according to their functionalities. OTN transports client signals into a G.709 frame, OTUk (k is for different line rates), that is transported by an OCh on a wavelength channel. Each wavelength channel carries its G.709 frame with its own frequency, there is no common clock for the different OTUk of the OTN. A trail is generated in an OTN network element that maps the client data into an ODUk and terminates it in another OTN network element that de-maps the client signal from the ODUk. Between the two OTN trail termination network elements, there can be 3R regenerators which perform complete regeneration of the pulse shape, clock recovery, and retiming within required jitter limits.

5.3.2 Lines Rates of OTN

Next we study mapping of the client signals of various rates in OTN protocol. Unlike the SONET standard, in OTN protocol different standardized OTN line rates are supported for the various signals. First the client signals are mapped in OPUk frames with few overheads which are then mapped into an ODUk frame, followed by mapping the ODUk in the OTUk frame. Here the subscript k stands for the bit rate of framing signal, where k = 1 is for 2.5 Gbps, k = 2 for 10 Gbps, k = 3 for 40 Gbps, and k = 4 for 100 Gbps. The ODUk signal is the server layer signal for client signals in the OPU frame. Switching is OTN is done at the ODU level. There are several standardized ODU and OTU rates as given in Tables 5.2 and 5.3. It can be observed that OTN is protocol agnostic, as it supports all types of payloads.

Here, ODUk ($k = 1/2/2e/3/3e2/4$) is the information structure which is mapped in the other information structure, OTUk (k = $1/2/2e/3/3e2/4$). The above ODUk information structures are defined in ITU-T Recommendation G.709.

As can be seen from Tables 5.2 and 5.3, the OTN frame carries all types of payloads and provides a transparent transport to these signals by wrapping them in the OTN frame. OTN transports both synchronous and asynchronous signals but it does not have to transport the network synchronization signals, since network synchronization can be transported within the payload, mainly by SDH/SONET client tributaries. Two types of mapping have been specified for the transport of CBR payload, e.g. SONET/SDH. The first one is asynchronous mapping, which is the most widely used, where the payload floats within the OTN frame. In this case, there is no frequency relationship between the payload and the OTN frame frequencies, thus simple free running oscillators can be used to generate the OTN frame. The second is synchronous mapping, where the timing used to generate the OTN frame is extracted from a CBR

Table 5.2 Standard ODU line rates.

Signal	Data rate (Gbps)	Signal protocol supported
ODU0	1.24416	A timing transparent transcoded (compressed) 1000 BASE-X signal or packets (Ethernet, MPLS, or IP) using generic framing procedure
ODU1	2.498775126	Two ODU0 or STS-48/STM-16 or packets (Ethernet, MPLS, or IP) using GFP
ODU2	10.037273924	Eight ODU0/four ODU1/or STS-192/ STM-64/WAN PHY (10GBASE-W)/packets (Ethernet, MPLS, or IP) using GFP
ODU2e	10.399525316	10 Gigabit Ethernet/compressed fiber channel 10GFC
ODU3	40.319218983	32 ODU0/16 ODU1/four ODU2/STS-768/ STM-256/compressed 40 Gigabit Ethernet/ packets (Ethernet, MPLS, or IP) using GFP
ODU3e2	41.785968559	Four ODU2e signals
ODU4	104.794445815	80 ODU0/40 ODU1/10 ODU2/two ODU3/100 Gigabit Ethernet
ODUflex (CBR)	239/238x (client bit rate)	Constant bit rate signal such as fiber channel, 8GFC, InfiniBand or Common Public Radio Interface
ODUflex (GFP)	Any configured rate	Packets (Ethernet, MPLS, or IP) using GFP

Table 5.3 Standard OTU line rates.

Signal	Data rate (Gbps)	Protocol signals supported
OTU1	2.66	SONET OC-48/SDH STM-16 signal
OTU2	10.70	OC-192, STM-64 or WAN physical layer (PHY) for 10 Gigabit Ethernet (10GBASE-W)
OTU2e	11.09	10 Gigabit Ethernet LAN PHY coming from IP/Ethernet switches and routers at full line rate of 10.3 Gbps
OTU3	43.01	OC-768 or STM-256 or a 40 Gigabit Ethernet
OTU3e2	44.58	Up to four OTU2e signals
OTU4	112	100 Gigabit Ethernet signal

client tributary, e.g. SDH/SONET. This specification allows for very simple implementation of timing in OTN switches compared with SONET/SDH.

Multiplexing in the OTN domain is shown in Table 5.2. There can be single-stage and double-stage ODU multiplexing. Double-stage multiplexing is typically used for mapping several small ODUs into one large ODU. The flexible multiplexing scheme allows mapping in one or more steps of small ODU data packets into a larger ODU container, maintaining a small data granularity while using a large transport capacity, viz. four ODU1s can be multiplexed to an ODU2. Up to 16 ODU1s or four ODU2s can be multiplexed to an ODU3. It is possible to mix ODU1s and ODU2s in an ODU3. G.709 defines the OPUk which can contain the entire SONET/SDH signal. Thus, the transport of such client signals in the OTN is bit transparent and also timing transparent. An operator can offer services at various bit rates (2.5G, 10G, …) independent of the bit rate per wavelength, using the multiplexing and inverse multiplexing features of the OTN.

For the OTN a Reed-Solomon 16 byte-interleaved FEC scheme is defined, which uses 4 × 256 bytes of check information per OTU frame. G.709 FEC implements a Reed-Solomon RS(255,239) code. Error correction of bursts up to 128 consecutive error bytes can be done. G.709 defines a stronger FEC for OTN that can result in up to 6.2 dB improvement in signal to noise ratio (SNR). The coding gain provided by the FEC can be used for extending the reach, decreasing power in each channel and thus increasing the number of channels in the DWDM system, by reducing launched power, extinction ratio, and noise figure, etc. for a given link. FEC has been proven to be effective in optical signal-to-noise ratio (OSNR) limited systems as well as in dispersion limited systems.

5.3.3 OTN Frame Structure

OTN is also called *digital wrapper technology* or *optical channel wrapper*. The G.709 OTN frame has payload and transport overheads that provides OAM

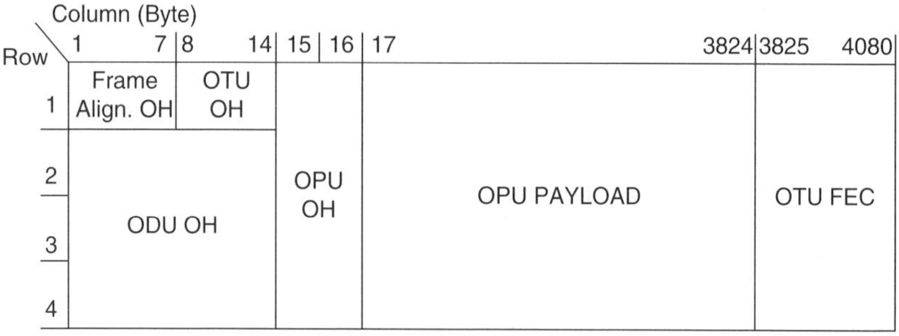

Figure 5.10 OTN frame.

capabilities and FEC. Bytes are provided for payload movement inside the OTN frame which can occur when line and client clock sources are asynchronous. In the case when synchronous clocking is supported by OTN, the payload is fixed relative to the overhead. The frame structure of the G.709 OTN frame is shown in Figure 5.10. It has three parts: overhead (OH), payload, and the FEC. There is a total of 4 rows and 4080 columns in 8-bit byte blocks and transmission of bytes is serially from right to left, from top to bottom, as in the case of SONET. OTN ensures that all 40 Gbps and 100 Gbps digital wrappers are fully packed to make maximum use of the network's available bandwidth. All the OTN signals at different data rates, as given in Table 5.1, have the same frame structure, but the frame period reduces as the data rate increases.

5.3.3.1 OTN Frame Structure Overheads
Figure 5.11 shows where and how the OHs for OPU, ODU, and OTU are added in the OTN frame in the different layers. The top layer is OPU, which has the clients' payload and OHs. The OPU frame structure has 4 rows and 3810 columns, as shown in Figure 5.10. Columns 15 and 16 are dedicated to OPU OH and columns 17 to 3824 of the OTN frame are dedicated to OPU payload. The OHs are there to support the adaptation of client signals for transport over an optical channel. There are bytes for identification of payload content such as SONET/SDH, ATM, or GFP. G 709 supports the virtual concatenation function link as in SONET, which groups multiple OPU interfaces onto a single OPU that runs at a higher rate.

The ODU layer provides end-to-end path supervision and supports TCM with six levels of TCM. The three main areas of the ODU frame, as shown in Figure 5.10 within the OTN frame, are the ODU overhead area, FEC area, and the OPU area. It has the largest number of overhead fields. It has fields for network operators for management functions, OHs for reporting fault type and their location for path level faults, alarm and path protection capabilities, etc. In the OTN frame structure, ODU information payload occupies four rows and columns from 15 to 3824 for its payload, which is the OPUk of (4×3810) bytes, and its own OHs occupy rows 2 to 4 and columns 1 to 14.

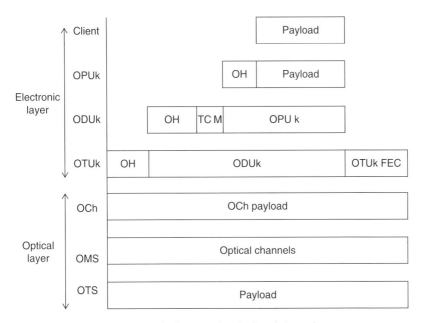

Figure 5.11 OTN layered hierarchy for optical multiplexed channels.

The OTU frame structure includes the FEC, multiple ODUs, and its own OHs. In the OTU layer, the OHs provide supervisory functions, optical section protection, and conditions the signal for transport between the optical channel termination points, where retiming, reshaping, and regeneration occur. The OTUk information structure is based on the ODU frame structure and extends it with 256 columns for the FEC, resulting in an octet-based block frame structure with 4 rows and 4080 columns of OTN frame. The 15 OH bytes occupies the top row from the first column, out of which the seven bytes are for frame alignment and the rest are the OTUk OHs for the various functions. The OTU FEC contains the RS(255,239) code. This is then carried by a single wavelength as an optical channel.

As multiple wavelengths are transported over the OTN, an overhead must be added to each wavelength channel to enable the management functionality of the OTN. Additional OH may be added to the OCh to enable the management of multiple wavelengths in the OTN. The OMS and the OTS are then constructed. The result is an optical channel comprising an OH section, a client signal, and an FEC segment, as Figure 5.11 shows.

5.3.4   OTN Switching

In OTN, different client services are mapped into the ODU frames, as shown in Table 5.2. OTN provides service-agnostic switching by switching them at ODU packet level. The ODU is the basic payload that is electronically groomed and switched within an OTN network. The actual switching is at the *digital wrapper* level, which is an ODU packet formed by multiplexing together more than one ODU frame. The digital wrapper encapsulates diverse data frames from different sources in a single entity so that they can be managed more easily. The ODU has frames of fixed sizes and uses bit rates ranging from 1 Gbps to 100 Gbps to match interfaces for a range of standards, such as Ethernet and SONET.

The traffic grooming in OTN is so done that it uses the large bandwidth of the optical wavelength channels very efficiently. It uses sub-lamda intermediate traffic grooming with an ODU switch to fully use the capacity of the OCh. Table 5.2 illustrates which service types can be mapped into an ODU packet and finally into the required transport OTU format. This type of grooming reduces wavelength usage around 30–40%.

5.3.4.1 OTN Switch

We are aware that OTN transparently transports the client signals which are mapped into OTU transport frames. The OTN switch which does the mapping and aggregation of signals has interfaces with them on the client side, and on the line side it interfaces with the WDM network, as shown in Figure 5.12. A large number of different service types can be processed by a single OTN switch simultaneously, such as single ODUk, dual-stage multiplexed ODUk, ODU flex, SONET and Ethernet. The key function that the OTN switch performs on the various client signals is the protocol processing of all the signals. A few of the more complex processes are the multiplexing and de-multiplexing

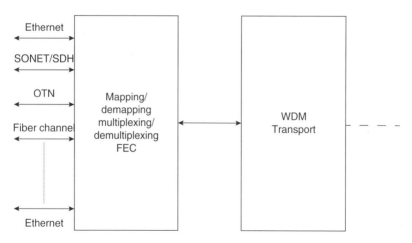

Figure 5.12 OTN switch interfacing with different user protocols on the client side and interfaced with WDM transport on the line side.

of OTN signals at the electronic level, mapping and de-mapping of non-OTN signals into and out of OTN signals, and FEC and packet processing in conjunction with mapping/de-mapping of packets into and out of OTN signals. The switch efficiently grooms the signals on a sub-wavelength level in the digital wrapper and then converts it to OCh for further multiplexing in optical domain. This increases network efficiency by enabling more effective use of bandwidth in a mixed network that carries synchronous and packet traffic. It thus minimizes the amount of traffic handled by client routers/switches and enables smaller and less costly routers to be used. The OTN switch fabrics are typically implemented using packet switch fabrics. In the case of higher bit rate (100 Gbps) OTN systems in the core metro, a single line card that supports many different protocols and data rates is used on the client side. The line card can also take several lower speed client signals and combine them into and separate them out of a single 100 Gbps optical transport signal.

The next issue is where in the network to deploy the OTN switches. When an OTN switch is deployed along with the cross-connect node, OTN switching enables transit traffic to bypass many of the network's IP layer routers and thus reduces the capacity requirement of the router. In an existing DWDM network with mesh topology, OTN switches can be deployed as a standalone network element operating as a cross-connect at a network node. The wavelengths of the DWDM system are optically de-multiplexed and individually connected to the OTN switch interface cards, as shown in Figure 5.13. In a newly deployed network where the OTN switch and the optical node can be integrated into a single network element in the mesh network, the composite element is managed by the network management system (NMS) as one NE featuring OTN switching, WDM switching, and WDM transmission.

5.3.5 Tandem Connection Monitoring

In SONET/SDH, connection monitoring can be done at the section, line, and path layers, but there is no provision for monitoring the connection when in a large network it passes through multiple domains of different operators. OTN enables path-layer monitoring at multiple user-defined endpoints and it is one of the powerful features of OTN. This is made possible with several TCM overheads. TCM is a layer between the line and path monitoring (PM). The user can monitor the connection in the various domains of the network when part of the connection passes through another carrier operator's network. In OTN, G.709 allows six TCMs, each for monitoring different parts of the connection segment in the multi-operator large network.

Figure 5.14 shows one such connection passing through three operators' domains. Here operator 1 needs to have operators 2 and 3 carry his signal. However, he also needs a way of monitoring the signal as it passes through their networks, which is made possible with TCMs. The different TCMs used

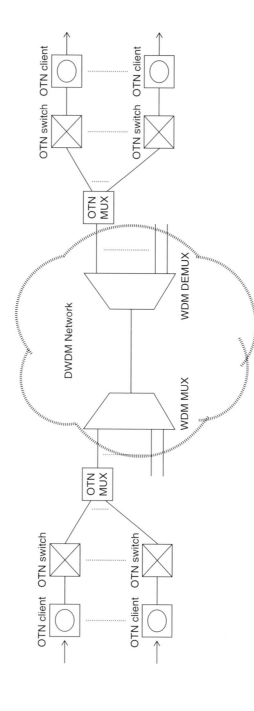

Figure 5.13 Deployment of OTN switch in the DWDM network.

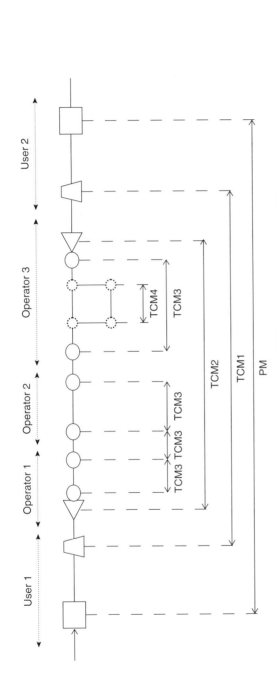

TCM4 – Protection supervision
TCM3 – Domain and domain interconnect supervision
TCM2 – Lead operator QoS supervision
TCM1 – User QoS supervision
PM – Path monitoring

☐ Path supervisory node   ◯ Domain end node

◻ User end node   ◌ Working and protection
                      path node

◁ Lead operator node

Figure 5.14 Tandem connection monitoring.

for monitoring, as shown, are TCM1, which is used by the *user* to monitor the QoS of the connection path; TCM2, which is used by the lead operator to monitor the end-to-end QoS; TCM3, which is used by the various domains for intra-domain monitoring; and TCM4, which is used for protection monitoring by the intermediate operators. Besides these, within the OTN overhead the PM bytes provide a critical function in monitoring end-to-end signal quality, enabling carriers to guarantee customer service-level agreements (SLAs).

TCM supports monitoring of ODUk connections for several network applications, such as monitoring OCh subnetwork connection protection switching, to determine not only signal disruption but also signal degrade conditions, monitoring an OCh tandem connection for the purpose of detecting a signal failure or signal degrade condition in a switched OCh connection, and to initiate automatic restoration of the connection during fault and error conditions in the network, verification of delivered QoS, etc.

5.4   Summary

Metropolitan networks occupy a strategic place in the overall network hierarchy, bridging end users to the core network's very high long-haul capacities. The presence of multiple customers with diverse bandwidth needs, together with traffic requirements that are constantly changing requires a compatible technology to deal with the network reconfigurations, varying network services, and load. Metro networks need to include scalability, capacity, transparency, and survivability. The legacy SONET/SDH have these characteristics. Meanwhile, metro networks are merging the optical and electronic domains, aggregating many user protocols onto large metro-core wavelength tributaries as OTN networks. The OTN standards were designed to provide better compatibility with a robust set of OAM features for WDM architectures, including TEM and FEC features and a standard mapping structure for multiplexing lower rate signals into higher-speed payloads of 40–100 Gbps. OTN is now being increasing, being used as switching and grooming large-core OTN switch nodes at major gateway locations. We have covered both SONET and OTN in sufficient detail considering their importance in metropolitan networks.

Problems

5.1   Which sublayer within the client layer (SONET or OTN) or optical layer would be responsible for handling the following functions?

    f)   A SONET path fails, and the traffic must be switched over to another path.

    g)   Layer where FEC added in OTN.

h) Many client streams are to be multiplexed onto a higher-speed stream and transmitted over a link.

i) A fiber fails, and SONET line terminals at the end of the link reroute all the traffic onto another fiber.

j) The error rate on a SONET link between regenerators is to be monitored.

k) Multiple wavelength channels multiplexed in OTN.

5.2 Why is OTN preferred over SONET/SDH for use in DWDM large area networks?

5.3 Consider a five-node UPSR with one as the hub node. Each of the four remote nodes have two STS-48 channels terminating at the hub node to carry traffic. How does the performance of the UPSR compare with BLSR/2 with the same ring speed in both cases?

5.4 Design a six-node ring network with one hub node and five remote nodes. Each remote node has two dedicated wavelengths which terminate at the hub node to carry traffic to/from the hub. Determine the cost of the following two networks. (i) The network uses ten channels in two bands, each with five channels. It provides band OADMs, which can drop one out of the two bands. Once a band is dropped, all five wavelengths in the band have to be regenerated. A band OADM costs $10 000, and a single-channel regenerator costs $5000. No optical amplifiers are required with this system. (ii) The second system uses ten channels without bands. It provides single-channel OADMs, which can drop any single wavelength. Each of this OADM costs $5000. For this system, two optical line amplifiers are required, each costing $10 000

5.5 Consider a WDM-SONET ring with five nodes. All the nodes are required to route one STS-48 SONET connection to every other node in the ring. (i) Determine the minimum number of wavelengths and number of SONET ADMs required for these connections. (ii) Give a solution, if instead of SONET-ADMs OADMs are used in the ring. How will the resources change?

5.6 Consider a six-node two-fiber bidirectional line switched ring. Ring 1 with fibers 1, 2, 3, 4, 5, and 6 has clockwise transmission. Ring 2 with fibers 7, 8, 9, 10, 11, and 12 has counter-clockwise transmission. If fiber 3 fails between nodes 3 and 4, show how the traffic will flow on the two fibers between station A connected to node-1 and station B connected at node 4. How will the traffic flow from station A to station C, which is connected to node 6?

5.7 A unidirectional ring network has four nodes, 1–2–3–4–1, each node having two SONET OC-12 connections to be transmitted to every other node. The SONET ADMs can multiplex four connections to be carried on a wavelength. Design traffic aggregation using minimum number of SONET ADMs. What will be minimum number of wavelengths required?

Draw a neat diagram and table to show the allocation of wavelength to different connections.

5.8 Consider next an optical metro ring network consisting of five nodes, 1–2–3–4–5–1, where node 1 is a hub node. The numbers of OC-12 connections which originate from hub to the remote nodes are: 3, 4, 4, 2 from 1–2, 1–3, 1–4 and 1–5, respectively. Design the rings so that the minimum number of ADMs and wavelengths are employed in the network.

References

1 Kartalopoulos, S.V. (1999). *Understanding, SONET/SDH and ATM: Communications Networks for the Next Millennium.* Wiley-Blackwell.

2 Ghani, N., Pan, J.-Y., and Cheng, X. (2002). Chapter 8 – Metropolitan optical networks. In: *Optical Fiber Telecommunications IV-B*, A Volume in Optics and Photonics, 4e (ed. I. Kaminow and T. Li), 329–403. Elsevier.

3 Sexton, M. and Reid, A. (1997). *Broadband Networking: ATM, SDH and SONET.* Boston: Artech House.

4 Siller, C.A. and Shafi, M. (eds.) (1996). *SONET/SDH: A Sourcebook of Synchronous Networking.* Los Alamitos, CA: IEEE Press.

5 Yuan, P., Gambiroza, V., and Knightly, E. (2004). The IEEE 802.17 media access protocol for high-speed metropolitan-area resilient packet rings. *IEEE Network* 18 (3): 8–15.

6 T. Nadeau, V. Sharma, and A. Gusmante, eds. , *"Series on Next-Generation Carrier Ethernet Transport Technologies"*, IEEE Communications Magazine, vol. 46, Mar. 2008.

7 van Helvoort, H. (2009). *The ComSoc Guide to Next Generation Optical Transport, SDH/SONET/OTN.* Wiley-IEEE Press.

8 O. Gerstel, R. Ramaswami, and G. H. Sasaki, *Cost effective traffic grooming in WDM rings, Proceedings of IEEE Infocom*, 1998.

9 Dutta, R. and Rouskas, G.N. (2002). On optimal traffic grooming in WDM rings. *IEEE J. Sel. Areas Commun.* 20 (1): 110–121.

10 ITU-T. *Recommendation G.709: Interfaces for the optical transport network (OTN)*, 2003.

11 Perros, H.G. (2005). *Connection-Oriented Networks: SONET/SDH, ATM, MPLS and Optical Networks.* Hoboken, NJ: Wiley.

12 https://www.fujitsu.com/us/Images/OTNNetworkBenefitswp.pdf

13 https://www.ciena.com/.../OTN-vs-SONETSDH-Comparing-the-differences-prx.html

14 https://www.itu.int/rec/T-REC-G.872

6

Wavelength Routed Wide Area Networks

6.1 Introduction

We discussed the shared-channel broadcast-and-select optical networks in an earlier chapter. We now move on to wavelength routed networks (WRNs) architecture where the optical media is not shared but the connection is directed from a source to its destination on a specific route, unlike in the case of B&S networks where the optical signal is broadcasted to all the connected nodes in the network. These are the second-generation optical networks, employing wavelength division multiplexing (WDM), and are used for wide-area networks (WANs) targeting nationwide and global coverage, and now also increasing being used in metro-networks. The optical nodes in the network form an arbitrary mesh topology with optical fiber carrying multiple orthogonal wavelengths, each carrying traffic with very high data rate of up to 100–200 Gbps. The optical nodes are routing nodes which route different wavelengths toward their destinations over the given physical fiber network.

Optical cross-connects (OXCs) have the capability to allow the wavelengths or *lightpaths* (LPs) traffic to bypass the node without an optical-electronic-optical (O-E-O) conversion and buffering at the intermediate nodes if the *lightpath* does not carry the traffic for the specific node. This process is known as *wavelength routing* and *bypassing*. A *lightpath* is defined as an *all-optical communication path* between the two client-node pairs, established by allocating the same wavelength throughout the physical route of the transmitted data on the fiber. It is uniquely identified by a physical path and a wavelength. Wavelength routed WDM networks have the potential to avoid the three major limitations of the B&S networks: one is wavelength reuse (discussed later), the second is the scalability to wide area core networks, and finally no power splitting loss at the node. The important characteristic which enables WRNs to span long distances is that the power is transmitted over the *lightpath* only to the relevant destination and not broadcasted to all the connected stations in the network. It thus provides a point-to-point connection between the

Optical WDM Networks: From Static to Elastic Networks, First Edition. Devi Chadha.
© 2019 John Wiley & Sons Ltd. Published 2019 by John Wiley & Sons Ltd.

source-destination pair nodes. Each optical network node (ONN) has multiple input-output ports connected to fibers, with each fiber carrying the optical connections by a large number of orthogonal WDM channels. Thus, the transparent optical network can have very high capacity in THz. Besides having high capacity, the transparent optical networks also have the other advantages of independence of protocol, data rates, format, survivability, etc. [1, 2].

The *lightpath* connection from the ingress client node to the egress client node in the WRN has to have three characteristics: (i) *wavelength continuity*, (ii) *distinct wavelength assignment*, and (iii) *wavelength reuse*. The requirement for a source-destination (s-d) connection to have the same wavelength on all the links along the selected route is known as the *wavelength continuity constraint*. This is required when there are no optical wavelength converters available at the nodes. Two lightpaths cannot be assigned the same wavelength on a fiber to avoid contention. This requirement is known as *distinct wavelength assignment constraint*. However, two lightpaths can use the same wavelength if they use disjoint sets of links. This property is known as *wavelength reuse* [3].

With all the advantages of transparency of data rates, protocols, and high capacity in the WRN network, they have few limitations as well. First, the scalability limitation. The number of connected user network nodes in a purely optical network, though large still cannot be increased indefinitely because of the resource limitations of number of wavelengths, optical transmitters and receivers, ports in the cross-connects, fibers, and the available spectrum within the network. Consequently, practically in any transparent WRN the number of connections cannot exceed a certain limit. The second limitation of the transparent WRN approach is *geographical reach*. A transparent optical connection accumulates noise, distortion, and cross-talk over long distances and eventually becomes unusable. Thus, signal regeneration, which is usually electronic, eventually becomes necessary, sacrificing the advantages of transparency in the network. The third limitation is the underutilization of spectrum. Many applications require sub-lamda connectivity among a large number of end systems, which leads to inefficient use of the available spectrum if full capacity of one λ-channel is allocated to each of these connections.

The above limitations in the transparent WRN can be overcome with the help of a multi-layered approach. An electronic layer of logical routed network (LRN) nodes is added on top of the transparent optical network layer. Typical examples of LRNs nodes are networks of Synchronous Optical NETworking (SONET)/synchronous digital hierarchy (SDH) digital cross-connects (DCSs) for synchronous traffic, networks of IP/optical transport network (OTN) routers for asynchronous traffic, and asynchronous transfer mode (ATM) networks carried on a SONET DCS layer. The DCSs, IP routers, etc. are electronic nodes making the composite network a *hybrid* network instead of a totally transparent WDM optical network. These electronic routers or switches do the aggregation of the lower rate traffic with time division multiplexing (TDM). The resultant high rate composite signal traffic can then, in turn, modulate a λ-channel which can be routed and switched through the optical network over the fiber. The topology formed by the set of optical network nodes connected with the set of *lightpaths* is called *virtual topology* (VT) or *logical topology* (LT) or *lightpath topology*. In the wide area WDM network, OXCs are the routing node in the physical topology (PT) interconnected with optical fiber links. The λ-channels or lightpaths (LPs) in the WDM system flow over the fiber links. The λ-channels carry the traffic of user data over the VT in the optical domain from one electronic node to another node. The OXC can bypass the LP by switching the input fiber to other output fiber without the electronic processing if the LP is not to be terminated at the node. But when the packets are to be moved from one LP to the other, then the optical signal is changed to the electronic domain and back to the optical domain onto the other LP. This kind of architecture therefore will have both single- and multiple-hops of the LPs in order to set up optical connection between all user pairs. Further, this architecture with multiple orthogonal wavelengths, has the advantage of carrying much higher aggregate data and has high connection capacity due to spatial reuse of wavelengths. Thus, it can support a large number of user connections with TDM using a limited number of wavelengths. Also, in case of dynamic traffic as the traffic matrix changes, the earlier set of LPs has to be taken off and the new set of LPs has to set up, forming a different VT to transport the traffic without any change in the PT [4–7].

Thus, with the *virtual topology* embedded in the *PT* architecture, a large number of users can be accommodated with full utilization of the network capacity and with limited number of wavelengths and other network infrastructure. Also, the architecture can respond with much higher speed with the use of relative strengths of both optics and electronics by incorporating transparent and high-speed optical switching in the physical layer, and grooming with electronic switching of the lower rate user signals in the logical layer. The LPs in the virtual topology are totally transparent in terms of data format and rate, protocols or services, hence the optical network can be used for all type of services, be it wavelength, IP, ATM, SONET, Ethernet with suitable interfaces.

In this chapter we discuss the design, analysis, and performance of these hybrid multi-layered optical networks with electronically switched overlays. After explaining how connections are established in the all-optical WRN with the basic constraints, the remainder of the chapter is organized as follows: in Section 6.2 we explain the architecture of hybrid wavelength routed WDM network followed by the *optical layer* concept in Section 6.3. The details of the WRNs with an overlay of logically routing nodes are given in Section 6.4 where we discuss the different sub-problems in the hybrid optical network. The analytical formulation of WRN with logical overlay and the optimal solution for the routing and wavelength assignment (RWA) problem with integer linear programming (ILP) is introduced in this section. Section 6.5 offers several heuristic problems with suboptimal solutions for the transparent optical network are discussed. We also give the performance parameters and the quality-of-service (QoS) issues of the network in this section. The virtual topology heuristics are discussed in Section 6.6 and Section 6.7 summarizes the chapter.

6.2 The Hybrid Wavelength Routed Network Architecture

We once again refer to Figure 1.1, in Chapter 1, for the geographical reach of the optical network in order to highlight how the user nodes get connected with each other through the mesh WRN WAN with the help of the access and metro networks for traffic flow over the λ-channel. Different users in the access network are connected to the metro area network with SONET switches, IP routers via the Ethernet, or directly to a *λ-channel* over the optical fiber infrastructure, depending on the capacity requirement of the traffic or the internet service provider (ISP) agreement. The logical switching nodes (LSNs) or the client-network nodes (IP routers or SONET switches) in the metropolitan area network (MAN) are mostly connected in a ring topology or sometimes in a mesh topology as well, and the users are connected in the star or tree topology of the access network to these electronic IP routers or the SONET switches. The LSNs process the user data electronically. The end node at the interface between the metro and the core WRN is equipped with a set of optical fixed or tunable transmitters and receivers which sends optically modulated data into, and receives data from the core, respectively. The network nodes or OXC in the mesh core network are interconnected in the mesh topology by multiple fiber WDM links and each fiber link is capable of carrying multiple signals simultaneously on a different wavelength. The lower rate data of the different users to be inputted at the particular end node are first aggregated with time multiplexing, then the composite high data rate signal modulates a laser diode in the transmitter module of the network-node to form the λ-channel.

The connection between a pair of users is called a virtual connection (VC). The traffic in each VC moves through the electronic LSNs on a *logical path* (LP) composed of a succession of interconnected *logical connections* (LCs) [8–9]. The LCs are formed by the routing functions performed by the LSNs so that the traffic associated with a given source destination VC follows a prescribed LP through the LRNs. The LCs are carried by a λ-channel on the fiber. Thus, the LT in the optical networks, formed with the *lightpaths* for a given traffic matrix, is important as it is the interface between the client layer and the physical layer to optimize the network resources, such as the optical transmitters, receivers, and fiber spectrum, by distributing the network processing load in the electronic

LSNs and the ONNs. Also, LT is important to interface the sub-lamda client traffic to lamda traffic on the lightpath.

In a WRN, circuit switching at any node is achieved by using an OXC which is capable of switching a *LP* from an input fiber to another output fiber. If there is no optical wavelength conversion in the OXC, as assumed in this chapter, the wavelength of the lightpath at the input does not change in the output fiber of the OXC. For setting up connections in this hybrid network architecture, both *single-hop* and *multi-hop* lightpath approaches are used, and it exploits the characteristics of both optical and electronic processing. A *lightpath* in this architecture can provide *single-hop* communication between any two client nodes, which could otherwise be far apart in the PT. However, by employing a limited number of wavelengths, it may not be possible to set up lightpaths between all the user pairs. As a result, multi-hopping of lightpaths by electrical conversion of the input optical wavelength and then again generating the output wavelength, known as O-E-O conversion, at intermediate optical nodes may be necessary for a connection to be established between distant source and destination pair. With such architecture, large numbers of users can be supported with a limited number of wavelengths with spatial reuse of wavelengths over diverse paths. In addition, in the case when the traffic pattern changes, the set of *lightpaths* forms a different *multi-hop* LT to support it. Hence, reconfiguration of the LT is required with multi-hop hybrid technology with minimal disruption to the network operations [10–11].

We explain the features of the WRN, such as the wavelength continuity, distinct wavelength assignment, and the reuse of a wavelength on disjoint paths, with the help of a five-node network in Figure 6.1. The nodes do not have optical wavelength converters. Each node has one fixed transmitter and receiver, either at wavelength λ_1 or wavelength λ_2, and each fiber can carry both λ_1 and λ_2 wavelengths. Assume that lightpaths are to be established one for each of the node pairs $(1 \rightarrow 3)$, $(2 \rightarrow 4)$, $(3 \rightarrow 5)$, $(4 \rightarrow 1)$, and $(5 \rightarrow 2)$. The figure shows a possible way of routing the five lightpaths l_{13}, l_{24}, l_{35}, l_{41}, and l_{52}, where l_{ij} is the lightpath with source client-node i and destination client-node j. Since l_{13} uses wavelength λ_2, l_{52} has to use λ_1 to avoid contention due to distinct wavelength constraint, because l_{13} and l_{52} share a link. Lightpath l_{35} can reuse λ_1 as l_{52} and l_{35} do not have a common link in their path. Lightpath l_{41} can use λ_2, as both l_{35} and l_{52}, which share links with l_{41}, use λ_1; therefore, the distinct wavelength constraint is not contradicted. Similarly, for other nodes and links we use the alternate two wavelengths, avoiding the two lightpaths to have the same wavelength. Also, the same wavelength is used for two different lightpaths over fiber-link disjoint paths and wavelengths are reused, thus helping increase the number of lightpaths established with a limited number of wavelengths. With λ-reuse, five lightpath connections are possible with two wavelengths. Thus, we can appreciate the importance of judiciously routing the lightpaths and assigning a wavelength to the desired connections. This is known as

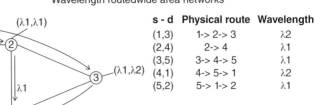

Wavelength routedwide area networks

s - d	Physical route	Wavelength
(1,3)	1-> 2-> 3	λ2
(2,4)	2-> 4	λ1
(3,5)	3-> 4-> 5	λ1
(4,1)	4-> 5-> 1	λ2
(5,2)	5-> 1-> 2	λ1

Figure 6.1 Wavelength routing and assignment.

wavelength routing and *assignment*, which minimizes the number of wavelengths in the given WDM network and can reduce the infrastructure cost for the transceivers in the case of a large area network.

To take it forward, we once again consider the five-node exemplar WDM network shown in Figure 6.2 with client network nodes (LSNs). These LSNs, which are electronic, are also connected to the optical network nodes of the WRN. The client traffic from the LSN is aggregated and then it modulates one of the wavelengths of the optical node transmitter which will carry the traffic over the fiber between the s-d client network node pairs. The fixed optical transmitter and receiver at each node are given in the inset in Figure 6.2. Any two network nodes (OXC) can be directly connected by a *lightpath* bypassing the intermediate network nodes if the receiving node has a receiver at that wavelength. This was shown in the case of connection from the node 3–5 with a lightpath at λ₁ in Figure 6.1. This is the *single-hop* connection. But if the two source-destination client nodes are not connected directly by a lightpath, they can also communicate using the *multi-hop* approach. For example, consider the case when node 5 in Figure 6.2 has a receiver at λ₂ instead of λ₁ wavelength. In this case, for the connection from node 3 to node 5, the intermediate node 4, which has a receiver at λ₁, receives the transmitted signal at λ₁, converts it to an electronic signal, and then in turn modulates its transmitter laser at λ₂ for onward transmission to node 5. This optical signal at λ₂ will be received by the receiver at node 5. There is an O-E-O conversion at the intermediate network node in this case. This is the *multi-hop* transmission to establish connection between the two nodes when the lightpath had to make two hops to reach its destination. In the hybrid optical network, the user/end nodes with sub-lamda traffic are connected to the core optical WRN through the SONET switches, IP router, etc., as shown in Figure 6.2. The end/user nodes communicate with each other using VC over the network, passing through a number of intermediate electronic LSN to connect the end nodes. Several of these VCs from different source-destination pairs are time-division multiplexed by the

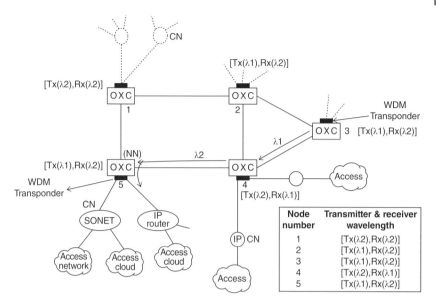

Figure 6.2 Wavelength routing and assignment with multi-hop connectivity.

electronic LSNs, converted to the optical domain to form a *λ-channel*, and transported over the underlying transparent WRN network. The ONN, i.e. OXC, with the help of LSNs performs the routing of these *lightpaths*, ensuring that the traffic associated with a given source-destination VC follows a prescribed *lightpath* through the network. Thus, the required connections in the network are made possible, both as single-hop and as multi-hop.

To summarize, upon arrival at an intermediate node, λ-channels from an inlet link are either bypassed or buffered and then forwarded out according to a given routing strategy. The buffering operations, which are currently implemented using electronic devices, limit the optical bandwidth utilization and result in an opto-electronic bottleneck problem. But the combination of a single-hop and a multi-hop system by employing an OXC *wavelength routed switch* results in a more efficient architecture. As the connections between any two source-destination nodes can span through different paths, potentially the network capacity increases as the number of nodes increase. Therefore, by properly designing *wavelength routing and assignment* schemes, a WRN can be made scalable and flexible. The WRN problem can be eased to a large extent by using wavelength conversion at each node. By adding a wavelength conversion optical device into an OXC, the switch becomes a *wavelength convertible routing switch*. By using a wavelength convertible routing switch, a wavelength in the network can be reused and the network utilization can be increased.

6.3 The Optical Layer in Wavelength Routing Networks

As introduced in Chapter 1, the optical network functional layered abstraction of second-generation WDM networks can be broadly segregated into three layers to handle the variety of protocols, data rates, and signal formats in the network: the *client/application layer*, the *logical layer*, and the *physical media layer*. The physical media layer was further divided into two sub-layers, with an *optical layer* being introduced between the *logical layer* and physical fiber layer in these networks, as shown in Figure 6.3. The *optical layer* serves the upper logical layer by providing *lightpaths* over the lower physical media (fiber) layer to the various electronic client networks in the client layer, such as the IP router, Ethernet, SONET or OTN switches. The payload of this electronic layer is passed to the optical layer as *lightpaths* which flow over the physical fiber media layer. Thus, the user traffic for the physical layer is now in terms of *lightpaths*. As per ITU-T Recommendation G.872 [12], the optical layer can be further decomposed into sub-layers, as discussed in detail in Chapter 1.

The optical layer provides client-independent and protocol-transparent services to the client networks. This is possible because the *lightpaths* are transparent to data rate, data format, or modulation schemes and therefore can carry a variety of client protocols. The *lightpaths* or *λ-channels* form an LT over the PT to carry the client traffic. While the PT is fixed once the fibers are laid, the virtual topology can be different from the PT and changes with the

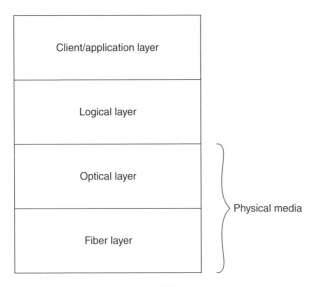

Figure 6.3 Layered structure of WRN.

client traffic. The advantages of the optical layer and the LT in a wavelength routing network are as follows:

- The LT of the optical layer is independent of the PT, hence the VT can be readily modified by reconfiguring the paths of the wavelengths in contrast to the fixed PT according to dynamic connection requirements. The lightpath can be automatically rerouted over alternate paths in case of failures, and reconfigured to an alternate LT for the connections, thus providing more reliability, flexibility, and robustness to the network, which is essential in large area networks with high connectivity. The fault recovery is possible at two levels now – at the VC level by rerouting traffic around a failed LC or LSN by assigning alternate paths for traffic flow between a given s-d node pair, and at the logical level by rerouting LCs supported by a cut fiber or failed ONN onto different fiber paths providing the optical layer protection.

- The optical layer also provides *grooming* of low data rate traffic with the help of client layer networks. A lightpath is a broadband pipe of bandwidth of 2.5 GHz and above. Many sub-lamda users do not require the full capacity, but large numbers of such users need to be connected to the network. Therefore the user traffic at the sub-lamda level is first aggregated to a higher rate electronically in the client networks, then these multiplexed aggregated user signals are put on the lightpath for better utilization of the optical spectrum.

The optical layer provides end-to-end networking of optical lightpaths for transparently conveying client data. In order to facilitate this, the OXC at the ONN establishes the *lightpaths* with the help of tunable transmitters and receivers at the network access terminals (NATs). The network node can be either static or reconfigurable routers. In the case of dedicated connections, frequently static routers are used, while in the case of dynamic random traffic and for demand connections, one uses reconfigurable wavelength selective cross-connects. These reconfigurable cross-connects along with the network management and control system reconfigure the connections as per demand and decide to accept or block the connection by establishing a route with the available selected λ-channel. This is the RWA problem, which we discuss in the next section.

6.4 Design of Wavelength Routed Network with Logical Routing Network Overlay

In the optical network with an electronic logical layer overlay over physical optical layer, the *PT* is formed by the OXC and the set of connecting physical fiber links. A physical link is bidirectional, realized by two unidirectional fiber links, or two orthogonal wavelengths propagating in opposite directions in a single fiber. A set of *lightpaths*, constituting λ-channels, is established over the PT, forming the *LT*, which is realized by a set of OXC interconnected by the *lightpaths*. The *lightpaths* are unidirectional. The message traffic from the user, in turn, is routed over the LT. If the two nodes are connected by a *lightpath* in an LT, a message can be transmitted with a single-hop in the optical form without requiring any O-E-O conversion at the intermediate nodes. A message goes through electronic processing at any node in three situations: (i) if the message has reached its destination node, (ii) the message is destined for some other node which does not have the receiver at the transmitted wavelength, hence needs processing for it to be converted to the required wavelength, and (iii) lastly, in the case of long reach destinations when the signal needs regeneration. Thus, to design an LT for a given PT, we not only have to do the RWA for each of the lightpaths, we also need to determine the LT to be embedded on the PT and then route the client packet/stream traffic over the LT. The PT is fixed once the fibers are laid between the nodes, but the LT can vary with the traffic request independent of the PT underneath.

With the background of the WRN network and its multilayered architecture explained in the earlier sections, we now discuss the end-to-end connectivity of the source node to its destination. Traffic from the client moves through the connected electronic LSNs on a logical path in the optical domain on a λ-channel over the fiber. The LSNs perform the routing function, ensuring that the traffic associated with a given s-d pair VC follows a prescribed logical path through the logical routing network. Each LC, in general, supports several multiplexed VCs, and each lightpath in turn can support multiple LCs. Thus, based on a traffic matrix, the data flows among the LSN connected to the network, forming the LRN. The selected LT along with LCs are embedded over the PT, therefore it is essential that the given physical infrastructure resources, such as fiber, network nodes, and the access NATs, should support the LT. The LRN should be able to provide alternate routing with changing dynamic traffic needs and traffic engineering strategies. The packet headers of the IP packets or the SONET frames have the destination information. These are read by the electronic LSN and the information is then used for routing, grooming, and fault detection/recovery purposes.

The design of the LRN over the WDM WRNs is complex and the optimized design of the WDM network requires design of LT and wavelength routing simultaneously for the given PT and traffic matrix. Ideally, design of a hybrid optical network should proceed simultaneously in both the logical and physical layers [10, 11, 13, 14]. But because of the complexity of the design problem, including the constraints of the logical and physical layers, it is helpful to separate it into four sub-problems: LT design, traffic routing over the LT on *lightpaths*, *lightpath* routing over the physical path, followed by wavelength assignment for the *lightpath*. Among these four, the first two are the LRN design issues and the other two are the *wavelength routing and assignment*

(RWA) sub-problems for the all-optical wavelength routing network. Each sub-problem is solved sequentially and the results obtained from one sub-problem are used in the next one to solve, and so on. We describe these sub-problems as follows:

- *LT design.* For a given PT, a possible LT is determined for the given traffic matrix. The LPs established by the LT should optimize certain objective functions with certain constraints, such as number of optical transmitters, receivers, and wavelengths per fiber available at a node, etc.
- *Traffic routing over the lightpath.* The problem concerns the routing of user traffic between the source-destination pairs over the LT according to the traffic matrix. The objective is to optimize the given performance parameters; blocking probability, latency, etc.
- *Lightpath route selection.* This problem determines a physical route for each of the *lightpaths* chosen by the LT. The chosen LP again needs to optimize the parameters, such as minimum number of hops to reduce delay, bandwidth efficiency, etc.
- *Lightpath wavelength assignment.* Assigning wavelength to the selected *lightpaths* with constraints of wavelength continuity and number of available wavelengths in the fiber, etc.

We will briefly introduce the ILP formulations for the analytical design of the optimized network and then give a few of the heuristic algorithms commonly employed in the next sections. For more details on the ILP formulations and resources on the heuristics algorithms, readers are advised to consult the references at the end of the chapter.

6.5 Analysis of WDM Wavelength Routing Networks

From the above discussion we can appreciate that the design of the WDM wavelength routing network with LRN overlay is quite complex. It is an optimization problem which has to consider a large number of parameters and constraints, leading to an NP-hard problem to solve. The problem can be analytically formulated as a mixed-integer linear programming (MILP) problem [4, 6, 15, 16] with inputs of traffic matrix, PT with number of wavelengths per fiber, and number of transmitters and receivers as the network nodes. It can be solved as the multi-commodity flow problem, where lightpaths and packet traffic are flows of (s-d) pair over the links. With these inputs, the objective is to design the LT to optimize certain objectives, be it maximizing throughput, minimizing traffic delay, or some other QoS parameters. The LT design has certain constraints within which it has to optimize the design. This problem is shown to be NP-hard and takes a very large amount of time and memory space to solve for a bigger network with dynamic traffic [15, 16].

Therefore heuristic solutions are developed which give a suboptimal solution but can be applied to dynamic large networks with sufficient accuracy.

In the following paragraphs, we mathematically formulate the problem of lightpath RWA in an all-optical network with the objectives and constraints defined.

The general problem of WRN design can be summarized as:

(i) For the given PT graph expressed as $G_p = (V,E_p)$, where subscript p refers to the physical graph, V is the set of network nodes, and E_p is a set of links or physical edges connecting the nodes. The links are bidirectional. The network nodes correspond to OXC and the edges correspond to optical fiber, either two different wavelengths transmitting in opposite directions or two fibers carrying lightpaths in the opposite direction for the bidirectionality. A node "i" with an input and output degree expressed as $D_{p,in}$ and $D_{p,out}$, respectively, provided by the $D_{p,in} \times D_{p,out}$ wavelength-routing switch or OXC. The in- and out-degree is provided by the number of fibers connected at the input and output of the switch. Here we are not considering a multi-fiber case but assume only a single fiber from each port.

(ii) Each fiber carries a fixed number of wavelength channels and each node has a fixed number of transmitters or tunable lasers and receivers or tunable filters.

(iii) The aim is to design a VT graph $G_v = (V,E_v)$, with a set of V nodes of the PT and E_v virtual edges corresponding to directed *lightpaths*. The in- and out-degree of the nodes are equal to the *lightpaths* or the transmitter and receivers connected, respectively. The *lightpaths* carry the traffic between the node pairs according to the given $N \times N$ traffic matrix T, where N is the number of network nodes. The traffic matrix element, $t_{i,j}$, gives the normalized traffic flow from node i to node j. The traffic is bidirectional but can be asymmetrical as well.

(iv) The LT has to be so designed that the wavelength assignment and routing of the *lightpaths* over the fiber optimize the desired metrics under the constraint of *wavelength continuity* and *distinct wavelength* on the common fiber link. The performance metrics can be either *throughput* or network *delay* or even minimizing the power or infrastructure cost of the network.

6.5.1 Optimization Problem Formulation of the WRN

The optimization problem using the principles of multi-commodity flow of (i) *lightpaths* over the fibers in the PT and (ii) packet or stream traffic flow over the LT is formulated using the following notations. It has been assumed that there are only single fiber links and no wavelength converters at the nodes:

- Subscripts s and d denote the source and destination node of packets, respectively.

- Subscripts i and j denote the originating and terminating nodes of a *light-path*, respectively.
- Subscripts m and n denote the endpoints of a physical link in a lightpath.

Given parameters are:

- Characteristics of the PT, P_{mn}, for m, n = 1,2,3......,N, where N is number of nodes, can be expressed as:

$P_{mn} = P_{nm} = 1,$	for direct fiber link between m and n
=0	otherwise

- For traffic matrix, T: The elements of the matrix, t_{sd}, denote average traffic flow (packets/second) from node (s-d), with $t_{ss} = 0$ for s, d = 1, 2, 3,...., N.
- For distance matrix, D: the elements, d_{mn}, denotes the distance from node $(m$-$n)$. It can also be used for propagation delay in time units, with

$d_{mn} = d_{nm} = 1$	as fiber links are bidirectional
$d_{mn} = 0$	if $P_{nm} = 0$

- Number of transmitters at node i = $Tx_i(Tx_i \geq 1)$.
- Number of receivers at node i = $Rx_i(Rx_i \geq 1)$.
- Maximum number of wavelengths per fiber = M
- Capacity of each channel = C (packets/second)

Variables used in problem formulation:

- In LT, as the lightpaths are unidirectional the variable: $V_{ij} \neq V_{ji}$,
  Also, if a lightpath exists between nodes i and j in the LT, then:
  $V_{ij} = 1$.
  Otherwise:
  $V_{ij} = 0$.
- Traffic routing variable $t_{ij,sd}$:
  $t_{ij,sd}$: is the traffic flowing from node s-d on link i-j on the LT V_{ij}.
- Physical routing variable will exist if both fiber and lightpath are present:
  $p_{ij,mn} = 1$ if $P_{mn} = 1$ and $V_{ij} = 1$.
- Wavelength color variable $c_{ij,k}$:

  $c_{ij,k} = 1$, if lightpath on $(i$-$j)$ is assigned color k, where k = 1, 2, 3,.......M
  and
  $c_{ij,k} = 0$ .... otherwise.

Objective function:

Among the varied objective functions of cost, throughput, etc., we express the following few functions mathematically:

- Minimizing network delay including propagation and queue delay:

$$\sum_{ij} \left[\sum_{sd} t_{ij,sd} \left(\sum_{mn} p_{mn,ij} d_{mn} + \frac{1}{C - \sum_{sd} t_{ij,sd}} \right) \right] \tag{6.1}$$

- Minimizing the maximum flow in a link: (min t_{max}):

$$\min \left[\max \left(\sum_{sd} t_{ij,sd} \right) \right] \equiv \max \frac{C}{\left[\max \left(\sum_{sd} t_{ij,sd} \right) \right]} \tag{6.2}$$

The objective in eq. (6.2) is to find a solution that minimizes the maximum flow on the lightpath. The smaller the maximum flow, the more aggregate throughput can be obtained by scaling up the flow on the maximally loaded connection.
Constraints:
The objectives have to be met under the following constraints:

- LT connection matrix, V_{ij} constraints are: the maximum in- and out-light-paths from any node in the network have to be less than the number of transmitters and receivers, respectively at the node:

$$\sum_{j} V_{ij} \leq Tx_i \tag{6.3}$$

$$\sum_{i} V_{ij} \leq Rx_{ij} \tag{6.4}$$

- *Lightpath* color:
The disjoint path constraint for wavelengths on a fiber is that not more than one *lightpath* on a fiber can have the same color:

$$\sum_{ij} p_{mn,ij} c_{ij,k} \leq 1 \, \text{for all} \, m,n,k \tag{6.5}$$

Continuity constraint of the same color wavelength for a lightpath is expressed as:

$$\sum_{k} c_{ij,k} = V_{ij} \tag{6.6}$$

- Constraints on the physical route variables, $p_{mn,ij}$, are given in Eqs. (6.7)–(6.11). These are basically the RWA problem constraints:

$$p_{mn,ij} \leq P_{mn} \tag{6.7}$$

$$p_{mn,ij} \leq V_{ij} \tag{6.8}$$

$$\sum_n p_{in,ij} = V_{ij} \tag{6.9}$$

$$\sum_m p_{mj,ij} = V_{ij} \tag{6.10}$$

$$\sum_m p_{mk,ij} = \sum_n p_{kn,ij} - \text{if } k \neq i, j \tag{6.11}$$

Equations (6.7) and (6.8) indicate that the $p_{mn,ij}$ variable will exist only when both fiber link and the corresponding lightpath exist. The next three Eqs. (6.9)–(6.11) are for the lightpath flow at any node in the PT: if node i is the source node in the PT then the net lightpath flow from i is equal to the total LP flow V_{ij}; if j happens to be the destination node in the PT then total LPs terminating from all the physical links is equal to V_{ij}; and finally, if the node happens to be neither the source nor destination but an intermediate node, then the difference between the in-coming and out-going lightpaths from the node is zero, or in other words no LPs are lost in the node, respectively.

- Constraints on the logical route traffic variables, $t_{ij,sd}$, are the following:

$$t_{ij} = \sum_{sd} t_{ij,sd} \text{ for all } i, j \tag{6.12}$$

$$t_{ij} \leq t_{\max} \text{ for all } i, j \tag{6.13}$$

$$t_{ij,sd} \leq t_{sd} V_{ij} \text{ for all } i, j, s, d \tag{6.14}$$

Equations (6.12)–(6.14) give the definition of $t_{ij,sd}$. Equation (6.12) explains that the packet flow between two nodes i and j in the LT is the sum from all the node pairs from source to destinations. Equation (6.13) is basically the congestion reduction in which the packet flow on any lightpath has to be constrained by the allowed maximum value.

The combined traffic flowing through a lightpath cannot exceed channel capacity. This constraint is expressed in Eq. (6.15).

$$\sum_{sd} t_{ij,sd} \leq V_{ij}^{*} C \tag{6.15}$$

The next three equations are for the packet flow conservation at any node in the LT. Equation (6.16) states that if node i is the source node in the LT, then the net flow from i is equal to the total traffic flow t_{sd}; if j happens to be the destination node in the LT, then total packets terminating from all the light-paths are equal to t_{sd}, which is for Eq. (6.17); and finally Eq. (6.18) expresses that if the node happens to be neither the source nor the destination but an intermediate node, then the difference between the in-coming and out-going packet traffic from the node is zero, or in other words no loss of packet in the node.

$$\sum_j t_{sj,sd} = t_{sd}$$ (6.16)

$$\sum_i t_{id,sd} = t_{sd}$$ (6.17)

$$\sum_i t_{ik,sd} = \sum_j t_{kj,sd} - \text{if } k \neq s,d$$ (6.18)

Cost functions considered for optimal design:
The cost functions which can be used for optimal design are:

(a) Minimize total number of *lightpaths*:

$$Minimize: \sum_{ij} V_{ij}$$ (6.19)

(b) Minimize amount of electronic switching or maximize single-hop connections:

$$Minimize: \sum_{ij,sd} t_{ij,sd} - \sum_{sd} t_{sd}$$ (6.20)

(c) Minimize the maximum number of *lightpaths* at a node:

$$Minimize_i : \left[\max\left(\sum_j V_{ji} \right) \right]$$ (6.21)

(a) and (c) are the LT design problem in minimizing the LPs terminating at a node. It is (b) where the transceiver cost or grooming cost comes into play.

The above graph construction algorithm is an ILP and is shown to be an NP-complete problem. The problem size in ILP formulation grows with the size of the network, traffic matrix size, number of variables, and constraints. For highly connected networks, with the required constraints the numbers of variables increase exponentially with the size of network, and the numbers of variables also grow exponentially with the size of the network. Since the formulations are computationally intensive for large networks, some approximation methods are proposed to solve the ILP for large networks. For example, the RWA problem can be decomposed into a number of sub-problems, which are solved independently. The ILP can be pruned by limiting to fewer alternate paths between (s-d) pairs to reduce the problem size. Further, suboptimal results can be obtained using *genetic*, *heuristic*, or relaxation methods such as *randomized rounding* [1, 17]. Once the lightpaths are chosen, the wavelength assignment can be done by employing *graph coloring* algorithms, which are, again, NP-complete problems.

6.6 Heuristic Solutions for WDM Wavelength Routing Networks

In this section, we will be giving the suboptimal heuristic solutions for WRN networks with single fiber connection from the different ports of the nodes. For optimally determining routes and assigning wavelengths for lightpath requests, ILP formulations can be used. But as the network size grows, the ILP solution takes much longer time. Therefore, some approximate solutions or heuristic algorithms are proposed for solving the RWA problem for large networks. The ILP solutions are used for small networks and also in static cases to obtain the optimum solution for the RWA problem. These solutions are then used for validations of the large networks solved heuristically.

In the PT, the ONNs or the OXCs in a WRN are connected in an arbitrary fashion with their neighboring nodes in the mesh topology with multiple input-output ports. There are multiple wavelengths on each fiber link. Any request which arrives at the client node or the router has more than one route to take; therefore, a good strategy is required for a network to select an appropriate route and a wavelength from source to the destination. The LT is to be formed with a set of select OXCs and set of select lightpaths. These lightpaths now form the traffic matrix of the given PT and so have to be routed over the fibers and assigned a wavelength to arrive at the best solution. There can be many solutions to this problem [18–24]. One solution can be to establish all the lightpaths using a minimum number of wavelengths and minimum blocking probability. Another objective can be to find a route with a minimum number of *hops*, which is *delay-oriented* optimization, or it can be a route which has maximum number of lightpaths that can be accommodated in a given network, in which case it is *utilization-oriented* optimization. There can also be a cost consideration, which includes the cost of switches, amount of bandwidth, connection durations, or can also be a power consumption optimization-oriented QoS consideration.

Depending on whether traffic requests are known a-priori and are static over time, or are dynamic in nature, the design problem can be classified into two categories: *static* and *dynamic*. In static RWA, all the lightpath requests between end-node pairs are known initially. For large transport networks, in the planning stage of the network the static design is based on an aggregate demand pattern and its future values. The objective in static RWA design is to assign routes and wavelengths to all demands to minimize the number of wavelengths used, or for the given number of wavelengths to maximize the demands met. Static RWA can be seen primarily as the case for wavelength routing networks for WAN distances. This is because typically in large area transport networks, the traffic demand between the end nodes remains almost fixed over hours or days or even months as dedicated traffic.

In the case of dynamic traffic, the lightpath requests between end nodes are assumed to arrive at random times and have random holding times. That is, lightpath requests are established on demand. Therefore, lightpath requirements may vary frequently over time. The objective now is to reduce the blocking of the requests or reduce the delay in the network with optimized network resources. There are many random distribution functions used to represent the traffic arrival and the connection time, viz. uniform, Poisson, exponential, or Pareto distribution for the bursty traffic. Unlike in the wide area core network with long links where we have almost a quasi-static traffic pattern, in data networks or small local area network (LAN), traffic or demands frequently change over time. Also, with changing traffic patterns as the high capacity demands are growing, wavelength routing networks are becoming common in metro and in LAN applications as well. Therefore, a combination of static and dynamic RWA is employed in the appropriate designing and functioning of the WRN networks.

6.6.1 Design Parameters, Performance Metrics, and QoS Issues

RWA algorithms provide a suitable route and wavelength to requested lightpath traffic. The RWA tries to optimize the required performance metrics for the given PT and infrastructure of the network, such as available spectrum, number of transmitters and receivers and node structures. The important performance parameters considered in the network are throughput, latency, blocking probability, spectral efficiency, i.e. providing a solution with a minimum number of wavelengths for the given demand, link utilization, load balancing, etc.

Throughput in the network is defined as the ratio of the average successful demands reaching the egress nodes to the total input demands at the ingress node. The input blocking probability of the connection request is defined as the probability of the connection not being successful due to non-availability of either the route or wavelength and hence has to be dropped. This will reduce the throughput. In the irregular mesh network topology with asymmetric traffic load, there is higher probability for congestion on some intermediate nodes. This congestion causes delay, inefficiency of network utilization, and complexity of network control. The congestion is higher if either the available number of alternate routes is less or the lightpaths are few. An efficient RWA has to take care of these limitations in order to improve the overall performance of the network.

Fairness of the routing algorithm is yet another important characteristic along with *efficiency* and *optimality*. An algorithm is said to be fair if it provides good performance to all the demands, i.e. it considers the average network performance of all possible routes in terms of either blocking or latency, instead of a select few. The RWA has to both be fair and still have good network performance. There is normally a conflict between finding an optimum solution and providing fairness because an optimum solution usually cannot be fair.

Next, we discuss the issue of QoS provisioning in WRN networks. In general, QoS in networks are the performance parameters, such as bit rates, bit error rates, and other characteristics that can be measured and guaranteed to clients in advance. Different types of applications in the network will have different QoS requirements. In the context of wavelength routing networks, QoS can be measured in terms of bit rate and call drops, i.e. the probability that a requested connection cannot find a route that can satisfy the QoS parameter requirement of the connection. However, the connection blocking probability will vary as a function of the different traffic and application characteristics. For example, a video stream can be admitted if we find a lightpath that satisfies the bandwidth and delay jitter requirements for it. Meanwhile, a file downloading request does not necessarily have these requirements. It does not have very strict delay and bandwidth requirements. Thus, the QoS performance of these two types of traffic in terms of blocking probability will be different, therefore the wavelength assignment and routing strategies are not the same. In the following sections we briefly describe the wavelength routing and assignment heuristic algorithms.

6.6.2 Route Selection Algorithms

For ease of computation, the design problem of a hybrid WRN network is separated into two parts which essentially are not independent, but we solve them individually to reduce complexity. The first part is determining the possible LT formed with a set of network nodes as vertices and connected with *lightpath* links to be embedded over the given PT and routing the given traffic matrix of the source-destination nodes over the LT. The second part is the lightpath-traffic routing and wavelength assignment or RWA over the PT. We look at the RWA part first in this section and will consider the LT topology formation in Section 6.7.

There can be a large number of wavelength routing algorithms for static as well as dynamic traffic [21–27]. As discussed earlier, many different cost factors can be taken into account, such as path length cost, power cost, and connection blocking, when assigning a route to a request. We discuss a few among the many good available path selection strategies. The most common strategy is the shortest path to minimize the cost metric of the path, be it distance, hop count, power cost or blocking, etc. The well-known two shortest-path algorithms are the *distance-vector* and the *link-state*. They are the breadth-first-search (BFS) algorithm and the Dijkstra algorithm (or the K-shortest path, K-SP), respectively. We explain the two algorithms in the following with examples.

6.6.2.1 Breadth-First-Search Algorithm

BFS is a *distance-vector* algorithm. In a distance-vector algorithm each node exchanges information about the entire network with neighboring nodes at regular intervals. The *distance-vector* algorithm determines the minimum-hop routes.

The BFS algorithm proceeds by considering all the links with a weight of unity. The objective here is to minimize the number of hops while searching for the shortest route. Therefore, it proceeds by considering all one-hop paths from the source first, then all two-hop paths from the source, and so on until the shortest path is found from source to destination. Each node stores information about the network in its routing table, which is periodically updated when a new destination is learned. If there are multiple shortest paths that are tied up, the BFS algorithm selects one randomly. In the case of fewer hops, lower cost or less wavelength contention is expected. The worst-case time complexity of this algorithm is $O(N^2)$, where N is the number of nodes in the network.

We explain the algorithm to find the hop length of the shortest path from the source node to the destination node by an example. Initially all the nodes are unmarked. A *queue* is used to keep sequence of nodes at various stages of the algorithm. To start with, the queue has only the source node and its hop-distance to all the rest of the nodes connected to it. In every next step, the node from the front of the queue is considered and the hop-distance vector to the connected edges to the node is updated. Thus, the queue and the hop-distance from the source are updated in each step as explained in the example. This procedure is repeated until the destination node appears at the front of the queue.

Example 6.1 Consider the five-node network given in Figure 6.4a along with the *directed* lightpaths connecting the nodes. The shortest distance in terms of hop count is to be determined between the source node 1 and the destination node 4. The contents of the nodes in the queue and the corresponding hop distance of the nodes from the source node are as given in Table 6.1.

In the first step, only the source node 1 is there in the queue. Nodes 2 and 5 are connected to the source and hence have a hop distance of 1, as $d(1)$ is 0.

Nodes 2 and 5 appear in the queue now. As node 2 is ahead in the queue, node 3 is considered which is connected to it. It is a two-hop distance away from the source node, as $d(2) = 1$. Node 2 is marked now and removed from the queue.

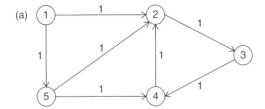

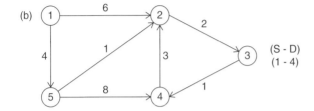

Figure 6.4 (a) Example of the BFS algorithm and (b) example of the Dijkstra algorithm.

Table 6.1 Nodes in the queue with hop distance from source.

Node number of the nodes in the queue	Hop distance from the source node 1
1	$d(2) = 1, d(5) = 1$
2, 5	$d(3) = 2$
5, 3	$d(4) = 2$
3, 4	$d(4) = 3$
4	

The next row has 5 and 3 in the queue. Node 5 has nodes 2 and 4 linked to it with one hop distance away. As node 2 is already marked and removed, the hop-distance column has only node 4, which is two hops away from the source, i.e. $d(4) = 2$ as $d(5) = 1$. Node 5 is now marked and removed from the queue in this step.

The next row takes 3 and 4 in the queue. Node 4, which is connected to node 3, has a hop distance $d(4) = 3$ from source through node 3. Node 3 is marked and removed.

The final entry in the queue is node 4. This is marked and removed from the queue. Thus, the shortest distance between nodes 1 and 4 in terms of hop count is 2 on the route 1-5-4.

The BFS algorithm finds the shortest path with the fewest number of hops. Fewer hops can potentially translate into lower cost or less wavelength contention. BFS can also work with negative link metrics. The basic flaw is the slow reaction to link or node failure because information comes only from neighboring nodes and it may be out of date. The algorithm has slow convergence.

6.6.2.2 Dijkstra Algorithm

In this section we explain the Dijkstra algorithm, which is used to find a shortest path from the source to the destination node. This is a *link-state* algorithm where each node sends information about its *neighborhood* to every other node in the network. Each link has a given weight or cost function attached to it, which may be length, traffic level, etc. The algorithm searches the shortest path which will minimize the cost function. The algorithm builds a shortest-path spanning tree for a node. Such a tree has a route to all possible destinations with no loops. The node running the algorithm is the *root* of its shortest-path spanning tree. As in the case of the BFS, the inputs to the algorithm are the network topology and the lightpaths giving the set of required s-d pair connections. The algorithm will have a maximum of N iterations, where N is the number of nodes, and with any iteration the shortest distance to a new node is achieved from the source node. If we assume $c(x, y)$ to be the cost of the link from x to y, at any iteration the value of distance $d(y)$ denotes the length of the best-known shortest path from source node to y-node traversing through only the nodes whose shortest distance has been found thus far. Starting with the source node, the cost function of each of the nodes is determined with respect to the source. All the links have the cost value $c(S, y)$ marked. For a node y, if there is no edge from source S to y, its distance $d(y)$ is considered to be infinity. The node with the least distance value from the source is selected and marked permanently. Now the distance value of the rest of the nodes is updated as $d(y) = \min [d(y), d(S) + c(S,y)]$. The same procedure is repeated until the destination node is reached. The worst-case time complexity of this algorithm is $O(N^2)$. We give an example to illustrate the algorithm.

Example 6.2 Consider the five-node network with link weights given in Figure 6.4b. The shortest distance is to be determined between the source node 1 and the destination node 4, as in the case of BFS algorithm. The sequence of iterations is given below in search of the shortest path from the source node:

Distance cost function from source route selected	
$d(1) = 0$	
$d(2) = 6, d(3) = \infty, d(4) = \infty, d(5) = 4$	$1 \rightarrow 5$ (5 selected as it is shortest distance from 1)
$d(2) = 5, d(3) = \infty, d(4) = 12$	$1 \rightarrow 5 \rightarrow 2$ (2 selected as shortest distance from 1)
$d(3) = 7, d(4) = \infty$	$1 \rightarrow 5 \rightarrow 2 \rightarrow 3$ (3 selected as shortest distance from 1)
$d(4) = 8$	$1 \rightarrow 5 \rightarrow 2 \rightarrow 3 \rightarrow 4$ (shortest path)

The Dijkstra algorithm is a "greedy" algorithm guaranteed to find the shortest path from source to destination, assuming a path exists. Greedy algorithms proceed by choosing the optimal option at each step without considering future steps. The shortest-path algorithm can be incorporated as part of a larger procedure to find K-SP. K-SP algorithms find the shortest path between the source and the destination, the second shortest path, and so on until the kth shortest path is found or until no more paths exist. These paths may not be completely disjointed from each other in all cases, i.e. the paths may have a few links in common. The K-SP algorithm allows several alternative routes for each lightpath. The shortest lightpath that cannot be set up due to constraints on routes and wavelengths is said to be blocked, so the corresponding network optimization problem is to minimize this blocking probability and select the next shortest path.

6.6.3 Heuristic Wavelength Routing Algorithms

A well-designed heuristic algorithm may give suboptimal solutions but are easily applicable to large-size networks. The run time of the RWA algorithm becomes very important in selecting an algorithm for a large network. In the case of network planning process or static traffic, the route and wavelength allocation for a lightpath with resource allocation to the route may require less than a few seconds to configure and in the case of dynamic real-time traffic load, it may be necessary to establish the new connection in less than 100 ms. Hence a fast heuristic RWA algorithm is very essential along with providing a near-optimal solution.

The route selection and wavelength assignment can be *centralized* or *distributed* controlled. Normally centralized control is suitable only for small-size networks, while distributed control is used in large-size and dynamic traffic real-time design where nodes have limited processing and memory capability. The routing algorithms can also be classified as *adaptive/dynamic* and *non-adaptive/static*. In adaptive routing (AR), routing and wavelength decisions may be changed when network topology or traffic load changes, while in non-adaptive the routing decisions are predetermined and not determined by the current situation of traffic or topology of the network. In practice, the routing algorithm should be adaptive and decentralized.

With the algorithms for selecting good routes either with SP, K-SP, or BFS criteria as discussed in the last section, we next describe, with examples, few of the algorithms which are commonly used for routing of the *lightpath* over the PT.

6.6.3.1 Fixed Routing

The simplest of the algorithms is to route the connection always on a fixed route (FR) for a source-destination (s-d) pair on some selected criterion discussed above. For example, the shortest-path route between all the s-d pairs can be calculated offline using either the Dijkstra or the BFS algorithm. The connection request can then be established using the predetermined routes.

This has the advantage of simplicity and less delay. However, it has the disadvantage of high blocking as wavelengths are tied with specific routes. It requires a large number of wavelengths to reduce blocking and cannot handle faults in the connections because of lack of alternate paths. In summary, fixed-path routing, including fixed shortest-path routing, is not recommended, except for the applications which require very stringent latency. The maximum latency in FR is $O(WL)$, where W is the number of wavelengths and L is the length of longest route of any node pair.

6.6.3.2 Fixed-Alternate Routing

In alternate fixed-path routing, the set of candidate paths is also generated prior to any demands being added to the network. In fixed-alternate routing (FAR), each node in the network is required to maintain a routing table that contains an ordered list of an M-alternate (disjoint) number of fixed routes to each destination node based on some criterion: shortest path, hop-count, etc. When a connection request arrives, the source node tries to establish the connection on all the possible predetermined routes in the routing table in order until a route with a valid wavelength assignment is found. If none is found, the connection request is blocked and then dropped. When there is a tie between two routes, selection is made randomly.

The advantage of FAR is once again the simplicity and also some degree of fault tolerance when the link fails. It also reduces the blocking significantly, but yes, with the cost of sacrificing the minimum delay as compared wth *fixed routing*. The latency is $O(MWL)$, where there are M fixed alternate routes. In order to reduce the cost, delay, and improving capacity utilization and likely lowering failure rate, the alternate path may drop the complete "disjoint" path restriction and may have a few common links between the alternate paths. We sacrifice diversity compared with the complete disjoint path case.

6.6.3.3 Adaptive Routing

Unlike in *fixed routing* and FAR, an adaptive routing (AR) scheme can be used for dynamic traffic. There is no predetermination of the paths – when the request arrives, the route from source to a destination node is chosen dynamically, depending on the network state determined by considering all the connections in progress. A cost function is assigned to all the links in the network. Links which have insufficient bandwidth available to carry the new connection are assigned a cost value of infinity, or in other words they are temporarily removed from the network. Similarly, unused links are assigned a cost of unity. Thus, we prune the topology based on the current state of the network. When a connection arrives, the shortest cost path between the s-d is determined. If multiple paths are of same cost, one is selected randomly. In the case when the node has a wavelength converter available, one can select that route, taking into account the wavelength conversion cost . The variety of cost functions can be used for AR. Typically, the cost functions are the number of hops/distance, or current congestion. The least-congestion route is the one which has a greater number of wavelengths available on the link. In shortest cost adaptive routing, a connection is blocked only when there is no route from the source node to the destination node in the network available. Adaptive routing can be either centrally controlled with the NMS (network management system) maintaining the complete network state information, or it can have distributed control, with each node maintaining and updating the current network state information as the connection request arrives and is stripped down. In practice, traffic forecast is used to assist in generating the first candidate path and then uses current network condition to select one of the available paths for a particular demand request.

AR requires extensive support from the control and management protocols to continuously update the routing tables at the node. The advantage of adaptive routing is low blocking probability and the disadvantage is high complexity and extra delay. The delay is due to examination of all links on all candidate paths before routing for the selection of the least-cost path. The latency complexity for the AR is $O(N^2 W)$.

6.6.3.4 Least Congested Path Routing

With the alternative routing approach, the least congested path (LCP) routing chooses the route with least congestion among the possible routes connecting a node pair. The congestion is suggested from the number of free wavelengths available on the entire route because the greater the number of free wavelengths, the less congested is the route. In LCP, the idea is to keep the *spare routes* as large as possible for a lightpath. When a new connection comes, LCP finds a path that least reduces the *spare route* set. In this way, the least congestion can be expected to be produced. LCP is computationally more complex because while selecting the LCP it has to consider all the links of all the possible candidate paths. The blocking probability improves, but at the cost of latency.

6.6.4 Wavelength Assignment Algorithms

Like the wavelength routing algorithms there are several *wavelength assignment* algorithms for both static and dynamic traffic. Wavelength assignment to the selected route is an integral part of the network planning process in networks with optical-bypass nodes. This is from the continuity property of optical-bypass elements, where a connection that traverses a node all-optically must enter and exit the node on the same wavelength. Therefore, effective wavelength assignment strategies must be used to ensure that there is no wavelength contention on any link on the route. In general, the approach for wavelength assignment algorithms is to assign wavelengths to the lightpath as they arrive one at a time, one after the other. In the static case we try to minimize the number of wavelengths, while in the dynamic case we assume the wavelengths are fixed so we try to minimize the blocking in the network. In the following we address the commonly used wavelength assignment algorithms.

6.6.4.1 Fixed Order First Fit

Each wavelength in the network is assigned an index from 1 to W, where W is the total number of wavelengths supported by the fiber. The algorithm searches for the free wavelength in a fixed order and the first free wavelength is selected and no correlation is required between the order in which a wavelength appears in the spectrum and the assigned index. The indexing order is basically used to potentially improve some parameter of the network performance. By doing this we try to pack all of the wavelengths presently being used into the lower end of the indexed space, and hence the higher end of the index space is then available continuously for longer paths with higher probability.

As it does not look for any other cost or any global search, the *first fit* algorithm is fast and can be applied in both centralized and distributed control. It has reasonably low blocking probability and fairness, has small computational overhead, and low complexity. The issue here is in the dynamic situation: when the connections which are established are later torn down, the indexing ordering does not guarantee the actual assignment order on a link. Thus, the indexing strategy of *first fit* may be viewed as a short-term means of potentially providing additional improvement in system performance. But with time, network growth and changes may cause the benefits to be lost.

6.6.4.2 Random Assignment

Each wavelength used in the network is indexed as in *first fit*. The random wavelength assignment algorithm starts with the search of all available wavelengths on the required route among the given number of wavelengths. Next, it randomly selects one among the available wavelengths on the route with a uniform probability. Thus the load gets uniformly distributed on all the wavelengths. This can also be used for both centralized and distributed control and does not require any global information. In comparison with *first fit*, the computation time involved is more as it has to search the complete set of available wavelengths for each route and is also computationally more complex.

6.6.4.3 Least Used Wavelength Assignment

Least-used (LU) wavelength assignment indicates spread or balance of load over all the wavelengths, i.e. when the demand arrives, the *least used* algorithm selects the wavelength which is least used in the network. This algorithm requires the actual or estimated network state information to know the usage of the entire wavelengths at any point in time to select the wavelength which is being used by minimum number of links. By selecting the least used wavelength, a shorter route can be determined that can leave many links free to be used by future connections for selecting wavelength. This algorithm is more useful for a centralized control scheme. Also, it requires more overhead for communication to fetch the control information and hence requires more storage and is expensive.

6.6.4.4   Most Used Wavelength Assignment

The most used (MU) wavelength selection criterion is in variance to *least used*. The algorithm in this case selects the wavelength which is most used on the links and hence tries to pack the wavelengths instead of spreading. The idea is to exhaust the wavelengths which are already in use so that maximum numbers of wavelengths are available for future connection on continuous links for the route on which each wavelength has already been assigned. Whenever the demand arrives, a wavelength order is established based on the number of fiber links. The wavelength that has been assigned to the most fiber links is given the lowest index. After the wavelengths have been indexed, the assignment proceeds as in *first fit*. The global network state information will be required, also as in the case of *least used*, to get the usage information of all the wavelengths. It also requires centralized control henceforth. This has longer overheads and is more costly than *least used*. It is more adaptive than *first fit* on the cost of more information.

Example 6.3 We once again consider the same five-node network example to explain all the RWA algorithms we have discussed. The network is given in Figure 6.5. It has five nodes and seven fiber links and each link can carry four wavelengths. Assume the request arrives for a connection from node 1 to node 4.

1) *Fixed routing.* Let the set of links on which each of the wavelengths is available be given as:

λ_1: l_1, l_2, l_4, l_6
λ_2: l_2, l_3, l_5, l_6, l_7
λ_3: l_3, l_6, l_7
λ_4: l_4, l_7

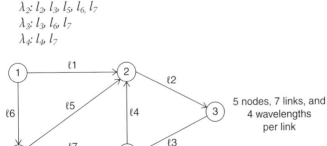

Figure 6.5 Example of *routing and wavelength* algorithms.

In the *fixed route* algorithm, the route assigned for a connection from node 1 to node 4 is $(l_6 - l_7)$. The set of wavelengths on the two links is:

l_6: $\lambda_1, \lambda_2, \lambda_3$
l_7: $\lambda_2, \lambda_3, \lambda_4$

The two links in the route have two wavelengths available, λ_2 and λ_3. λ_2 is available on five links, while λ_3 is available on three links only. Once the route is selected, the choice the available wavelengths is made on the wavelength selection algorithm used. For *most used* we will select λ_2 and for *least used* λ_3 is selected. In the case of *random selection* any one can be chosen and in the *first fit* case the lower indexed wavelength λ_2 will be selected.

a) *Fixed alternate routing.* Continuing with the same example but with the wavelengths available on the different links in this case are as given below:

λ_1: l_1, l_2, l_5, l_6
λ_2: l_2, l_3, l_5, l_7
λ_3: l_1, l_2, l_3, l_5, l_7
λ_4: l_1, l_4, l_6

The two predetermined alternate routes available in this case for a connection from node 1 to node 4 are $(l_6 - l_7)$ and $(l_1 - l_2 - l_3)$.
The set of wavelengths on the two links of the first route $(l_6 - l_7)$ is now:

l_6: λ_1, λ_4
l_7: λ_2, λ_3

Each of the two links on this route has two wavelengths available, but no common wavelength is there for the continuity constraint to be satisfied, hence the route will be blocked.
On the second route $(l_1 - l_2 - l_3)$, the set of available wavelengths is:

l_1: $\lambda_1, \lambda_3, \lambda_4$
l_2: $\lambda_1, \lambda_2, \lambda_3$
l_3: λ_2, λ_3

Studying the available wavelengths on the three links of the route we find that it has only one common wavelength, λ_3, on all the three links. Hence there is no choice but to select this wavelength for all the wavelength selection schemes. Thus, we find with FAR we reduced the blocking of the connection with the alternate path available.

a) *Adaptive routing.* Consider once again the same five-node, seven-link, and four-wavelength per fiber link example. The wavelengths available on the different links in this case are as given below:

λ_1: l_2, l_3, l_5, l_6
λ_2: l_1, l_2, l_4, l_6
λ_3: l_3, l_4, l_5, l_7
λ_4: l_2, l_5, l_6, l_3

There is a total of three alternate routes available in this case for a connection from node 1 to node 4. These are $(l_6 - l_7)$, $(l_1 - l_2 - l_3)$, and $(l_6 - l_5 - l_2 - l_3)$. The set of wavelengths on the two links of the first route $(l_6 - l_7)$ is now:

l_6: $\lambda_1 \lambda_2 \lambda_4$
l_7: λ_3

On the two links in this route we find no common wavelength, hence the continuity constraint is not fulfilled; therefore, the route will be blocked for the connection from node 1 to node 4.

On the second route $(l_1 - l_2 - l_3)$, the set of available wavelengths is:

l_1: λ_2
l_2: $\lambda_1 \lambda_2 \lambda_4$
l_3: $\lambda_1 \lambda_3 \lambda_4$

Once again on this also there is no common wavelength available for the connection to be carried on the route and hence the connection request is blocked.

On the third route from $(l_6 - l_5 - l_2 - l_3)$ there are four links involved and the wavelengths available on these links are:

l_6: $\lambda_1 \lambda_2 \lambda_4$
l_5: $\lambda_1 \lambda_3 \lambda_4$
l_2: $\lambda_1 \lambda_2 \lambda_4$
l_3: $\lambda_1 \lambda_3 \lambda_4$

We find that λ_1 wavelength is available on all the four links connecting nodes 1–4. Thus, with alternate routing with more alternate paths, the blocking can be reduced further.

6.6.5 Joint Routing-Wavelength Assignment Algorithm

All the algorithms discussed above select the route and wavelength independently. We can also have joint wavelength and route selection algorithms, where the wavelength and the route are selected simultaneously. In this case a cost function is assigned to each wavelength-route pair. The cost function can be hop count, distance, congestion, reliability, or some other metric. For example, the cost for the K-shortest routes for all the node pairs in the network is calculated offline. When any demand is received, the search for the appropriate wavelength-route pair is calculated and assigned. After each demand or periodically, the cost function weight needs to be updated before the new demand is received. Therefore, the residual capacity over all wavelengths and over K-alternate routes can be maximized. Simulations show that the joint routing-wavelength assignment algorithm can achieve much better blocking performance compared with other RWA algorithms where wavelength and routes are assigned independently, but at the cost of more complexity, computational time, and overheads involved. The performance, however, depends on the modeling of the cost functions and their weights.

6.7 Logical Topology Design Heuristics

At the optical layer in the WRN network, *lightpaths* are established between a subset of network node pairs to form an LT according to the given user stream/ packet traffic matrix of the network, which can be static or dynamic. The topology design problem includes designing and optimizing the LT so as to optimize the network performance metrics, such as delay, throughput, cost or network congestion, network resources, and traffic. As we do not have lightpaths between all the source-destination pairs, the LT problem becomes difficult to solve by ILP for large area network and dynamic traffic. Therefore, heuristic solutions are given which can give near-optimum solutions.

In the case of LT design, the network performance parameters to be taken into account are reducing maximum traffic on the LP. By reducing maximum traffic, the congestion of LP reduces and we manage to minimize the average weighted number of hops, which in turn reduces O-E-O conversion and hence reduces delay and makes efficient use of resources to support traffic.

The LT formation problem can be decomposed into two subproblems: first, to determine a heuristic for good LT design, and second, to find the solution of the multi-commodity client traffic flow over the LP. The LSN does traffic grooming, i.e. aggregating the low-rate traffic of the user to high-rate traffic by time multiplexing. Therefore, grooming can reduce network design cost while increasing its operational efficiency. In the first subproblem, attempts are made to find an LT that maximizes either the single-hop traffic to reduce delay or throughput by reducing the blocking probability. The LT can be either regular or irregular/ mesh in architecture. The regular LT networks are well studied. The prescribed symmetrical LTs are the ShuffleNet, Hypercube, deBruijn, or Kautz digraph [28–30]. The routing on these networks is relatively simple and efficient. Therefore, mapping a regular topology network onto the WRN can simplify the process of routing. In the case of mesh irregular topologies though, the performance is much better, but the routing optimization problem is complex. In the following sections, we give a few of the LT heuristics used in irregular topologies.

6.7.1 Logical Topology Design Algorithm with Congestion Minimization

For the given PT and traffic matrix, the number of transmitters and receivers at each node and the number of wavelengths per fiber are known. The objective is to design the LT so as to minimize congestion in the network [31]. Congestion in a network is defined as the maximum offered traffic on any lightpath. The offered traffic on the lightpath is the sum of the traffic flows between all those source-destination node pairs that use this lightpath. The congestion is to be kept to a minimum in order to support for future traffic growth. The algorithm therefore maximizes the single-hop traffic in order to reduce congestion. The following sequence of steps is carried out for the LT design problem in the algorithm:

1) The node pairs are arranged in decreasing order of traffic.
2) The node pair with maximum traffic is selected first. The lightpath is established between the two nodes if the physical path is available over fibers between the nodes along with a wavelength, transmitter, and receiver at the source and destination nodes, respectively.
3) If lightpath is established, then the traffic for the node pair is updated by subtracting the traffic of the next highest node pair at the top of the list.
4) The lightpath is established for the next highest traffic node pair at the top of the list and similarly updated as in 3.
5) If a lightpath cannot be established between any node pair, the traffic associated with it is set to zero and the next node pair on the list is taken.
6) When all node pairs with nonzero traffic have been considered, the procedure stops.
7) If few transmitters and receivers at few nodes are available, the algorithm creates lightpaths between such nodes to use all the transmitters and receivers.

While designing the LT in the algorithm, the RWA problems are also solved simultaneously by selecting the shortest path route or least congested route and free available wavelength is assigned from a pack of *most used* or *least used* wavelengths.

6.7.2 Logical Topology Design Algorithm with Delay Minimization

The criterion selected in this algorithm is to design an LT with minimum traffic delay between every node pair which gives least average delay for the traffic in the network. The algorithm starts with making a lightpath for all the links between the nodes in the PT. Thus, the LT replicates the PT. Then it makes the new light-path according to the previous algorithms to minimize congestion. Therefore, more transmitters and receivers are required at the nodes to set up the enhanced lightpaths required to reduce the delay in the network. This is a greedy approach to reduce the propagation delay of the traffic flow in the designed LT.

For initial mapping, the nodes are mapped in decreasing order of the *degree* of nodes in the PT. In order to reduce propagation delay, the adjacent nodes in the PT are also kept adjacent as far as possible in the LT, which is later refined in a greedy fashion. Next the nodes are considered in reverse order of the initial mapping. The algorithm then checks whether swapping a node with some other node reduces the average delay. If it does, then swapping is done; otherwise the same connection is maintained. Once all the nodes are processed, the problem of LT design is said to be complete.

6.7.3 Logical Topology Design Algorithm with Link Elimination and Matching

Link elimination with matching heuristics tries to minimize both single-hop and multi-hop traffic in order to minimize congestion with better utilization of the physical infrastructure [32]. The algorithm starts with a fully connected topology as in the previous algorithm and eliminates lightly loaded links until a topology with the desired nodal degree is obtained:

- The PT with N nodes, traffic matrix, and the number of transmitters and receivers at each node are given.
- A complete bipartite graph is divided into two partitions, each containing all the nodes of the network. Edges with weights as the traffic demands between the corresponding node pairs from one partition to the other partition are made. This is a completely connected graph in LT (each node with (N–1) degree), which is realized with shortest paths at the physical level.
- There is no restriction on the number of wavelengths per link initially. If there is traffic with multiple shortest paths on the route, the route which has minimum wavelength congestion on any physical link of the path is chosen.
- Now from the complete bipartite graph, least congested edges are selected via minimum perfect matching and they are eliminated one by one. The traffic that was routed over those edges is rerouted over the remaining edges in the network via minimum congestion routes.

- This process is repeated until the degrees of all the nodes is reduced to Δ. Note that the degree of each of the nodes in the bipartite graph decreases by one every iteration. Thus, this phase of the algorithm is repeated for N-Δ times.

6.7.4 Simulated Annealing-Based LT Heuristics

As discussed earlier, LT design is an NP-hard problem. Many heuristic algorithms are therefore designed as discussed above, such as a sequential algorithm, a degree-descending algorithm, etc. Some other algorithms which can perform optimization, such as genetic algorithm (GA) and stimulated annealing, are also used in LT design. The GA is used successfully to improve the performance over the heuristic algorithms. GAs are a part of evolutionary computing. The GA is initialized with a set of solutions, represented by chromosomes. GA is based on the principles of natural selection – surviving of fit individuals and losing of non-fit individuals [33, 34]. In the case of optical networks, GA improves performance in terms of the average number of wavelengths required. GA is a heuristic, adaptive approach for deciding topological problems in network planning. But GA has the disadvantage of long running times of a few hours.

Another routing algorithm is the simulated annealing (SA) algorithm to improve the performance of solving the LT problem and optimizing the result. SA originated in the annealing processes found in thermodynamics and metallurgy [34, 35]. SA is used to approximate the solution of very large combinatorial optimization problems, e.g. NP-hard problems. It is based on the analogy between the annealing of solids and solving optimization problems. In SA, the value of an objective function [34] that we want to minimize is analogous to the energy in a thermodynamic system. The most important aim of SA is to avoid getting trapped in a local minimum and obtain a globally better solution by employing the so-called annealing schedule or cooling schedule, which specifies how rapidly the temperature is lowered from high to low values. SA is a Monte Carlo approach for minimizing multivariate functions. The algorithm of the SA is an approach that integrates most of the local search algorithms. An essential feature of SA is that it can climb out from a local minimum, since it can accept worse neighbors at the next step. Such an acceptance happens with a probability that is smaller if the neighbor quality is worse. The probability of the acceptance can be presented as follows:

$$\Pr(accept) = \begin{cases} 1, \ldots\ldots\ldots\ldots if \, \Delta \leq 0 \\ e^{-\Delta/T}, \ldots\ldots\ldots\ldots\ldots if \, \Delta > 0 \end{cases} \tag{6.22}$$

where Δ is the cost change and T is a control parameter that is called temperature. There are four problems for the initializing of the algorithm – defining the initial temperature, defining the cooling schedule, defining the number of iterations on each temperature step and stop criterion.

In the SA process to determine the LT, the algorithm starts with an initial random configuration for the LT. The average message delay on the LT is calculated for this mapping. The mapping is refined over several iterations. Node-exchange operations are used to arrive at neighboring configurations. In a node-exchange operation, adjacent nodes in the LT are examined for swapping and the neighboring configurations which give lower average latency than the current solution are accepted automatically. Solutions which are worse than the current one are accepted with a certain probability, as in Eq. (6.22), which is determined by a system control parameter. The probability with which these failed configurations are chosen, however, decreases as the algorithm progresses in time so as to simulate the cooling process. The probability of acceptance is based on a negative exponential factor and is inversely proportional to the difference between the current solution and the best solution obtained so far.

6.8   Summary

In this chapter we have studied the WRN WANs with nationwide or global coverage. In an all-optical WRN, by setting *lightpath* through OXC nodes, a large area can be covered using single-hop access without any O-E-O conversion. Each intermediate node between the source-destination pair provides a circuit-switched optical bypass to support the lightpath. Sub-wavelength groomed traffic from logical network nodes with high bit rate can also be sent over a wavelength channel to the end nodes with multi-hop connections through intermediate O-E-O nodes to reach the destination. An efficient wavelength routing and assignment algorithm is required to find an optimum and scalable solution for the large-size network. The ILP formation to obtain the optimum solution is considered in the chapter. Also, several RWA heuristics algorithms have been discussed to obtain solutions for large-size networks with dynamic traffic.

Problems

6.1   Determine the possible logical topologies for a six-node Star physical network topology and show how packet from node 1 will be routed to node 3. Repeat the same for the ring topology as well.

6.2 (a) What are the sub-problems in the design of wavelength routed multi-layered optical networks. What are the performance metrics of these networks? In the two-layer optical networks explain how with the above sub-problems the message is transmitted in the upper application layer.

(b) What is the in-degree and out-degree in virtual topology and what is it limited by? If the physical in/out degree is given by D_{in}^P and D_{out}^P, and W wavelengths are available on each fiber link, what can be the maximum value of the in-degree and out-degree in virtual topology.

(c) Discuss the constraints, objectives in the RWA and the sub-problems for the solution of wavelength routed networks.

6.3 There are four logical connections which have to be transported on an optical WDM network. Each connection has to be first FEC coded, the RF subcarrier modulated and finally the LD intensity modulated before being transmitted over the fiber. In which sub-layer and functional block will these operations be carried out for a point-to-point connection?

6.4 (a) The physical and virtual topology for a four-port network is shown in Figure 6.6. For a message to be sent from 1 to 2 and another from 1 to 3, what will you use at node 1, 2, 3, and 4 for the two messages: OXC, OLT, or OADM?

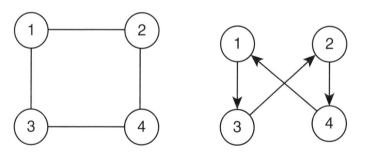

Figure 6.6 Four-node network: (a) physical topology; (b) virtual topology.

(b) Consider the given physical topology in Figure 6.7, assuming each node to have three transmitters (wavelengths: λ_1, λ_2, and λ_3) and three receivers. The nodes are connected by a pair of fibers for the bi-directional routes. Write down the possible s-d pairs, their LPs, routes, and possible wavelength assignment. There is no wavelength converter at any node.

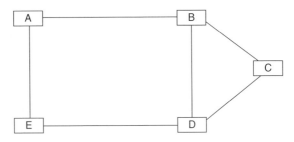

Figure 6.7 Physical topology of a six-node network.

6.5 (a) What are the objective functions and constraints in VT design problems?

(b) In a wide area network the *message delay* is dominated by ____ delay in real time application, and by ____ delay in offline application. How can one reduce the two types of delays?

(c) Where are the following algorithms used in wavelength routing networks:

- Annealing
- Graph coloring
- Randomized rounding

(d) What is the advantage/disadvantage of stacking IP over ATM over SONET over WDM as compared to IP/WDM?

6.6 What are the factors limiting throughput in the following types of networks: (a) a star coupler-based LAN; (b) an all optical wavelength-routed WAN; (c) a WAN consisting of electronic switches joined by point-to-point WDM links?

6.7 Write a computer program for a five-node network as in Figure 6.5 to study dynamic routing and wavelength assignment in WRN. Study the various algorithms discussed in this chapter with Poisson-distributed connections with λ arrival rate and μ as the service rate. Each connection is destined randomly to different nodes with equal probability. Calculate the blocking probability of the network.

6.8 (a) What is the least congested path routing scheme? What are its advantages or disadvantages?

(b) For the given network (Figure 6.8) for the node pair (0–4) use the *Least Congested Path Routing algorithm* to select the route and an appropriate wavelength. Give a reason for the selection of the route and wavelength. Assume all lengths are of equal lengths and the set of links on which each of the wavelengths is available is given by:

$\lambda_0:L_0,L_2;\ \lambda_1:L_0,L_1,L_2,L_3,L_5;\ \lambda_2:L_1,L_2,L_5;\ \lambda_3:L_0,L_1,L_2,L_4,L_5$

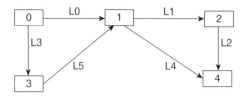

Figure 6.8 A five-node topology.

6.9 For the five-node optical network given in Figure 6.8 with uniform traffic flow on all the links, formulate the complete ILP with the objective of minimizing the delay in the network. Write all the constraints and variables giving explanations qualitatively.

6.10 (a) A WRN network has eight nodes. Six access stations (A–F) are connected to the different nodes as given below. Traffic matrix for the six stations and the connection matrix of the eight nodes are as given. Neatly show the light-path connections for the given traffic matrix using minimum number of wavelengths in the network.

References

1 Mukherjee, B. (2006). *Optical WDM Networks*. New York, USA: Springer Publications.
2 Stern, T.E., Ellinas, G., and Bala, K. (2008). *Multiwavelength Optical Networks-Architecture, Design and Control*, 2e. Cambridge University Press.
3 P. Roorda, C.-Y. Lu, and T. Boutlier, "Benefits of All-optical Routing in Transport Networks", In *OFC'95 Technical Digest*, pp. 164–165, 1995.
4 Cahn, R. (1998). *Wide Area Network Design: Concepts and Tools for Optimization*. San Francisco: Morgan Kaufmann.
5 Chlamtac, I., Ganz, A., and Karmi, G. (1992). Lightpath communications: an approach to high-bandwidth optical WAN's. *IEEE Trans. Commun.* 40 (7): 1171–1182.
6 Mukherjee, B., Banerjee, D., Ramamurthy, S., and Mukherjee, A. (1996). Some principles for designing a wide-area optical network. *IEEE/ACM Trans. Networking* 4 (5): 684–696.
7 Ramaswami, R. and Sivarajan, K.N. (1996). Design of logical topologies for wavelength-routed optical networks. *IEEE JSAC/JLT Spec. Issue Opt. Networks* 14 (5): 840–851.
8 Ganz, A. and Wang, X. (1994). Efficient algorithm for virtual topology design in multihop lightwave networks. *IEEE/ACM Trans. Networking* 2 (3): 217–225.
9 Labourdette, J.-F.P. and Acampora, A.S. (1991). Logically rearrangeable multihop lightwave networks. *IEEE Trans. Commun.* 39 (8): 1223–1230.

10 S. Banerjee and C. Chen, "*Designof Wavelength-Routed Optical Networksfor Circuit switched Traffic*", *In Proceedings of the IEEE Global Telecommunications Conference (Globe- com)*, London, UK, November 1996.

11 Baroni, S. and Bayvel, P. (1997). Wavelength requirements in arbitrary connected wavelength-routed optical networks. *IEEE/OSA J. Lightwave Technol.* 15 (2): 242–252.

12 *G.872: Architecture of optical transport networks – ITU* https://www.itu.int/rec/T-REC-G.872.

13 Chiu, A.L. and Modiano, E.H. (2000). Traffic grooming algorithms for reducing electronic multiplexing costs in WDM ring networks. *IEEE/OSA J. Lightwave Technol.* 18: 2–12.

14 T. Cinkler, D. Marx, C. P. Larsen, and D. Fogaras,"Heuristic algorithms for joint configuration of the optical and electrical layer in multi-hop wavelength routing networks", In *Proceedings of IEEE Infocom*, 2000.

15 Garey, M.R. and Johnson, D.S. (1979). *Computers and Intractability: A Guide to NP- Completeness*. New York: W. H. Freeman.

16 Bienstock, D. and Gunluk, O. (1995). Computational experience with a difficult mixed-integer multicommodity flow problem. *Math. Program.* 68: 213–237.

17 Raghavan, P. and Thonlpson, C.D. (1987). Randomized rounding: a technique for provably good algorithms and algorithmic proofs. *Combinatorica* 7 (4): 365–374.

18 Chatterjee, B.C., Sarma, N., Sahu, P.P., and Oki, E. (2017). *Routing and Wavelength Assignment for WDM-Based Optical Networks; Quality-of-Service and Fault Resilience*. Springers.

19 Kershenbaum, A. (1993). *Telecommunications Network Design Algorithms*. New York: McGraw-Hill.

20 A. Sridharan and K. N. Sivarajan, "Blocking in all-optical Networks", In *Proceedings of IEEE Infocom*, 2000.

21 Yen, J.Y. (1971). Finding the K-shortest loop-less paths in a network. *Manag. Sci.* 17 (11): 712–716.

22 A. Narula-Tam and E. Modiano, "Dynamic load balancing for WDM-based Packet Networks", In *Proceedings of IEEE Infocom*, 2000.

23 Zhang, Z. and Acampora, A.S. (1995). A heuristic wavelength assignment algorithm for multihop WDM networks with wavelength routing and wavelength reuse. *IEEE/ACM Trans. Networking* 3 (3): 281–288.

24 Ramaswami, R. and Sivarajan, K.N. (1995). Routing and wavelength assignment in all-optical networks. *IEEE/ACM Trans. Networking* 3 (5): 489–500.

25 Rayward-Smith, V.J., Osman, I.H., Reevesand, C.R., and Smith, G.D. (1996). *Modern Heuristic Search Methods*. New York: Wiley.

26 Reeves, C.R. (1993). *Modern Heuristic Techniques for Combinatorial Problems*. New York: Wiley.

27 S. Subramaniamand, R. Barry, "Wavelength Assignment in Fixed routing WDM networks", *In Proc. of the IEEE Int'l Conf. Commun. (ICC)*, Montreal, Canada, June 1997.

28 Bermond, J.-C., Dawes, R., and Ergincan, F.O. (1997). De Bruijn and Kautz Bus Networks. *Networks* 30 (3): 205–218.

29 Hluchyj, M.G. and Karol, M.J. (1991). ShuffleNet: an application of generalized perfect shuffles to multi-hop lightwave networks. *IEEE/OSA J. Lightwave Technol.* 9 (10): 1386–1397.

30 K. N. Sivarajan and R. Ramaswami,"Multi-hop Lightwave Networks based on de Bruijn graphs", *In Proc. of the IEEE Conf. on Computer Communications (Info-com)*, pp. 1001–1011, Miami, FL, April 1991.

31 Ramaswami, R. and Sivarajan, K.N. (1996). Design of logical topologies for wavelength routed optical networks. *IEEE J. Sel. Areas Commun.* 14 (5): 840–851.

32 Banerjee, S., Yoo, J., and Chen, C. (1997). Design of wavelength routed optical networks for packet switched traffic. *IEEE/OSA J. Lightwave Technol.* 15 (9): 1636–1646.

33 Chardaire P., A. Kapsalis, J.W. Mann, V.J. Rayward-Smith and G.D. Smith, "Applications of Genetic Algorithms in Telecommunications", *Proc. of the 2nd Inter. Workshop on App. of Neural Net. to Telecomm.*, pp. 290–299, 1995.

34 Kirkpatrick, S., Gelatt, C.D. Jr., and Vecchi, M.P. (1993). Optimization by simulated annealing. *Science* 220: 671–680.

35 Aleksandar Tsenov, "*Simulated Annealing and Genetic Algorithm in Telecommunications Network Planning*", World Academy of Sc., Eng. and Tech. Int. J. of Elect. and Comm. Eng., vol. 2, No. 7, 2008. https://waset.org/.../simulated-annealing-and-genetic-algorithm-in-telecommunications...

7

Network Control and Management

7.1 Introduction

In earlier chapters, we have discussed the access, metro, and large area optical networks in detail with regard to their architecture, performance, and the routing and switching functions. We have also discussed various devices and network resources, such as logical nodes, access nodes, optical switching nodes, optical amplifiers (OAs), links, optical sources and detectors in the transport or data plane, both for the optical and logical layers which make up these networks. Also, we studied the network design issues for the access, metro, and core area networks. What we have not yet considered is, using these resources and devices, how does the optical network operate? How are the connections actually made for the data signals to be transmitted over them, monitored, and taken off, so that the network is made available for future connections? How can survivability be increased or the network made more energy efficient? These are the network control and management functions which are an integral part in any network functioning. Today optical transport networks support several client networks: Internet Protocol core networks, cellular networks, point-to-point private-lines, enterprise networks, and others. The larger networks may have many domains with equipment from different vendors having different control or management systems that do not interoperate with each other. These networks are highly managed and there is hierarchy among network controllers. The connection provisioning process is highly automated in the electronic higher network layers, but in the optical layer it is a combination of automated and manual steps, hence it is time consuming. Today the traffic is very high and dynamic, and therefore expectations from the optical networks are that they should be able to efficiently handle resources and quickly reconfigure as fast as possible with traffic demands. The ITU-T proposed the automatically switched optical network (ASON) which fulfills few of these requirements of optical networks, such as automatic end-to-end fast connection provisioning, support of different clients, higher

Optical WDM Networks: From Static to Elastic Networks, First Edition. Devi Chadha.
© 2019 John Wiley & Sons Ltd. Published 2019 by John Wiley & Sons Ltd.

reliability, scalability, and provision of different levels of quality of service (QoS). But it still has certain limitations.

Functionally, a communication network can be viewed as being composed of a data forwarding or transport plane and networking plane, which is the control and the management plane. The data plane is directly responsible for physically carrying the data across the network, whereas the management plane and control plane are responsible for the connection management, performance monitoring, and other network management operations. Network control is hence an indispensable part of a communication network operation. It is responsible for ensuring best continuous functioning, detection-restoration of faults, and security of the network.

The management plane performs management functions not only for the transport plane but also for the control plane and the system as a whole. It handles *accounting* and *safety* features in the network. The management plane is usually less automated and works on a longer time scale without change: hours, weeks, or months. The safety or s*ecurity management* involves protecting the network resources and data on the network by controlling and authenticating access to the network, data encryption, etc. The *accounting management* issues involve keeping track of the cost of users' utilization of the network and the various maintenance costs involved in managing the network. The control plane is more dynamic, performs in real time, and for shorter spans of time. It is responsible for implementing configuration, performance monitoring, and fault detection-restoration of the network. The *configuration management* deals with a set of functions associated with management of connections, resources, and adaptation in the network. *Connections* as part of configuration management deals with setting up connections, keeping track of them, and taking them down when they are not needed anymore. It also involves maintaining state information about the network and the current status of connections. *Resource management* accounts for tracking the resources, network topology, and notifying their state in the network, and *adaptation function* is required at the input and output of the optical network to convert client signals to signals that are compatible with the data layer, and vice versa. *Performance management* is responsible for monitoring and managing the various parameters that measure the performance of the network. It is linked with fault management, which deals with fault detection and identification, reporting and restoration of the faulty connection in the network. We therefore find that network management and control become more crucial in the case of optical networks as they are vast and carry heavy traffic of many varieties, hence any fault or underperformance of the network can cause a loss of a large volume of data.

Earlier optical networks, though, could offer large capacity but were quite inflexible. Most of their limitations were due to the fact that they were operated with complex and slow network management systems (NMSs). There were long connection provisioning times as some functions were either manual or semi-automated, inefficient resource utilization, and difficult interoperability between client networks and between networks belonging to different operators. With the present-day, highly dynamic, and multimedia traffic demands, flexibility in the control plane is essential for dynamic provisioning of the connections. Presently, most of these network functions are done in the electronic domain, but they need to move to the optical domain.

In this chapter, we explain the control and management functions of an optical network, which is a fairly broad and rapidly evolving subject. There are a number of control plane protocol standards being developed by various international bodies: the main organizations working in this area are the ITU-T, IETF (International Engineering Task Force) and the OIF (Optical Internetworking Forum). The International Telecommunication Union (ITU)'s approach to unified control architecture for the wide area OTNs is the ASON [1]. ASON defines a set of requirements that must be satisfied for automated control of network operations in a more abstract way. But the control protocols are proprietary to the vendors and are not standardized, therefore the network management functions have many interoperability difficulties. Large networks which are divided into many independent domains are managed by different service providers and equipped with network elements (NEs) of different vendors, so it becomes essential to have a standardized approach for signaling and routing for its proper functioning. The IETF, meanwhile, has developed a set of new protocols for the automation of the control plane. It has developed GMPLS (generalized multi-protocol label switching) [2, 3]. GMPLS is considered to be a framework for providing concrete implementation of the ASON requirements with explicit specification of a set of protocols to handle routing, link management, and signaling which are concerned with the smooth operation of the network. In IP packet networks, a standardized control plane based on MPLS (multi-protocol label switching) has been developed. GMPLS is working in the direction of standardization of the control plane protocols derived from MPLS for the optical network.

In Section 7.2 we give the details of an NMS architecture of an optical network and Section 7.3 outlines the logical frame structure of current automatic switched optical network with interaction of the control layers of client and the optical layers. Section 7.4 looks at the functionalities of control and management planes. Though control plane reliability is a critical subject in the operation of optical networks, we will be discussing that in more detail in a separate chapter. In Section 7.5 we will discuss in brief the GMPLS control protocols in optical networks, though for a more comprehensive treatment of GMPLS the reader is referred to texts completely devoted to the subject [2, 3].

7.2 NMS Architecture of Optical Transport Network

A basic NMS architecture of an optical transport network is shown in Figure 7.1. The client network nodes get interconnected with the wavelength division multiplexing (WDM) optical network through the optical network nodes. When data are to be sent over the network, the optical switching nodes are required to switch to the correct ports to facilitate the appropriate path for the *lightpath* (LP) to carry the signals to their destination node. Therefore all the optical network components, including optical line terminals (OLTs), optical add/drop multiplexers, optical amplifiers (OAs), and optical cross-connects (OXCs), along with the *lightpaths*, need to be managed for the purpose. After solving the wavelength routing and assignment (RWA) problem, the connection requests are provisioned and set up in the network by the NMS. There are fundamentally two different options for the network provisioning operations: either they are set up locally on each *ingress* node in a *distributed* manner, or the requests are forwarded to a *centralized* entity for path computation operations. Since in a distributed scenario, connection requests are provisioned by their respective ingress nodes, it requires the availability of global network state information on each network node to perform RWA locally. Distributed provisioning has advantages in terms of network scalability and resilience to network failures, as path computation responsibility is now with multiple nodes in the network and no single failure can bring the whole network down. However, for effective deployment in a dynamic scenario, the network state information available with

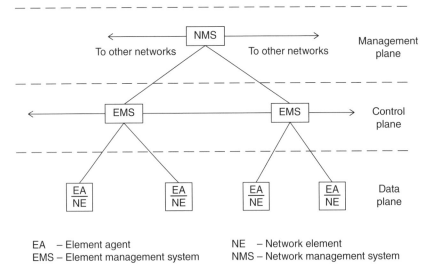

EA   – Element agent   NE   – Network element
EMS – Element management system   NMS – Network management system

Figure 7.1 Architecture of distributed network management system.

different network nodes has to be updated periodically. Dissemination of network status updates from each node is required at much higher speed because if not fast enough it will result in less optimal path computation decisions. The requirement of higher rate of network status update will increase the control overheads (OHs). Compared with distributed provisioning, in a centralized provisioning approach only a single entity is involved in path computation operations and state information is not to be conveyed to every other network node. Centralized control also eliminates the need for complicated routing protocols for disseminating network status information, which results in much lower control overheads. There is a more optimal use of network resources in centralized network provision as compared with a distributed approach. Hence, a centralized system is cheaper and less complicated, but it has limitations with scalability of a large network. It becomes very difficult for a single centralized NMS to manage the many functions of the entire large network. Hence distributed control is more appropriate in such cases.

The all-optical automatic switched network has three functional planes. The lowest plane, referred to as the transport or data forwarding plane, represents network resources. The functions of the transport plane are generation, transport, and termination of signals which are transferred on network connections. It provides transfer of data and few control and management information signals. The control plane performs the call and connection control functions. The functions of the ASON control plane are automated based on networking intelligence that includes discovery, routing, and signaling. The management plane performs management functions for the transport plane, the control plane, and the system as a whole, as well as coordinating operation of all the planes. These management functions are quasi-static and are related to the management of network elements and networks services, and are less automated than those of the control plane.

To carry out distributed control, each optical network element in the transport layer is managed by its own *element management system* (EMS) (Figure 7.1) in the control plane. The information of the performance parameters and the attributes of the elements to be managed are made available in an abstracted form in a built-in *element agent* (EA) in the NE. EA is a software application residing in a microprocessor in each network element. This information is also made available in the EMS and the NMS through the communication network for signaling, known as the DCN (data communication network). With this information, the network can be monitored and configured automatically by the management system. The EMS can make queries with the EA, and the EA also can initiate a message by sending a *notification* message in the form of alarms. EMS provides an interface to the NMS or OSS (operation support system), typically called a *northbound* interface, using protocols such as *simple network management protocol* (SNMP) [4, 5], *common object request broker* (CORBA), or extensible markup language (XML).

Three types of connections can be made in the ASON: permanent, switched, and soft-permanent. The permanent connection is set up by the network operator via the management plane and is an equivalent to a traditional leased line. The switched connections, involving the control plane, are set up within seconds. They enable services such as bandwidth on demand (BoD), etc. The soft-permanent connections, triggered by the management plane but set up within the network by the control plane, does connection provisioning and may normally support *traffic engineering* (TE) or dynamically establishing failed connections. The control plane of an optical network provides very fast connection provisioning, rapid failure recovery, and has TE features. TE provides service differentiations, QoS assurance, traffic monitoring, etc. It adaptively maps the traffic flow onto the physical topology of a network allocating various network resources, such as fiber links, wavelengths, and time-slots in order to balance the traffic load to improve the throughput of the network. This information is then used in the routing and switching the LP over the optical infrastructure. In other words, traffic engineering essentially signifies the ability to accommodate traffic wherever the capacity exists in order to enhance the throughput.

7.3 Logical Architecture of Automatic Switched Optical Network

Figure 7.2 shows the logical architecture of an ASON with the three planes and their signaling interfaces. The transport plane, responsible for data forwarding, contains a large number of forwarding switches, optical or digital, which are connected to each other via a physical interface. The control plane, which is responsible for resource and connection management, consists of a series of OCCs (optical connection controllers) interconnected via NNIs (network to network interfaces). The functions of OCCs are to carry out network topology discovery, signaling, routing, wavelength assignment, connection set-up and tear-down, connection protection and restoration, TE, etc. of the control plane. The internal network-to-network interface (I-NNI) is a bidirectional signaling interface between the adjacent OCC in the same domain providing signaling and routing, and external network-to-network interface (E-NNI) is a bidirectional signaling interface between control plane entities belonging to different domains for the inter-domain under different administrators. Connection control interface (CCI) is the interface between the network signaling element and the transport network element (NE), and the function of the protocol running across CCI is to support add/drop of connections and port status monitoring of switches. The management plane is responsible for managing the control plane, which includes configuration management of the control plane resources,

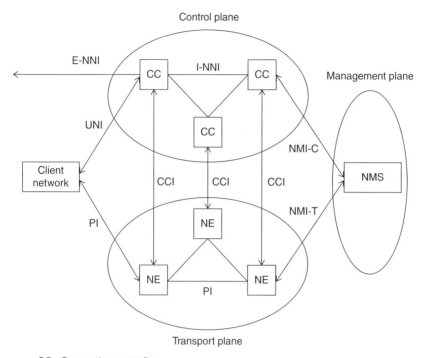

CC: Connection controller
CCI: Connection controller interface
E-NNI: External network-network interface
I-NNI: Internal network-network interface
NE: Network element
NMI-C: Network management interface-control plane
NMI-T: Network management interface-transport plane
PI: Physical interface
UNI: User network interface

Figure 7.2 Management and functional architecture.

transport resource in control plane, and policy. The network management entity in the management plane is connected to an OCC in the control plane via the NMI-C (network management interface for control plane) and to the transport switches via NMI-T (network management interface for the transport network). The user networks are connected to the transport plane via a physical interface, while it communicates with the control plane via a UNI (user network interface). This is a bidirectional signaling interface which carries out the functions of connection creation, deletion, and modification; neighbor discovery function; and service discovery function that allows the client network to discover the services that are available from the transport network. The UNI enables the client networks to dynamically establish connections across the optical network.

The different interfaces are defined by the information flow between control plane entities. For example, the connection service messages are done by UNI, E-NNI, and I-NNI; authentication and connection admission control messages are carried by UNI and E-NNI; endpoint name and address messages are again carried by UNI; reachability information is by E-NNI; topology information and network resource control information by I-NNI, etc. These control interfaces should be reliable, scalable, and efficient and should be fast enough to provide connections to be set up and taken down.

7.3.1   Interaction Between the Client Control Layer and the Optical Control Layer

The different client layers of the optical network – Synchronous Optical NETworking (SONET)/synchronous digital hierarchy (SDH), IP, OTN – are served by the optical layer for provisioning of the connections appropriately with best possible performance and efficiency. The interaction between the logical switches and the optical routers to obtain end-to-end connectivity can happen as an *overlay*, a *peer* or through *hybrid* models as mentioned below and shown in Figure 7.3.

- *Peer model.* This integrates the control layers of the client network and optical transport planes and is managed with a common NMS. As shown in Figure 7.3a, the NMS of the client network has full knowledge of the transport plane. All the network elements in both planes know the network topology and resources. Thus, there is better coordination for connections, performance monitoring, fault handling among different network elements, and better network resource utilization with faster failure discovery. The client layer digital routers have full topology awareness of the optical layer and can therefore control optical layer connections directly. The main drawback in the peer model is scalability. As the network becomes large, the number of states and control information increases and becomes very high to be handled by its *unified control plane.* Also, the management authority of the transport layer has to share the information of the network resources with clients, which the service provider does not want. It also makes the network more insecure. The other drawback is that optical layer elements have significantly different constraints with respect to routing and protection, compared with the electronic client layers. Therefore, the electronic logical routers have to use the equivalent routing constraints of the optical layer. These have to be determined by path computation engines (PCEs) residing in the electronic logical routers.
- *Overlay model.* In the overlay architecture, the client and optical control layers are completely separated for the purpose of control. The optical layer is totally opaque to the client network NMS, as shown in Figure 7.3b. The client network equipment has no knowledge of optical system resources and topology, and hence does not participate in the connection provisioning

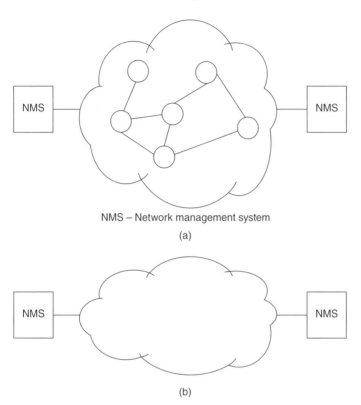

NMS – Network management system

(a)

(b)

Figure 7.3 (a) Peer model, and (b) overlay model.

process, just sends a request to the optical layer though the UNI for a connection. Within the optical layer, different sub-networks can interoperate through a standardized NNI. As control systems and protocols are proprietary to the vendors, they differ from system to system. Full privacy of the optical layer is maintained as the details of the optical network topology are hidden from the client layer through the UNI. It becomes very convenient to use the overlay model for a large transport network with several client layers and many independent sub-networks operated by different administrators. In this scenario the optical layer, client layers, and sub-networks can grow and evolve independently. The overlay model also has the ease of providing various services, such as optical virtual private networks (VPNs), transport bandwidth brokering, leasing transport facilities, and others.

The overlay model has a few disadvantages. It does not use the network resources efficiently as it completely separates the state and control information at the boundaries of the transport and client layers. It requires two separate but interoperable control mechanisms. One mechanism operates within the *core* optical network and the other acts as an interface between the core and *edge* components.

- *Hybrid model.* The hybrid approach combines the peer and overlay models and has good features of both. Some of the client logical nodes are peers to the optical switches in the transport network and thus share the same control plane, whereas a few others have their own control plane and interface with the transport network with a UNI. We also have another model called an *augmented model*, where the IP layer (client) operates as a separate control plane but has the concise routing, addressing, and state information of the optical layer. There is a secure intermediate intelligent controller between the two layers which has abstracted information of specific clients and optical layer topology and state information. The controller can use this information to request and release LPs based on specific policies, such as specific service-level agreement (SLA) and QoS. These models require more privacy and closer coupling between the IP and optical layers compared with the overlay models.

7.4 Functions of Management and Control Plane

After the details of the architecture of the automated management and control system of the optical network, we next discuss its different functionalities in further detail. As mentioned above, the management plane performs management functions for all the planes and coordinates their operation. Although each plane is autonomous, some interactions need to occur because all planes operate on a common underlying infrastructure. In centralized control the controller has global knowledge about the network, hence optimization is easier. Centralized control can have easier coordination among multiple client and vendor technologies and network layers. But with the increase in network capacity, in both size and traffic volume and dynamicity, the NMS has become more distributed. A few functions, such as accounting and security, are carried out by the centralized management while the functions which require fast response time of the order of milliseconds, such as fault discovery and restoration, topology and network discovery, route computation and establishment of the connections, are done by a distributed control plane.

To offer the features discussed above, the different functions that the control plane has to carry out are discussed below. These include discovery functions, routing, signaling as well as protection and restoration schemes.

7.4.1 Discovery or State Information

In ASON there is automatic discovery of the state of the network which results in fast configuration management, fault diagnosis, and other management functions. There are three groups of discovery functions: *neighbor discovery*, *resource discovery*, and *service discovery*.

- The neighbor discovery function determines the state of local links connecting to all neighbors. Automated neighbor discovery is essential for acquiring the topological information needed for setting up signaling and data paths in a network with dynamic load. This kind of discovery is required to detect and maintain node connectivity between adjacent network elements which would otherwise require manual interconnection among the network elements. The neighbor discovery protocol determines not only the neighboring nodes and ports for each interconnection but also the types of links, such as fiber, wavelength, or logical, plus the length of each link and its physical parameters, e.g. bandwidth, optical characteristics, etc.

  In order to carry out neighbor discovery, in the case of distributed control mechanisms, each node periodically transmits *Hello* messages. The message contains a unique link ID associated with the transmitting node. When the other node receives a *Hello* message from the neighboring node, it returns a new *Hello* message on that link with both the local and remote IDs. This is the neighbor discovery protocol. With the neighbor discovery procedure, each network element automatically learns its connectivity to other network elements. This connectivity information is spread throughout the network from node to node in a distributed system, so that each node acquires the information to construct its own view of the network topology. In the case of a centralized system, each network element periodically sends its own node and link interface information to a central manager to build the network topology.

  In an optical network, for neighbor discovery, the adjacency of the neighbor involves not only the physical layer but also the layer adjacency and control adjacency. Layer adjacency is required to know the endpoints of a logical link that terminates at the given layer. This is used for building layered network topology to support routing. The control adjacency involves two control entities associated with neighboring transport plane network elements. This is required for identifying the endpoints of the link connection which are needed for connection management.

- The resource discovery function has a larger scope. It keeps track of the system resources' availability, such as bandwidth, time multiplexing capability, number of fiber ports, etc. It allows every node to discover network topology and resources. This discovery determines the available resources, capabilities of various network elements, and how the resources are protected. Resource discovery can be achieved through either manual provisioning or automated procedures. Resource management in the network is done with the help of the *resource protocols*. They determine how the network resources are discovered, updated, and maintained in the link-state databases.

- Finally, the service discovery function is responsible for verifying and exchanging service capabilities of the network, such as the class of service, the grade of service supported by different administrative domains, and the services supported over a trail or link.

7.4.2 Routing

We have discussed the various wavelength routing algorithms, for example the link state or the distance vector-based approach, in Chapter 6. In this section, we now discuss the routing functions of the control plane. Routing is used to select paths for establishing connections through the network. When a connection request is made to the network, the network needs to find a route and obtain resources along the route to support this connection. This can be done by applying a routing algorithm on the topology database of the network. The routing algorithm has to consider the various constraints imposed by the network, such as wavelength conversion ability and the bandwidth available on each link along the route. The control protocol for route computation determines how the route of a connection request is selected according to TE policies of the carrier, which decides how to allocate network resources to a given request if multiple routes are available. Also, in an all-optical network, the transmission impairments accumulate along the optical paths and have to be taken into account while calculating the route. This information has to decimate throughout the network. Based on this information, a control agent can determine a feasible route for a desired connection. With the help of signaling of the control plane, to be discussed in the next section, the data plane performs information transmission, switching, and routing to appropriate ports with the help of OXC and other optical components.

7.4.3 Signaling

Signaling involves transporting control messages between different layers and entities which are communicating through the network's control plane. Signaling protocols are used for fault recovery and connection management, which involves creating, maintaining, restoring, and releasing connections. For the circuit switching case, this process involves reserving the resources required for the connection and setting the actual switches along the link for the connection. The source/destination node sends control signals to each of the intermediate nodes along the path for making a connection. A command for connection provisioning is issued by a client network in the case of a switched connection, or by an NMS in the case of a soft-permanent connection. Then the provisioning command triggers a signaling exchange and the connection is provisioned from a port of the source node to the other port in the OXC on to the destination node, with an explicit route assigned from the source along the path which requires configuring the OXCs with proper settings. The signaling protocol messages are carried on a signaling network that may be carried with a data transport network or is on a separate control network, as we will discuss in the next section. Such protocols are essential to enable fast provisioning or fast recovery after failures.

7.4.3.1 Signaling Network

Optical networks need to have robust signaling networks. A fast, reliable, and efficient signaling network is essential between the NEs and the EMS, and also between EMSs because of the high agility and different demand functionalities of the dynamically switched optical networks. With the failure of a single signaling channel there can be simultaneous failure of multiple high data-rate channels, resulting in a large amount of data being lost. Performance monitoring of the network parameters will also be affected due to the failure of the signaling network [6–8]. This will not only provide degraded QoS to the clients but is equally important for detecting faults. Fault management in the network requires detecting problems in the network, alerting the systems appropriately, and then restoring them. The performance monitoring of the optical layer, fault and configuration management is done by the control and management plane by communicating through the DCN. The signaling control channels can be outside the optical layer or they can be inside the optical layer, either in-band or outside the data band. In the following we discuss the signaling network further in detail:

- The control signaling channel network outside the optical layer is part of the DCN. The DCN links the EAs and EMS and also the different EMSs. The EMS is usually connected to one or more EAs and communicates with neighboring EMSs (Figure 7.1). The DCN is supported by the transmission control protocol (TCP)/IP or open systems interconnection (OSI) as well as on dedicated leased lines.
- For very fast signaling we have the signaling channels in the optical plane. In the out-of-data band we have the optical supervisory channel (OSC) on a different wavelength. This is available for WDM line equipment, such as the optical transmission section (OTS) and optical multiplex section (OMS) layers of the OTN protocol. Optical amplifiers are managed using this approach. The OSC on a separate wavelength is used for fast signaling of critical information in real time to support optical path trace, defect indicators, etc.
- Another approach in the optical plane for signaling is the in-band *pilot tone*, also known as the *subcarrier modulated signaling*. In this case, the optical wavelength is subcarrier modulated with an RF signal of 1–2 MHz which carries the pilot tone. Pilot tones carry overhead information within the subnetwork for fast signaling. The overheads have small modulation depth (5–10%) of the pilot signal at the rate of a few Kbps to carry the signaling

overhead information. By carefully choosing the pilot and subcarrier frequency, the data is relatively unaffected. At intermediate locations, a small fraction of the optical power can be tapped off and the pilot tones extracted. The advantages of the pilot tone approach are that it is relatively inexpensive and it allows monitoring of the overheads in transparent networks without requiring knowledge of the actual protocol or bit rate of the signal. Pilot tone is used in the optical channel (OCh) layer of the OTN or wherever the signal is available in the electronic domain, such as the OLT or 3R locations. It can be inserted or modified at the transmitter or at a regenerator. The limitation with this is that the *pilot tone* will be available only after the O-E conversion, i.e. at the edges of the network but not within the transparent optical subnetwork. At the boundaries of each subnetwork, the signal is regenerated (3R) by converting it into the electrical domain and back.

In-band signaling is very fast and is used in real-time exchange of information or where very fast switching is needed for protection purposes, etc. It has a disadvantage, however: there will no control signal available in the network when the control channels are brought down for maintenance purposes while the data links are still in use.

7.4.3.2 Alarm Management System

For fault detection and correction the signaling is done with the *alarm management* system in the control plane. When a link fails, the node downstream of the failed link detects it and generates a *defect condition*. If the defect persists for a certain fixed time period, typically a few seconds, the node generates an alarm so that fault correction can take place. Many times it can happen that a single failure event may cause multiple alarms to be generated in the network, leading to incorrect actions taken in response to the failed condition. In such a situation, the management system reports the single *root-cause* alarm. Multiple alarm suppression is accomplished by using a set of special signals, called the *forward defect indicator* (FDI) and the *backward defect indicator* (BDI). Immediately upon detecting a defect, the node inserts an FDI signal downstream to the next node. The FDI signal propagates rapidly and nodes further downstream receive the FDI and suppress their alarms. A node detecting a defect also sends a BDI signal upstream to the previous node to notify the node failure. If this previous node did not receive the FDI, it then knows that the link to the next node downstream has failed.

7.4.3.3 Resource Reservation Signaling Protocol

Resource reservation signaling protocols fall into two categories: *parallel resource reservation* and *hop-by-hop* resource reservation. In the case of parallel resource reservation, the control scheme reserves wavelengths on multiple links in parallel. As an example, as in the *link-state parallel* reservation every

Example 7.1 A long all-optical link connection has two optical amplifiers (OAs) connected between the end network access terminals (NATs). (i) Show the different signaling schemes used in different segments of the optical layer. (ii) If a fiber cut develops between the first NAT and the first in-line OA, how do FDI and BDI flow in the link and how are they carried along the link?

Solution
Figure 7.4 gives the signaling and FDI and the BDI signals for the alarm suppression in the network.

(i)   The different types of optical layer overheads for monitoring and signaling are: the OSC as an out of data band optical signaling technique. The OSC on a different wavelength will be used hop by hop in the link from one NE to the next, as shown in Figure 7.4. It can be monitored in the optical domain in between at nodes 2, 3, and 4. The pilot tone can be inserted by a transmitter at the NAT1 as an in-band signal and can also be monitored at NEs until it is terminated at a receiver in OLT2. The rate-preserving OHs used end to end through the link and removed at the NAT2 receiver.

(ii) When there is a link cut between NAT1 and the optical amplifier, as shown in Figure 7.4, the amplifier detects the cut. It immediately sends an FDI signal downstream indicating that all channels in the fiber have failed and also a BDI signal is sent upstream to NAT1. Both FDI and BDI are sent over the OSC. The OA in the downstream sends the FDI to the OA, which in turn sends to NAT2 on the OSC. At the NAT2 all the wavelengths are de-multiplexed and the NAT2 then generates FDI signals for each failed connection and sends them downstream to the ultimate destination of each connection as part of the overhead. Finally, the only node that issues an alarm is the first OA.

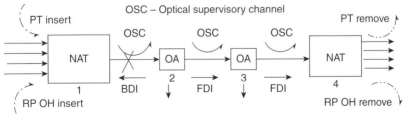

Figure 7.4 Example of optical signaling and the BDI and FDI generated in the link.

node has global state information of network topology and wavelengths on each link, the source node reserves the desired wavelength on each link in the route by sending a separate control signal to every node in the route. Each node receives a *reservation request* message for a specified wavelength and then will send either a positive or a negative acknowledgment back to the source. On receiving positive acknowledgments from the nodes, the source establishes the lightpath with the destination; it thus calculates an optimal route to a destination on a given wavelength. The advantage of a parallel reservation scheme is that it reduces the lightpath establishment time by parallel processing the reservation requests but with the disadvantage of requiring global knowledge.

Hop-by-hop resource reservation can be of two types: *forward path* or *backward path*. At each intermediate node, the control message is processed before being forwarded to the next node. When the control message reaches the destination, it is processed and is sent back toward the source node. The actual reservation of link resources may be performed either while the control message is traveling in the forward direction toward the destination, or while the control message is traveling in the backward direction toward the source. If the wavelength is blocked, the source node may select a different wavelength and reattempt the connection. The limitation of this approach is of high latency.

There are also other aggressive reservation schemes in order to maximize the likelihood of establishing a lightpath. These schemes have the drawback of network resources being over-reserved for a short period of time, which leads to blocking of subsequent connection requests and lower network utilization. To prevent the over-reservation of resources altogether, *backward reservation* schemes are used: the source node sends control packets to the destination without reserving any resources but collects information about wavelength usage along paths. The destination will then utilize this information to decide on a route and a wavelength. The destination sends a reservation message to the source node along the chosen route and this reservation message will reserve the appropriate network resources along the way.

7.4.4 Performance Monitoring

The performance of the network parameters has to be periodically monitored in order to comply with the service guarantees assured by the service provider to the clients and also for fault management in the network. Monitoring the status of the link can be accomplished with continuity check messages by keeping statistics on packet losses and BER (bit error rate). If performance falls below an acceptable threshold, as an early warning a failure indication signal can be sent to the other end of the link, which can allow a switchover to an alternate link before the link fails. Performance monitoring must be able to detect the transmission failures which affect the signal quality and reduce the QoS, though it may not disrupt the network services totally. If a certain parameter is being monitored and its value falls outside its preset range, the network equipment should generate an alarm.

The key parameter associated with optical layer performance is the BER, which can be monitored when the signal is available in the electrical domain after detection. For client nodes SONET/SDH or OTN service over the optical layer, BER can be made available from the framing overhead parity check bytes, which can provide the performance guarantee within the optical layer. But in the optical domain the performance parameters can be measured with optical power meters, optical spectral analyzer (OSA), optical time domain reflectometer (OTDR), or pilot tones.

In the case of an optical transparent WDM optical network, transmission effects associated with the impairments in various transmission links and optical components need to be monitored continuously in each lightpath during normal operation. This is important as the transparency in optical networks may introduce various transmission impairments, such as dispersion, crosstalk, and many other miscellaneous linear and non-linear impairments, etc. But there are many difficulties in providing continuous control and monitoring of all supported lightpaths simultaneously in a WDM system. One needs to monitor optical signal-to-noise ratio (OSNR) in the optical domain for each wavelength channel in the core of an optical network. The lightpaths are transparent to data rates, protocols, and signal format and the optical performance measurements, which are typically limited to optical power, OSNR. These measurements do not directly relate to latency, BER, or other QoS issues used by carriers. Normally, a small part of the optical power from a desired channel is tapped via a low-loss tap to a measurement device such as a power-averaging receiver.

7.4.5 Fault Management

Fault management functions include prevention, detection and correction of the faults in networks. The function of fault management in optical networks is to detect the location of the fault in the network and to rectify it quickly by separating the failed link or node. It then has to restore the traffic either by providing an alternate route or by replacing it with the standby component. The detection and rectification of the fault need to be done automatically within 50–60 milliseconds. The network does this by alerting the control plane appropriately through alarms and by rapid notifications, as discussed earlier.

Different types of signaling are provided in order to effectively carry out fault management in the optical networks. In protecting all-optical networks, supervisory techniques are required to detect failures [9, 10]. These techniques are categorized in two groups. In the first group, statistical analysis of the transmitted data is done. The data analyzed is the optical power, spectrum using OSAs, and BER. The second group of techniques uses probe signals, such as the pilot tones and OTDR signals for diagnostics. Fault management functions need to be fast and accurate, and need to function over an extended period of time.

As a link failure can potentially lead to the loss of a large amount of data, we need to develop appropriate protection and restoration schemes which minimize the data loss when a link failure occurs. The higher layers of protocols have their own procedures to recover from link failures. However, the recovery time for upper layers may be significantly higher, of the order of seconds, whereas the fault-recovery time in the optical layer would be of the order of milliseconds. Also, survivability at the optical layer can provide protection to higher-layer protocols that may not have built-in fault recovery. Restoring service in the event of failures is an autonomous control function and so has to be a distributed application as the network restoration has to be fast in order not to lose the data.

Details of fault tolerance and survivability in the optical network will be discussed in the next chapter.

7.4.6   Security, Accounting Management, and Policing

Security, policing, and accounting are the functions handled by the management plane of the network. Security and safety management handles the issues of limiting the maximum power launched by the laser transmitter and limiting the laser emission below the maximum allowed under the standard safety limit [11]. Using shuttered optical connectors to avoid any leakage and have some automatic shutdown of equipment in case of any leakages or emission also come under the safety measures. Protecting users' data from bugs and unauthorized people are the security features provided by the NMS.

The accounting management function is responsible for the billing functions. Another function is to monitor actual service being utilized by the user, such as the bit rate, QoS, etc. In order to do so the system needs to monitor the power levels of signals in the network. The network also needs to monitor the thresholds of the parameters set for each user signal. Basically, the services provided need to be monitored according to the tariff being charged from the clients.

7.5   Generalized Multi-Protocol Label Switching

We have discussed the unified control architecture of ASON in the previous sections. As described earlier, most of the network control functions are automated in ASON. The software application of the distributed control plane resides in the physical network equipment in the transport layer, thus the network configuration functions can be easily automated. But the control protocols are proprietary to the vendors and are not standardized; therefore, the network management functions have many interoperability difficulties. It is important to have standardized control protocols for signaling and routing in a network. The ITU endeavor in this direction was to standardize the existing protocols, such as PNNI (private network-to-network interface) for signaling, which is similar to open shortest path First (OSPF) used for IP routing, and many such others. The IETF developed a new suite of protocols for the automation of the control plane, the GMPLS protocols [3, 12–14]. In IP networks, a standardized control plane based on MPLS is now in place. MPLS not only provides a faster method for forwarding IP packets with QoS but also offers a variety of new functionalities and services such as VPN, TE, etc. The success of MPLS in packet-switched network was a motivation for applying the concept of MPLS to other kinds of transport technologies. MPLS is an open-ended technology and hence can be extended to future undefined networks as well. GMPLS is the work in the direction of standardization of the control plane signaling protocols derived from MPLS for optical networks.

GMPLS provides a general control plane which can support any kind of underlying transport technologies. The benefits of having a general control plane for all types of network devices is that it alleviates the complexity in managing and controlling network devices, due to which network resource management and service provisioning of end-to-end TE paths are automated. It also has the ability to automate network operations, such as providing protection and restoration rapidly and automatically across network and domain boundaries involving interconnected dissimilar networks. The basic label switching features of MPLS have been moved down to the physical layers of general classes of networks, such as labels for time slots switching in time division multiplexing (TDM) SONET/SDH networks, wavelengths and space switching in ports or fibers for the wavelength routing transparent optical network. In GMPLS the idea of a *label* is generalized to identify any traffic flow. With the data and other multimedia services demanding high speed, the control plane needs to dynamically manage traffic demands and balance the network load on different data flows, such as fiber links, wavelengths, and time-slot at the switching nodes so that none of these components is over- or underused. GMPLS facilitates this TE process of adaptively mapping traffic flows onto the physical topology of a network and allocating resources to these flows. With TE, resources can be reserved and routes predetermined for the optical network. The client network devices that connect to the optical network can request the connection setup in real time as needed and thus can provide the service of BoD, QoS, etc. Once the connection parameters are made available to the ingress node, the network control plane determines the optical paths across the network according to the parameters that the user provides. Thus, dynamically, the control plane can choose a route and a wavelength that maximizes the probability of setting up a given connection, while at the same time minimizing any blocking for future connections [15–17].

7.5.1 Interfaces in GMPLS

Multiple types of interfaces are required to make GMPLS flexible enough to provide all types of services, from electronic switching to all-optical network switching. The five main interfaces supported by GMPLS are the *packet switching capable* (PSC) interface to recognize packet boundaries and forward the packets based on the IP header; layer 2 *switch-capable* (L2SC) interfaces to recognize frame and cell headers and forward data based on the content of the frame or cell header, such as the asynchronous transfer mode (ATM) label switched router forwards on its virtual path indicator (VPI)/virtual channel identifier (VCI) value or Ethernet routers forward their data based on the media access control (MAC) header. *Time-division multiplexing-capable* (TDMC) interfaces forward the data based on the time slot, for example, SDH digital cross-connect (DCS), add drop multiplexer (ADM), or OTN switches' interfaces implement the digital wrapper etc., *lambda switch-capable* (LSC) interfaces for wavelength-based control of optical add-drop multiplexers (OADMs) and OXCs, operating at the granularity of the single wavelength and *fiber-switch-capable* (FSC) interfaces forward the signal from incoming fibers to the outgoing fibers for physical fiber switching systems.

7.5.2 GMPLS Control Plane Functions and Services

GMPLS has a distributed control plane. It has to support both the packet switched services and the non-packet-switched services. How does GMPLS support these services in various devices that switch in different domains? In the case of the IP MPLS network, a label contains an arbitrary integer value which represents FEC (forwarding equivalence class) of a packet. In this case a logical quantity, i.e. an arbitrary integer representing FEC, is switched. By contrast, in optical network devices a physical quantity, such as a wavelength, a time-slot, or space switching within a fiber bundle, is switched. Furthermore, network devices in optical networks cannot examine individual packets passing through them. Therefore, labels in optical networks should represent the physical quantities, i.e. a specific wavelength, fiber, or particular time-slot in these network devices. Thus, conventional MPLS labels are not applicable to optical network devices. For this reason, format of a label is redesigned in GMPLS. The new format of a label is referred to as *generalized label* or G-label. G-label supports the diversity of switching identities. For example, an LSP (label switched path) can specify the label for a packet, wavelength, fiber as the switching type which specify the switch capability such as packet switch, wavelength switch, time-slot switch, etc. and the payload identifier to identify the type of payload such as ATM, Ethernet, etc.

Besides G-label, another important feature in GMPLS that enables it to support the various kinds of transport network devices is the complete separation of control plane and data planes. In contrast to the packet-switched IP networks, which have a unified data and control plane, in optical networks control plane messages cannot be delivered in the same channel as the data channel. For this reason, the control plane needs to be separated out of the data plane. The control traffic or signaling may be transported over a dedicated optical channel, via separated links or even digitally over separated networks, such as DCN.

The OXCs operate as LSRs (*label switching routers*). In GMPLS the labels (time slot, wavelength, or space) are distributed to the OXCs which are used for label swapping. The label-swapping operation corresponds to the cross-connect settings for the resultant path, analogous to LSP in IP routers. In optical networks for a circuit-switched connection the label swapping is required only during connection setup time, for which the labels for the path are requested via messages sent downstream from source node through intermediate nodes to destination node. The destination node responds with a message for the upstream path and similar messages are sent upstream to the remaining nodes propagating the label back to the source node, after which the communication starts.

Next, we have to see how an LSP is established in a GMPLS control network. An end-to-end optical trail, the generalized label switched path (G-LSP), is created between *ingress* and *egress* nodes and label forwarding is done at each intermediate router according to the lookup tables. Thus, GMPLS uses *label switching* along the path. There are high-speed LSRs which switch the traffic within the GMPLS domain with the *label distribution protocol* (LDP). The sequence of labels and the LSR routing tables define the virtual paths for the packets. GMPLS supports TE by allowing the ingress node to specify the route that a G-LSP will take by using explicit *lightpath* routing. This path is set up before the data transmission, similar to circuit switching. *Path* and *Resv* messages are used to reserve the resources along the session path. First, a source sends a *Path* message to the destination indicating the traffic characteristics of the session. The routers that process this signaling message, along with the data path, also create the *Path state* for the session in their databases and forward the message to the next hop. Based on the *Path* message received and the traffic characteristics specified in the message, the destination sends back a *Resv* message along the reverse path. Each router on the reverse path that processes the *Resv* message creates a local *Resv state*. Thus, resources are reserved for the traffic flow from the source to the destination.

In GMPLS there is an LSP hierarchy. This is a concept of aggregating the lower-speed LSPs into higher-speed LSPs. The hierarchy of LSPs is based on physical link granularity. As shown in Figure 7.5, packet links PSC-LSP have the lowest granularity. They are nested in a TDM link: the TDM-LSP. A set of TDM-LSPs is aggregated within a lightpath link, LSC-LSP, and finally, a group of LSC-LSPs is nested within a fiber.

The rule of LSP establishment in GMPLS is *tunneling*, that is, the LSP in the higher hierarchy must be established first in order to act as a virtual link so that the LSPs in the lower hierarchy can be tunneled through it. For example, the ingress node LSR sends a *Path/Label Request* message destined for the egress LSR through the intermediate LSR. The intermediate LSR will check whether there are any existing higher-order LSPs between it and the next LSR which are able to carry the lower-order requested LSP by the ingress node. If there is a higher-order LSP that is capable of tunneling the lower-order LSP, this higher-order LSP will be used as a link to transport a *Path/Label* message to egress LSR. When egress LSR receives the message, it will reply with a *Resv/Label Mapping* message, including a G-label that may contain several generalized labels. Then, when ingress LSR receives the G-label with a *Resv/Label Mapping* message, it will send a *resource ReSerVation Protocol* (*RSVP*)/*Path* message to establish an LSP with destination LSR. But if no higher-hierarchy LSPs exist to tunnel, the intermediate LSR will again send a *Path/Label Request* message destined for the next intermediate LSR now to establish a higher-order LSP which can carry the requested LSP. Once the next LSR receives this message, it will check whether the next intermediate LSR has any higher-order LSPs that can carry lower-hierarchy *LSP*. If not, this process of establishing higher-order LSP will be performed recursively until all higher-order LSPs have been established. Once all higher-order LSP have been established and egress LSR has received a *Path/Label* message from the ingress LSR, it will reply with a *Resv/Label Mapping* message including a G-label that may contain several generalized labels. Then when ingress LSR receives a G-label, it will connect and start sending the traffic to the egress LSR by sending *RSVP/Path* messages. This is illustrated in Figure 7.6 where the lower-order LSP5 from the ingress node is tunneled through the higher-order LSPs from LSP4(OC-12) to LSP1(OC-192) through the intermediate LSRs to reach the egress node on the other end of the link.

The advantage of the GMPLS LSP protocol is that the *Path/Label Request* message from ingress to egress are bidirectional and include the TE

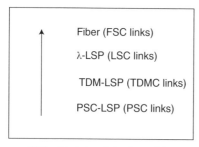

Fiber (FSC links)

λ-LSP (LSC links)

TDM-LSP (TDMC links)

PSC-LSP (PSC links)

PSC – Packet switched capable
TDMC – TDM capable
LSC – Lambda switched capable
FSC – Fiber switched capable

Figure 7.5 The hierarchy of LSPs in GMPLS.

parameters, local protection, etc. With bidirectionality of LSP, network latency is reduced, memory consumption for the LSP setting-up process is reduced, and the time required to configure the switching fabric on the upstream node is reduced.

7.5.3 GMPLS Protocol Suite

We now consider the GMPLS protocol suite which carries out different control functions. The full suite of RFC is given in [18–25]. Protocols in GMPLS are categorized into three groups based on their functions as follows:

- *Signaling protocols.* In a network, signaling protocols are required to exchange control information among nodes to set up LPs. GMPLS signaling is accordingly a means for conveying the necessary provisioning information, i.e. selected port, channel, etc., from one node to the next along the chosen path. The LSPs are either control-driven or data-driven, i.e. circuit switched, established prior to data transmission or upon detection of a certain flow, respectively. Suitable signaling protocols for the GMPLS control plane include RSVP and *Constraint based Routing-Label Distribution Protocol* (CR-LDP). Each data packet encapsulates and carries the labels during their journey from source to destination. The signaling protocols are closely integrated with the routing and wavelength assignment protocols. Through *RSVP* message QoS is introduced in a routing protocol for defined traffic flows by reserving the necessary link and node resources for the session. These signaling protocols are responsible for carrying messages for all the connection management actions to set up, modify, or remove the G-LSPs. These signaling protocols therefore have new requirements for a photonic network, heterogeneous networking elements to support provisioning, and restoration of end-to-end optical trails. The extensions to the *RSVP-TE* signaling protocols must support the generalized label concept for establishing LSPs at any architectural layer in a network, be it the logical or the optical layer. TE reserves different resources and predetermines the routes for a packet-switched network and thus provides virtual links or *tunnels* through the network to connect source destinations nodes. Packets injected into the ingress of an established tunnel are label-switched so that they automatically follow the tunnel to its egress. Thus, a virtual circuit switched path is made.
  CR-LDP uses TCP sessions between nodes in order to provide a hop-by-hop reliable distribution of control messages, indicating the route and the required traffic parameters for the route. Each intermediate node reserves the required resources, allocates a label, and sets up its forwarding table before backward signaling to the previous node. To implement a signaling protocol, a DCN is required that is used to transport signaling messages between the network nodes. DCN normally provides network layer connectivity between the control agents in an optical network.

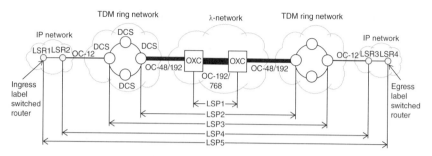

Figure 7.6 LSP establishment in GMPLS.

For protection and restoration, provision for a path protection is made prior to failure event occurring by GMPLS signaling. So, when a failure occurs and is detected at the source by *Notify* message, the source node with RSVP-TE signaling protocol activates the protection path that was precomputed and reserved for the connection that failed. The message is used in RSVP-TE to notify a session endpoint when a failure occurs. Intermediate nodes do not process the message but rather forward it to the appropriate endpoints of the affected connection. On reception of the *Notify* message, the session end-point identifies the failed connection using the LSP ID and sends an *Ack* message to the node that generated the *Notify* message.

- *Link management protocols.* LMP (link management protocol), which plays an important role in a GMPLS network, has two main functions. The first is to ensure the health of a link by isolating a single or multiple faults in the optical domain. The second is to provide physical connectivity of the link between two adjacent nodes with the neighbor discovery procedure discussed earlier.

LMP provides control-channel management, link-connectivity verification, link-property correlation, and fault isolation [26, 27]. *Control-channel management* establishes and maintains connectivity between adjacent nodes using a *keep-alive* protocol. LMP automatically generates and maintains associations between links and labels for use in label swapping. Monitoring of the control channel is carried out by the exchange of *hello* messages at specified time intervals to ensure that the LMP session is operational. *Link verification* verifies the physical connectivity between nodes, thereby detecting loss of connections or misrouting of fiber connections. *Link-property correlation* performs the verification of similarity of the TE requirements between links of adjacent nodes in the case of link bundling. GMPLS requires a way to verify that all TE requirements are similar between links of adjacent nodes in the case of link bundling. *Fault isolation* pinpoints failures in both electronic and optical links without regard to the data format traversing the link. In transparent networks, fault isolation is particularly important as a failure propagates downstream across multiple nodes and it is difficult to isolate its location. LMP determines the link where the failure has occurred by *backtracking* upstream until it finds a data link that has not failed. After LMP is executed and all node-port associations are obtained, the network topology is created automatically either by a centralized management system or in a distributed fashion. The network is then ready for connection provisioning, which is the function of the routing protocols discussed in the next section.

For topology management, similar to techniques used in IP networks, the nodes in the optical network have to maintain a database of the network topology resources available in the network and used to support the traffic. In order to do so the nodes periodically, or in the event of changes, flood the updated information to all the network nodes. A routing and topology management protocol is then used to represent the optical layer topology and update it periodically. To monitor the continuity or discover current neighbors, a *hello* message is sent periodically. *Ping* or *loopback* messages are sent from one end of a connection and are returned by the other end to verify the connectivity of connection. *Trace route* or *link trace* messages are sent from one end of the connection to discover the path of the connection. LMP is used to communicate the cross-connect information between the network elements for establishing connectivity.

- *Routing protocols.* The two most prominent IP routing protocols are extended Open Shortest Path First-Traffic Engineering (OSPF-TE) and extended Intermediate System-Intermediate System-Traffic Engineering (IS-IS-TE), which include attributes of the links such as available bandwidth, delay, etc. The functions of these protocols in GMPLS networks are to automatically discover the network topology, disseminate link status, and advertise resource availability. In addition to computing routes for carrying the *working* traffic, the algorithm may have to compute *protection* or *backup* routes for the connection, which are used in the event of failures. The Bellman-Ford algorithm and Dijkstra's algorithm are the most commonly used routing algorithms.

The path computation algorithm is a small part of the complete routing problem from the network control point of view. In GMPLS, link-state and topology information originates from neighbor discovery procedures using LMP. If the availability of a shortest path tree in each node is there, then the tree can be used for explicit routing when that node is an ingress node. OSPF supports hierarchical routing where a large network is divided into smaller areas that are all interconnected through a backbone area. The flooding of detailed link-state information is now confined to single areas. This reduces the scalability problems associated with the flooding process and reduces the size of the database that each node must keep. OSPF neighbors achieve *adjacency*, first, when they have exchanged and agreed on key parameters, second, when they have provided a copy of their databases to their neighbors, and third, when they have checked that these databases are consistent and complete. Once adjacency is achieved, the OSPF neighbors can proceed with their routing computation. In large networks, as there are several independent administrative domains, routing has to be done through several of these domains. Protocols that deal with routes within a domain are the interior gateway protocols (IGP) – OSPF is an example. Similarly, routing protocols that deal with inter-domain routing are called exterior gateway protocols (EGP). The basic EGP used in the Internet is the Border Gateway Protocol (BGP).

7.5.4 Path Computation Element

As defined by IETF [28–30], a path computation element (PCE) is an entity that is capable of computing a network path or route based on a network graph and applying computational constraints during the computation. A PCE is needed in many situations – situations when the computation of a path requires a large amount of calculations to be performed – and hence, it may not be desirable for some routers to perform the path computation by themselves. It is also used in situations when a node responsible for computing the path does not have complete visibility of the network to the destination or the situation when traffic engineering data (TED) is not available at a node. It can also be used when the node has a limited control plane or routing capability. For example, if there is a legacy device in the optical network which does not have the control plane or routing capability but has to integrate with GMPLS, then a PCE can be used for path computation.

We explain in brief the path computation function provided by PCE in the GMPLS network. In order to calculate paths, the PCE needs knowledge of the available network resources and the network topology. This information is stored in the TED. The TED can be built with information distributed by a routing protocol, like OSPF-TE or ISIS-TE. When the PCE is used as a standalone device and performs path computation as a separate service in the network, it computes the path with the given network constraints. It then sends a request to the ingress router, specifying the LSP route that it wants to set up. The LSP is then signaled all across the network, as explained earlier, and finally the LSP gets established.

A PCE node can also be deployed as a composite PCE node, or as an external PCE node, depending on the requirements of a provider and the available resources. A PCE as a composite node can be seen as a router that also implements the PCE functionality. This node uses the routing protocol in the control plane to give and receive TE information, and constructs the TED based on this information. When the ingress node receives a request to set up an LSP, it sends a request to the PCE, which communicates with the TED to calculate the requested path and then sends the response to the ingress node. After that, it can start signaling to establish the LSP.

The PCE can also be external to the network element that makes the request. In this case, when the ingress network node receives a path initiating request, it makes a path computation request to the external PCE. The PCE consults the TED and calculates the requested path, which is sent as a response to the ingress network node. After that, it starts signaling to set up the service. In the case of external PCE, it also creates and populates its own TED. In this case, each node that wants to communicate with the PCE to request an LSP needs to implement a path computation client (PCC). The PCC can connect to the PCE by means of the path computation engine communication protocol (PCEP) over a TCP connection. The steps are as follows: the ingress node sends a *path request* to the PCE. The PCE analyzes the TED to know the actual topology and network constraints and computes a path and sends it as a *path response*. After that, the ingress node sends an *RSVP* path message to the egress node, containing the specification of the required traffic. When it reaches the destination, it sends a *Resv* message as a reply. The resources are reserved in each hop through which the message passes. Finally, the message arrives at the ingress node, which finishes its resource allocation. After the LSP is set, all the involved nodes update their information about used resources and then flood a link state update (LSU) informing the situation as a LSU message. The LSP setup with PCE is given in Figure 7.7.

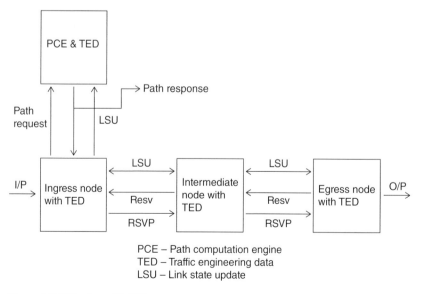

PCE – Path computation engine
TED – Traffic engineering data
LSU – Link state update

Figure 7.7 LSP setup with PCE.

7.6 Summary

A well-designed NMS is very important to ensure smooth and reliable operation of the network. It also improves efficiency and reduces the operating cost of the network. The current wide-area mesh optical networks require a robust and flexible optical control plane to manage network resources, provisioning and maintaining network connections across multiple control domains automatically. The implementation of a control plane requires information transfer among control, transport, and management entities. To achieve this, fast signaling channels are required between network elements to exchange real-time control information to manage all the supported connections and to perform other control functions. The chapter gives details of ASON's NMS architecture, its functional details and protocols for automatic configuration, signaling, performance monitoring, and fault detection and management issues.

The success of MPLS in packet-switched networks leads to the development of GMPLS which is able to operate on various kinds of traffic flows and client networks types in the optical network. As the control plane and the data plane in GMPLS networks are completely separated from each other, it offers many benefits. GMPLS protocols and the various interfaces to handle all types of devices have been discussed. Certain legacy optical network devices, which do not have the control plane or routing capability of GMPLS but use PCE for integration with the network, has also been described.

Problems

7.1 (a) What are the different categories of signaling protocols? State the advantages and disadvantages of each.

(b) Differentiate between the link-state and distributed-routing approaches for connection management.

7.2 If multiple NATs are connected to an all-optical OXC, without any O-E-O conversions at the ports, how does it communicate with other such OXCs in the network and what performance parameters does it monitor? Instead, if the OXC has an electronic switch core with O-E conversions at its ports, how would it communicate with other such OXCs in the network and what performance parameters can it monitor?

7.3 Identify the sub-layer in the optical network which performs the following functions. Also explain how these functions are performed.

(a) Setting up and taking down lightpaths in the network.

(b) Detecting a fiber cable cut in a WDM line.

(c) Lightpath failure detection.

(d) BER detection in a lightpath.

(e) Monitoring and changing the digital wrapper overhead in a lightpath.

7.4 How is fault recovery and reconfiguration controlled and managed in the user, logical, and physical layers in the case of multilayer system?

7.5 (a) How is QoS achieved with a GMPLS control plane in the optical network?

(b) How does GMPLS support both packet-switched and circuit-switched services in various devices that switch in different domains?

(c) What is the use of PCE in GMPLS?

7.6 What functions do alarms have in the control and monitoring system of an optical network? What type of signaling is it part of: DCN or optical?

7.7 In the network given in Figure 7.8, the numbers on the links represent the distance function. A lightpath has to be reserved between the nodes A and E. Compute the time needed to reserve all the lightpaths using: (i) parallel reservation; and (ii) hop-by-hop forward reservation. The different delay times are: message processing time at a node is 10 μs; OXC setup time is 100 μs; and propagation delay per km is 5 μs. What are the advantages and disadvantages of the two schemes?

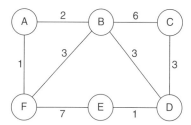

Figure 7.8 Six-node network topology.

References

1 *ASON-Automatically Switched Optical Network – ITU*; https://www.itu.int/dms_pub/itu-t/oth/0B/04/T0B040000150005PDFE.pdf
2 Farrel, A. and Bryskin, I. (2005). *GMPLS: Architecture and Applications*, Morgan Kaufmann Series in Networking. Elsevier Science.
3 Yamanaka, N., Shiomoto, K., and Oki, E. (2005). *GMPLS Technologies: Broadband Backbone Networks and Systems*. CRC Press.

4 Subramanian, M. (2000). *Network Management: Principles and Practice*. Reading, MA: Addison-Wesley.

5 Udupa, D.K. (1999). *TMN Telecommunications Management Network*. New York: McGraw-Hill.

6 Ramaswami, R. and Sivarajan, K.N. (2001). *Optical Networks, A Practical Perspective*. Academic Press.

7 Medard, M., Chinn, S.R., and Saengudomlert, P. (2001). Node wrappers for QoS monitoring in transparent optical nodes. *J. High Speed Networks* 10 (4): 247–268.

8 Larsen, C.P. and Andersson, P.O. (2000). Signal quality monitoring in optical networks. *Opt. Networks Mag.* 1 (4): 17–23.

9 Rejeb, R., Leeson, M.S., Machuca, C.M., and Tomkos, I. (2010). Control and management issues in all-optical networks. *J. Networks* 5 (2): 132–138.

10 Doverspike, R.D. and Yates, J. (2012). Optical network management and control. *Proc. IEEE* 100 (5): 1092–1104.

11 Medard, M., Marquis, D., Barry, R.A., and Finn, S.G. (1997). Security issues in all-optical networks. *IEEE Network* 11 (3): 42–48.

12 de Dios, O.G., Casellas, R., and Paolucci, F. (2016). GMPLS control plan. In: *Elastic Optical Network: Architectures, Technologies, and Control* (ed. V. Lopez and L. Velasco), 189–215. Springer Publishing.

13 Internet Engineering Task Force. *Generalized Multi-Protocol Label Switching (GMPLS) Architecture*, 2004.

14 Roebuck, K. (2011). *GMPLS/ASON: High-Impact Strategies – What You Need to Know: Definitions, Adoptions, Impact, Benefits, Maturity, Vendors*. Lightning Source.

15 Awduche, D., Berger, L., Gan, D. et al. (2001). *RSVP-TE: Extensions to RSVP for LSP Tunnels*. Internet Engineering Task Force.

16 Bradon, R., Zhang, L., Berson, S. et al. (1997). *Resource Reservation Protocol— Version 1 Functional Specification*. Internet Engineering Task Force.

17 Palmieri, F. (2018). GMPLS control plane services in the next-generation optical internet. *Internet Protocol J.* 11 (3, available at http://www.cisco.com/c/en/us/about/press/internet-protocol…/113-gmpls.html, viewed in March 2018).

18 Mannie, E. (2018). *Generalized Multi-Protocol Label Switching (GMPLS) Architecture"*, RFC 3945. Internet Engineering Task Force (IETF), available at http://tools.ietf.org/html/rfc3945.

19 International Engineering Consortium, *"Generalized Multiprotocol Label Switching (GMPLS)"*, available at: http://www.iec.org/online/tutorials/gmpls/index.asp, viewed March 2018

20 Kompella, K. and Rekhter, Y. (2018). *OSPF Extensions in Support of Generalized Multi-Protocol Label Switching (GMPLS)"*, RFC 4203. Internet Engineering Task Force (IETF), available at: http://tools.ietf.org/html/rfc4203.

21 Kompella, K. and Rekhter, Y. (2018). *IS-IS Extensions in Support of Generalized Multi-Protocol Label Switching (GMPLS)*", RFC 5307. Internet Engineering Task Force (IETF), available at: http://tools.ietf.org/html/rfc5307.

22 Lang, J.P., Rekhter, Y., and Papadimitriou, D. (2018). *RSVP-TE Extensions in Support of End-to- End Generalized Multi-Protocol Label Switching (GMPLS) Recovery*", RFC 4872. Internet Engineering Task Force (IETF), available at: http://tools.ietf.org/html/rfc4872.

23 Berger, L., Bryskin, I., Papadimitriou, D., and Farrel, A. (2018). *GMPLS Segment Recovery*", RFC 4873. Internet Engineering Task Force (IETF), available at: http://tools.ietf.org/html/rfc4873.

24 Ashwood-Smith, P. and Berger, L. (2018). *Generalized Multi-Protocol Label Switching (GMPLS) Signaling*", RFC 3472. Internet Engineering Task Force (IETF), available at: http://tools.ietf.org/html/rfc3472.

25 Papadimitriou, D. and Farrel, A. (2018). *Generalized MPLS (GMPLS) RSVP-TE Signaling Extensions in Support of Calls*", RFC 4974. Internet Engineering Task Force (IETF), available at: http://tools.ietf.org/html/rfc4974.

26 Fedyk, D., Aboul-Magd, O., Brungard, D. et al. (2018). *A Transport Network View of the Link Management Protocol (LMP)*", RFC 4394. Internet Engineering Task Force (IETF), available at: http://tools.ietf.org/html/rfc4394.

27 Berger, L. (2018). *Generalized Multi-Protocol Label Switching (GMPLS) Signaling Functional Description*", RFC 3471. Internet Engineering Task Force (IETF), available at: http://tools.ietf.org/html/rfc3471.

28 Farrel, A., Vasseur, J.-P., and Ash, J. (2018). *A Path Computation Element (PCE)-Based Architecture*", RFC 4655. Internet Engineering Task Force (IETF), available at: http://tools.ietf.org/html/rfc4655, viewed March 2018.

29 Ash, K. and Roux, J. (2018). *Path Computation Element (PCE) Communication Protocol*", RFC 4657. Internet Engineering Task Force (IETF), available at: http://www.ietf.org/rfc/rfc4657.txt.

30 Dasgupta, S., de Oliveira, J.C., and Vasseur, J.-P. (2007.). Path-computation-element-based architecture for interdomain MPLS/GMPLS traffic engineering: overview and performance. *IEEE Network* 21 (4,).

8

Impairment Management and Survivability

8.1 Introduction

Today's high data rate DWDM transport networks are built on the paradigm that optical links provide a highly reliable communication infrastructure with end-to-end bit error rates (BERs) below 10^{-15} and typical system availability exceeding five nines. These stringent quality of service (QoS) requirements are based on commercial carriers' requirement to provide reliable end-to-end connectivity to most of their customers. Faults occur in networks, both because of non-ideal characteristics and impairments in hardware or firmware, or due to disruption of the signal flow in any part of the network due to link or node failures. The QoS requirements of the network can be met by good infrastructure and design, accompanied by impairment awareness and high fault tolerance, giving the network high survivability and availability.

The trend in the optical network is toward using a large number of carrier wavelengths carrying high data rates, spectrally efficient advanced modulation schemes, significantly increased transmission distance between repeaters with the use of optical line amplifiers, all adding up to an increase in the data rate by multifolds. The performance of optical networks operating at ultra-high data rates and long transmission distances strongly depends on the impairments introduced into the fiber and the network elements (NEs).

To achieve the desired availability of 99.999% of time, the network needs to be *survivable*. It should be able to continue to provide services in the presence of faults or failures in the network and needs to restore the required QoS in tens of milliseconds. This can be achieved by providing some redundant capacity within the network so that the network automatically reroutes the traffic around the failure using this redundant capacity. The most common cause of failure in optical networks is fiber cut, followed by node failures due to the failure of active components in the nodes. Failures in networks can also happen due to natural catastrophic events. Malfunctioning in the network software can cause many serious issues in the network functioning. Several

Optical WDM Networks: From Static to Elastic Networks, First Edition. Devi Chadha.
© 2019 John Wiley & Sons Ltd. Published 2019 by John Wiley & Sons Ltd.

protection schemes need to be used to achieve high survivability in the network providing redundancy, signaling and alarms, using proper software design with protection from malware, etc. Another important issue is restoration time. With today's high data rate in Gbps, the restoration time of the network after a failure needs to be of the order of tens of milliseconds. Protection is therefore usually implemented in a distributed manner without requiring centralized control in the network. This ensures fast restoration of service after a failure without losing a large amount of data.

Considering the above, we will first discuss the causes of impairments in the optical network and how we take them into account in the design, followed by the survivability issues in optical networks. In Section 8.2 we discuss the various impairments in the optical network, their evaluation criteria followed by impairment awareness and compensation techniques used in the networks. Section 8.3 consists of the basic concepts behind the survivability in optical networks. In Section 8.4, both protection and restoration schemes are discussed for achieving increased availability. In Section 8.5, survivability in the multilayer wavelength division multiplexing (WDM) networks is explained. Many of the protection techniques used in today's telecommunication networks were developed for use in Synchronous Optical NETwork (SONET) and synchronous digital hierarchy (SDH) networks, and we will explore these techniques. Next, we will look in detail at protection functions in the optical layer and then discuss how protection functions in the different layers of the network can work together.

8.2   Impairments in Optical Networks

If the physical layer of the network has desired ideal characteristics of links and nodes in the network along with advanced control and management functions, then it can have ideal performance as well. But practically all the devices have limitations and are not ideal. Signal impairments in the network are due to its linear and nonlinear scattering and dispersion characteristics. These impairments lead to QoS issues in the network. In order to achieve very high physical-layer performance, considerable power margins have to be provided, taking into consideration signal distortions arising from various signal impairments and the hardware aspects, component aging, as well as partial subsystem failure. In long-haul fiber-optic transport networks, system margins are usually specified as the difference between the signal to noise Ratio (SNR) that is delivered at the receiver input under normal operating conditions and the SNR that is required at the receiver in the absence of impairments to guarantee the target BER performance. This SNR margin allocation ensures operation better than the required BER over the entire life of the system. Having said that, the cost of a network also increases with the increase in system margins. It is expected that with more advanced dynamic and high capacity networks the tradeoff between design complexity and system flexibility may result in overly high system margins and an unrealistic outage probability, which in turn will increase the cost of the transport network infrastructure. Therefore, instead of increasing system margins, one may think of introducing impairment-aware cross-layer control to network management as an important alternative solution to the problem.

Signal impairments which can reduce the SNR margin are the chromatic dispersion (CD), polarization mode dispersion (PMD), impairments due to fast power transients, or nonlinear effects, e.g. self-phase modulation (SPM), cross-phase modulation (XPM) or four-wave-mixing (FWM), etc. among WDM channels [1]. Degradations in delivered OSNR (optical signal to noise ratio) to the detector can arise from optical loss of fiber, connectors, splices, aging, and partial failure of system components as well. Fortunately, there are techniques in place to take care of several of these impairment issues in the network design when deploying optical connections, such as increasing signal power, dispersion management, using coherent receivers and advanced modulation formats, etc.

In the following sections, we start with an introduction to various impairments in the optical network followed by the various techniques that are popular to make networks more resilient and to maintain the required QoS.

8.2.1 Impairments in Transparent Optical Networks

In an all-optical network, with optical amplifiers, multiplexers and de-multiplexers, switches, and the other optical devices, and no O-E-O conversions or repeaters used at the intermediate nodes of a connection, cost to build the network is much lower. Such a transparent optical WDM network is the right candidate for building the next-generation backbone network at low cost. A transparent optical network in which a data signal that is transmitted remains in the optical domain for the entire lightpath has several advantages. However, the quality of an optical signal degrades as it travels through the several optical components along the lightpath. In an *opaque WDM* network, data transmission occurs over point-to-point links so that the signal is regenerated at every intermediate node along a lightpath with O-E-O conversion and it is cleaned of impairments and can be transmitted further ahead. The operating expenses of such a point-to-point system are quite high due to the large number of regenerators. In a *translucent* network, as a compromise, where the regeneration functionality is employed at some nodes only instead of at all nodes, the cost could be reduced but at the expense of losing the transparency of the signal at the regenerating nodes [2].

The objective of considering physical impairments when establishing optical connections is to assure good signal quality and avoid potential degradations of QoS. QoS is typically defined as the performance specification of a communication system and is usually represented quantitatively by its performance parameters. In the transparent domain QoS is applied to the physical layer of the transparent network and is quantified with physical-layer parameters, such as OSNR, optical power, or end-to-end latency. There are two main alternatives to obtain information about physical impairments: modeling and real-time monitoring. There are different impairments models available in literature. Through these models the affected performance parameters for the dominant impairments in high-speed networks, such as CD/PMD, fiber non-linearities, etc., can be taken into account. With the recent advances in optical monitoring techniques it is possible to obtain optical performance monitoring (OPM) parameters (e.g. optical power per channel, wavelength drift, channel OSNR, and in-band OSNR and others) in milliseconds by extracting a portion of WDM signals by tapping optical fibers.

The transmission impairments induced by non-ideal transmission components can be classified into two categories: linear and nonlinear. Some important linear impairments are amplifier noise, PMD, group velocity dispersion (GVD) in the fiber, component cross-talk, etc., and the nonlinear-impairments are FWM, SPM, XPM, scattering, etc. The linear effects are independent of signal power in the fiber. Their effects on end-to-end lightpath might be estimated from link parameters, and hence they can be handled as a constraint on routing. The nonlinear effects are significantly more complex as they depend on input transmitted power. Both the linear and the nonlinear impairments start affecting network performance with the increase in bit rate and transmission length. Therefore, in the case of high bit rate long-haul systems we need to have O-E-O electrical regeneration of signal in order to reshape, retime, reamplify, and then relaunch the signal in the system for further transmission. In order to do so, the composite WDM signal is to be fully de-multiplexed, amplified, and multiplexed again, which is neither cost effective nor efficient. Meanwhile, in transparent optical networks with optical amplifiers only reamplification takes place, but the signal remains entirely in the optical domain. This has the disadvantage of the impairments increasingly deteriorating the signal quality with distance. Also, amplifiers at the same time add their own complement of amplified spontaneous emission (ASE) noise. Amplifier noise is a severe problem in system design, reducing the OSNR of the received signal and eventually the BER performance.

8.2.2 Evaluation Criteria of Signal Quality

Impairments in the network cause distortion in the output signal and reduce the signal quality. There are several criteria which could be used to evaluate the signal quality of a lightpath in an optical network. The most common and important criterion used for the performance evaluation is the BER requirement of the network. The BER is normally between 10^{-9} and 10^{-12} for acceptable performance in fiber communication system in long-distance communications. The minimum power at the receiver has to be greater than the *receiver sensitivity* for a given BER. BER can be improved by increasing the input power but at the expense of increasing nonlinearities in the fiber. If we increase the transmit power to a higher level, the fiber nonlinearities coming into play deteriorate the BER at the receiver output. BER is considered an important figure of merit for WDM networks as most network designs adhere to the requisite BER quality at the output. Power budget and power margins are important issues as well in the design of optical system, which eventually also control the BER. It is worth noting that BER is one of the major parameters of the signal quality because it captures all the impairments along the path; however, in a transparent domain BER can only be measured in the electronic domain after the receiver. In the optical domain BER cannot be measured, it can only be estimated by using OSNR as proposed in [3].

In a dynamic system, calculating BER instantaneously is a difficult task; instead the quantity measured is the Q-factor, which provides a qualitative description of the receiver performance. The Q-factor suggests the minimum SNR required to obtain a specific BER for a given signal. The higher the value of the Q-factor, the better the BER. Mathematically, the Q-factor of an optical signal is expressed as:

$$Q = \frac{I_1 - I_0}{\sigma_1 + \sigma_0} \tag{8.1}$$

In Eq. (8.1), I_1 is the value of the current for 1-bit, I_0 is the value of the 0-bit current, σ_1 is the standard deviation of the noise for 1-bit signal, and σ_0 is the standard deviation of the 0-bit noise. The relationship of Q-factor to BER in the case of OOK (on-off keying) modulation is as:

$$BER = \frac{1}{2} erfc\left(\frac{Q}{\sqrt{2}}\right) \tag{8.2}$$

BER calculations are quite difficult as the values we are looking for are in the range from 10^{-9} to 10^{-15}. Yet the Q-factor analysis is comparatively easy. The next question is how to dynamically calculate Q for transparent networks. One can obtain the value of Q from OSNR, which is also a practically measurable quantity in optical domain for a given network. The relationship of OSNR with Q-factor can be obtained as:

$$Q_{dB} = 20\log\sqrt{OSNR}\sqrt{\frac{B_o}{B_e}} \tag{8.3}$$

where B_0 is the optical bandwidth of the photodetector and B_e is the electrical bandwidth of the receiver filter. For practical designs OSNR (dB) is greater than Q (dB) by at least 1–2 dB. Hence, if a lightpath is not satisfying the OSNR requirement, it will not be able to satisfy the BER requirement either.

Each impairment in the physical layer of the network results in the deterioration of the quality of the received signal and hence causes power penalty to the system. Therefore, for the same desired BER, in the presence of any impairment, a higher signal power will be required at the receiver. Qualitatively, power penalty can be considered as the net extra power required to pump up the signal so that it reaches the receiver while maintaining the minimum BER requirement of the system. Typically, the power penalty for most networks is in the range of 2–3 dB. The International Telecommunications Union (ITU-G957) standard specifies this penalty to be less than 2 dB. When the increased power is launched, initially the Q factor of the optical system increases with the increase in launched power, reaches a peak value, and then decreases with a further increase in power because of the onset of the nonlinear effects.

The power penalty due to any impairment is expressed as the increase in power required to maintain the same BER in the presence of impairment with respect to the power without impairment. In decibel units, power penalty is defined as:

$$PP = -10\log_{10}\frac{Increased-Power}{Original-Power} \tag{8.4}$$

There are several mechanisms which degrade the receiver sensitivity, viz. finite extinction ratio of the signal, pulse broadening induced by fiber dispersion, timing jitter of electronic circuits, fiber impairments, etc. The power penalty can therefore be also defined as:

$$PP = -10\log\left(\frac{S'/N'}{S/N}\right) \tag{8.5}$$

where S'/N' and S/N denote the received SNR at the receiver with and without impairments, respectively, assuming an optimized threshold setting. When estimating network power budget, power penalties for all different impairments have to be taken into account. We need to keep in mind though that not all impairments are uncoupled in a system but to get the net power penalty due to various impairments present at any point of time, we normally consider them to be uncoupled. In practice, the power penalty due to each impairment is calculated one at a time assuming that the rest of the system is ideal.

8.2.3 Optical System Impairments

Factors affecting the performance of the optical system are the linear and nonlinear impairments. Linear impairments appear in the system when the power transmitted is not high and hence they are independent of the signal power. But in the DWDM system with large number of channels multiplexed in the fiber, the high-power transmission leads to nonlinearities and limiting the system performance. We will be discussing both these impairments in the following sub-sections.

8.2.3.1 Linear Impairments

Initially, fiber attenuation was considered the biggest factor in limiting the length of an optical channel. However, as data rates grew group-velocity dispersion and PMD became important considerations.

Linear Dispersion One of the most important linear impairments in optical fiber is dispersion. The main contributions to dispersion are CD and PMD. The CD in fiber, which is due to both material and waveguide dispersion, is deterministic in nature, and can be compensated in the optical links, especially in reasonably linear systems. CD is taken into account in terms of GVD [1, 2]. The group-velocity dispersion parameter, D, is defined as:

$$D = -\frac{2\pi c}{\lambda^2} \frac{d^2\beta}{d\omega^2} \tag{8.6}$$

where λ is the free space wavelength of light and $d\omega/d\beta$ is the group velocity in the fiber. The parameter of $|D|$ has units of ps/nm. km, defined as the temporal pulse spreading per unit line-width of the source per unit distance travelled. GVD induces pulse broadening, chirping, and compression in the signal. In the case of pulse broadening, the pulse spreads into the time slots of the other pulses. This not only causes ISI (inter-symbol-interference), but it also introduces a power penalty, which can cause degradation of the system's SNR and thus lowers its power content. Therefore, when we are considering dispersion-limited systems, we must consider a power penalty due to dispersion. For Gaussian pulses, in the case when source bandwidth dominates, the peak power is reduced by the pulse broadening factor, b_f, as:

$$b_f^2 = 1 + \left(\frac{DL\sigma_\lambda}{\sigma_0}\right)^2 \tag{8.7}$$

where σ_λ is the spectral width of the source and σ_0 is the pulse width. The power penalty due to pulse spreading is given by:

$$PP_D = -10\log_{10}\left(b_f\right) = -5\log_{10}\left[1 + \left(DL\sigma_\lambda / \sigma_0\right)^2\right] \tag{8.8}$$

In the case of a narrowband source and unchirped Gaussian pulses, the power penalty is:

$$PP_D = -5\log_{10}\left[1 + \left(\beta_2 L / 2\sigma_0^2\right)^2\right] \tag{8.9a}$$

where L is the link length and

$$\beta_2 = d^2\beta / d\omega^2 \tag{8.9b}$$

Techniques are available to compensate dispersion. Dispersion flattened or dispersion shifted fibers (DSFs) can be used in the system to compensate for dispersion. Further dispersion-compensating fibers can be placed at strategic locations in a network so that we can reshape the broadened pulse as desired. Fiber Bragg gratings (FBGs)-based dispersion compensators can also be used for reducing the pulse spreading [1].

Frequency Chirping   Intensity modulation of the laser sources causes phase modulation due to the carrier-induced changes in the refractive index of the laser material. Such optical pulses with a time-dependent phase shift are called *chirped pulses*. Chirping of optical pulses affects pulse broadening. For chirped Gaussian pulses with chirping factor of C, the pulse broadening factor is:

$$b_f^2 = \left(1 + \frac{C\beta_2 L}{2\sigma_0^2}\right)^2 + \left(\frac{\beta_2 L}{2\sigma_0^2}\right)^2 \tag{8.10}$$

Power penalty due to dispersion in the chirping pulse is expressed as:

$$PP_D = -5\log_{10}\left[\left(1 + \frac{C\beta_2 L}{2\sigma_0^2}\right)^2 + \left(\frac{\beta_2 L}{2\sigma_0^2}\right)^2\right] \tag{8.11}$$

Penalty can be quite large when $\beta_2 C > 0$. This is the case for directly modulated distributed feedback (DFB) lasers ($C > -4$) operating near 1.55 μm when $\beta_2 < 0$. To keep penalty below 0.1 dB, $|\beta_2|B^2 L < 0.002$ is required, where B is the bit rate of the signal.

For standard fibers $B^2 L$ is limited to 100 (Gb/s)^2-km. System performance can be improved by ensuring that $\beta_2 C < 0$. Theoretically, the chirp-induced power penalty is difficult to calculate, but it can be approximated to a 0.5 dB margin while considering system design.

Polarization Mode Dispersion   Besides CD, the other kind of dispersion in the linear region is the PMD, which is due to waves of different polarization states traveling with different velocity in the fiber leading to dispersion. The asymmetry of the fiber causes *walkover* effect, which in turn causes the pulse to spread in time. A measure of PMD is the differential group delay (DGD), which is the time difference in multiple spectral components at multiple speeds over a given length of fiber. The mean DGD can be calculated as:

$$DGD = \zeta x \sqrt{L} \tag{8.12}$$

where ζ is the PMD coefficient. PMD is a stochastic time-varying effect that can be modeled statistically but is more difficult to compensate in practice.

Though GVD is a primary cause of concern for bit rates higher than 2.5 Gbps, PMD starts influencing system performance at high-speed transmissions above 10 Gbps. As the channel bit rate increases to 10 Gbps and beyond, PMD becomes one of the most critical limiting problems for data transmission in a high-speed network. PMD strongly affects the transparent transmission length, which is expressed as below:

$$Bx\sqrt{\sum_{k=1}^{M}\zeta^2(k)xL(k)} \leq a \tag{8.13}$$

where B is the bit rate, $\zeta(k)$ is the fiber PMD parameter in the kth span of the transparent lightpath which consists of M fiber spans, and $L(k)$ is the fiber length of kth span. The parameter a which represents the fractional pulse broadening should typically be less than 10% of a bit period for which the PMD can be tolerated. The typical system margin for PMD is 1 dB for general long haul, but it depends on the transmission length.

Finite Extinction Ratio Extinction ratio of a binary signal is defined as, $r_{ex} = P_0/P_1$, where P_0 and P_1 are the power level of the 0 and 1 bit, respectively. A finite extinction ratio increases the BER, hence the input power needs to increase in order to compensate. The power penalty in this case can be obtained by calculating the Q parameter of the received signal, which can be expressed as:

$$Q = \left(\frac{1-r_{ex}}{1+r_{ex}}\right)\frac{2R_d P_{rec}}{\sigma_1+\sigma_0} \tag{8.14}$$

where R_d is the responsivity of PIN (p-type, intrinsic and n-type) detector, P_{rec}, is the received average power, σ_1 and σ_0 are the standard deviation of received noise when 1 and 0 bit are transmitted. In the case of thermal noise limit, $\sigma_1 = \sigma_0 = \sigma_T$ and the received power for a finite extinction ratio can be expressed as:

$$P_{rec}(r_{ex}) = \left(\frac{1+r_{ex}}{1-r_{ex}}\right)\frac{\sigma_T Q}{R_d} \tag{8.15}$$

Hence the power penalty for extinction coefficient can be obtained as:

$$PP_{ex} = -10\log_{10}\left(\frac{P_{rec}(r_{ex})}{P_{rec}(0)}\right) = -10\log_{10}\left(\frac{1+r_{ex}}{1-r_{ex}}\right) \tag{8.16}$$

Non-ideal Optical Components Transmission impairments due to non-ideal optical components in the physical layer may significantly affect the network performance. Cross-talk between the transmitted signals is due to the imperfections in the optical components. Some of the transmitted power will leak out to the unintended port, thus corrupting the signal passing through that port. Cross-talk is defined as the ratio of leaked signal to signal power in the port considered. If one estimates system performance by simple power addition adding a random binary cross-talk power to the signal power, it is easily shown that the system power penalty in decibels should be no larger than for the cross-talk ratio of ε:

$$PP_{Xer} = -10\log(1-\varepsilon) \tag{8.17}$$

This will indeed be the case if the signal and cross-talk differ in optical frequency by much more than the receiver's radio frequency (RF) bandwidth, or if their polarization states are orthogonal. This is known as inter-channel cross-talk. An example of such a case is a multiplexer in which a small part of power from one channel gets leaked into other port of a wavelength and cause inter-channel cross-talk. As the leaked power can be separated with an optical filter, the cross-talk can be taken care of. However, practical systems will in general need to provide good performance even when the cross-talk's frequency and polarization state are matched to the signal, which is known as intra-channel cross-talk. In this case, one must consider interference of the underlying fields. The field incident at the photodetector is then:

$$E(t) = Ed_s(t)\cos\left[\omega_0(t)+\varphi_s(t)\right] + \sqrt{\varepsilon}\,Ed_x(t)\cos\left[\omega_0(t)+\varphi_x(t)\right] \tag{8.18}$$

where E is the signal-field amplitude, normalized so that $(\frac{1}{2}E^2)$ is the signal power. $d_s(t)$ and $d_x(t)$ are random binary functions with values of 0 or 1, representing the signal and cross-talk bitstreams and ϕ_s and ϕ_x are the phase of the signal and cross-talk signals. In the practical system, the region of interest is that where $\varepsilon << 1$, in which case the detected power is approximately:

$$P(t) = \frac{1}{2}E^2\left[d_s(t)+2\sqrt{\varepsilon}d_s(t)d_x(t)\cos\left(\varphi_s(t)-\varphi_x(t)\right)\right] \tag{8.19}$$

The second term inside the brackets of (8.19) represents random power due to phase fluctuations of the interfering waves. For worst-case design the power penalty is given as:

$$PP_{Xra} = -10\log(1-2\sqrt{\varepsilon}) \tag{8.20}$$

Practical examples of inter- and intra-channel cross-talk are shown in Figures 8.1a,b, respectively. If there are N interfering channels, each with average received power $\varepsilon_i P$, then ε and $\sqrt{\varepsilon}$ are given in (8.17) and (8.20), respectively by:

$$\varepsilon = \sum_{i=1}^{N} \varepsilon_i \text{ and } \sqrt{\varepsilon} = \sum_{i=1}^{N} \sqrt{\varepsilon_i}$$

Example 8.1 Consider the WDM link shown in Figure 8.1a, where the de-multiplexer (DEMUX) has three input wavelengths: λ_1, λ_2, and λ_3. The DEMUX introduces inter-channel cross-talk from the adjacent channels which is ($-30\,$dB) below the desired channel. Compute the maximum cross-talk in the case of λ_2 channel at the output of the DEMUX.

Solution
The worst effected is the λ_2 channel as it has cross-talk from both λ_1 and λ_3. Therefore, the total cross-talk is: $2\varepsilon = 0.001 \times 2 = 0.002$.

The power penalty $= -10 \log (1-0.002) = 0.00869\,$dB

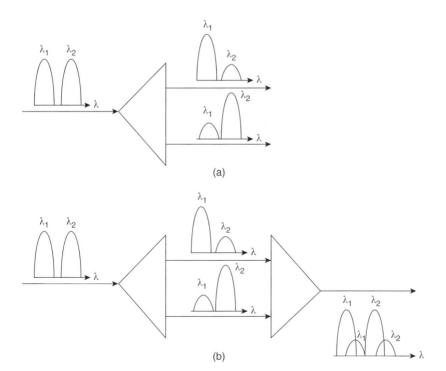

Figure 8.1 Examples of (a) inter-channel cross-talk, and (b) intra-channel cross-talk.

8.2.3.2 Impairments Due to Nonlinearities

Nonlinear impairments can impose significant limitations on both high-speed channel provisioning and highly dense DWDM transport systems. Also, while considering nonlinear impairments in wavelength routed networks (WRNs), detailed knowledge of the entire physical infrastructure and current loading of each optical signal on every link is required. There is no generic analytical model of nonlinear impairments available that allows handling these effects as constraints in impairment-aware routing and assignment (RWA) in the networks. For these reasons, the effects of nonlinear impairments are handled by increasing the required OSNR level at the receiving end by a link budget margin. However, this approximation may lead to either an overestimation or an underestimation of the real impact of the nonlinear impairment effect.

The different nonlinear impairments of concern are due to nonlinear phase and nonlinear scattering effects. The refractive index of the fiber has a strong nonlinear component that depends on the power level of the signal, which is expressed as:

$$n' = n + n_2 \left(\frac{P}{A_{eff}} \right) \tag{8.21}$$

where is n_2 the *nonlinear-index coefficient*, P is the optical power, and A_{eff} is the effective mode area of the fiber. The numerical value of n_2 is about 2.6×10^{-20} m^2 W^{-1} for silica fibers and varies somewhat with dopants used. Because of this relatively small value, the nonlinear part of the refractive index is quite small but the high power level in WDM system affects the long fiber lengths system considerably. In particular, it leads to the phenomena of SPM and XPMs [1, 3].

Self-Phase Modulation Nonlinearity in the refractive index of the fiber produces a nonlinear phase shift denoted by φ_{NL} as shown in Eq. (8.22).

$$\varphi_{NL} = \gamma P_{in} \frac{\left(1 - e^{\alpha L}\right)}{\alpha} \tag{8.22}$$

In Eq. (8.22), γ is the nonlinear coefficient that depends on the nonlinear refractive index, expressed as:

$$\gamma = \frac{\omega_0 n_2}{c A_{eff}} \tag{8.23}$$

Further, φ_{NL} depends on the effective length of the link given by the third term in Eq. (8.22), and input power, P_{in}, which is time varying. The nonlinear phase shift induced by the time varying optical pulse, P_{in}, now also varies with time. The result is that the time varying pulse at frequency, ω_0 would have time varying phase components in the range of $\omega_0 t \pm \varphi_{NL}$, where φ_{NL} is time varying. The result is pulse spreading, which is a result of the power dependence on the induced phase shift. Therefore, to keep a check on the maximum phase shift that a pulse can have, it is imperative to set a threshold to the maximum input power. This nonlinear phase shift is SPM.

GVD and SPM work in quite the opposite way with the input power. In other words, there is a tradeoff involved: we need more power to take care of the dispersion-induced power penalty, but this additional power leads to fiber nonlinear effects, which creates more spread.

Example 8.2 Calculate the power launched into a 40 km long single mode fiber for which the self-phase modulation induced nonlinear phase shift becomes 180^0. Assume $\lambda = 1550\,nm$, $A_{eff} = 40\,\mu m^2$, $\alpha = 0.2$, and $n_2 = 2.6 \times 10^{-20}\,m^2\,W^{-1}$.

Solution
SFS phase shift, $\varphi_{NL} = (\omega_0 n_2/cA_{eff})P_{in}(1 - e^{-\alpha L})/\alpha = (\omega_0 n_2/cA_{eff})P_{in}/\alpha$, ignoring the term $e^{-\alpha L}$

$$\pi = (2\pi/1550 \times 10^{-9}) \times (2.6 \times 10^{-20})/(40 \times 10^{-12})(P_{in}/4.6 \times 10^{-5})$$

$$P_{in} = 55\,mW$$

Cross-Phase Modulation   We have considered nonlinear refractive index of the fiber in the case of single channel in SPM. But in the case of a WDM system, the parallel channels have an effect on each other. Two or more channels have nonlinear effects on each other called the XPM. In multichannel WDM systems, XPM causes intensity-based modulation to adjacent frequency channels. XPM causes phase fluctuations in pulse propagation due to the neighboring adjacent channels. Furthermore, if adjacent channels are traveling at the same bit rate, XPM effects are more pronounced. One way to avoid XPM is by carefully selecting bit rates for adjacent channels that are not equal to each other. XPM has more impact on certain types of modulation formats. Typically, frequency-shift keying (FSK) and phase-shift keying (PSK) have a more pronounced impact than on-off keying (OOK) NRZ and RZ coded signals. The induced phase shift is due to the *walkover* effect, whereby two pulses at different bit rates or with different group velocities walk across each other. The total phase shift depends on the net power of all the channels and on the bit rate of the channels. Maximum phase shift is produced when two "1" bits walk across each other due to the high power in both the bits. The phase shift due to XPM in the jth channel is given as:

$$\varphi_j^{NL} = \gamma \left[\frac{1 - e^{\alpha L}}{\alpha} \right] \left[P_j + 2 \sum_{k \neq j}^{W} P_k \right] \tag{8.24}$$

where W is the total number of channels and P_k is the power of the kth channel.

The maximum phase shift (with all channels with 1 bit) is given as:

$$\varphi_{\max}^{NL} = \left(\frac{\gamma}{\alpha}\right)[2W-1]P_j \tag{8.25}$$

Four-Wave Mixing   FWM is a third-order nonlinearity in optical links that can be compared to the intermodulation distortion in standard electrical systems. FWM has very detrimental cross-talk effect for equally spaced WDM systems and at high powers. When three optical channels at frequencies ω_i, ω_j, and ω_k travel such that they are close to the zero dispersion wavelength, they have mixing due to fiber nonlinearity and produce a fourth signal whose frequency is expressed as:

$$\omega_{ijk} = \omega_i \pm \omega_j \pm \omega_k \tag{8.26}$$

This ω_{ijk} optical frequency, when it falls close or on the signal wavelength channel frequency, can mix with the WDM channels, causing severe cross-talk. For W wavelengths in a fiber, the number of FWM channels produced is expressed as:

$$N = \frac{W^2}{2}(W-1) \tag{8.27}$$

Not all these combinations have phase-match with the WDM channels, hence those FWM channels do not cause interference or cross-talk. When generated frequency combinations become nearly phase-matched with the channel wavelengths for the multichannel system, they are most troublesome. These combinations are of the form ($\omega_{ijk} = \omega_i + \omega_j - \omega_k$). The degenerate FWM process for which ($\omega_i = \omega_j$) is the dominant process and impacts the system performance most.

Figure 8.2a shows the effects of FWM in equally spaced systems. Three equally spaced channels generated nine FWM signals, out of which few fall along the signals with same phase. Figure 8.2b shows the same considerations for unequally spaced systems. Unequal spaced channels generated nine FWM signals; none of the generated signals falls on top of the original signals.

The solution for minimizing FWM is to use unequal channel spacing in such a way that the generated wavelength does not interfere with the signal channel. Using NZ-DSF also minimizes the effect of FWM.

Raman Stimulated Scattering   When high pump optical signal passes through the silica fiber, nonlinear scattering takes place by the silica molecules. This is known as spontaneous Raman scattering. The high-power pump photons passing through the fiber give up their energy and create other photons of reduced energy along with vibrational energy absorbed by the silica molecules.

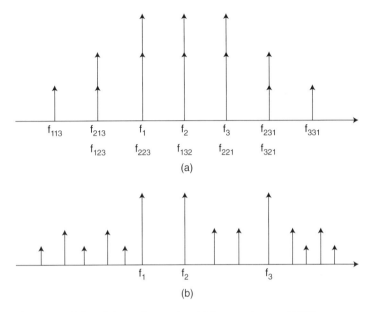

Figure 8.2 (a) Equal channel spaced, and (b) unequal spaced FWM.

If a signal is propagating through the fiber with a radial frequency ωs, then the generated photons by Raman scattering are at the lower frequency ($\omega p - \omega s$), where ωp is the high-power pump frequency. The Raman scattering process becomes stimulated Raman scattering (SRS) if the pump power exceeds a threshold value. SRS can occur in both the forward and backward directions in optical fibers. The backward SRS causes loss but the forward can cause cross-talk in the propagating signal. SRS power threshold is about 570 mW at 1.55 μm and the SRS bandwidth is a very large in THz. As channel powers in optical communication systems are typically below 10 mW, SRS is not a limiting factor for single-channel lightwave systems. However, it affects the performance of WDM systems considerably because of its large bandwidth. Raman scattering has a gain bandwidth of 125 nm, with peak occurring at a frequency shift of 13 THz. The total WDM channels, which have spacing of 50/100 GHz, hence get affected when fall in the 125 nm Raman gain bandwidth.

The Raman-induced cross-talk degrades the system performance and is of considerable concern for WDM systems. In order to calculate the worst-case effect of Raman scattering in a WDM system, power depletion due to Raman scattering from the highest-frequency (ω_1) channel to the lower frequency channels is obtained with a simple model. The maximum power depletion for a M-channel WDM system will occur when all channels carry simultaneously bit "1." The total power depletion from the highest frequency channel to the lower channels is approximately given by [4]:

$$P_R = \frac{1}{2} \frac{M(M-1)}{2} g_R \Phi P_{ch} L_{eff} / A_{eff} \qquad (8.28)$$

where g_R is the peak Raman gain coefficient with the value of 6×10^{-14} m W^{-1} at 1.55 μm, P_{ch} is the peak power in each channel, A_{eff} and L_{eff} are the effective area and the effective interaction length of the fiber and Φ is the ratio of WDM channel spacing to the gain bandwidth of SRS which has a value of 125 nm. The power penalty due to Raman scattering is then expressed as:

$$PP_R = -10\log(1 - P_R) \qquad (8.29)$$

Example 8.3 Consider a WDM system of four channels with 100 GHz spacing starting from 1569.59 nm (191.00 THz). It is an unamplified thermal noise limited system of 100 km length. Each channel carries a power of 10 mW. Calculate the power penalty due to Raman scattering in the system with α = 0.2 dB km^{-1}, A_{eff} = 50 μm^2, g_R = 6 × 10^{-14} m W^{-1}, and Raman gain bandwidth of 125 nm.

Solution

$L_{eff} = (1 - e^{-\alpha L})/\alpha = 1/\alpha = 21.7$ km;

P_R (from eq. 8.28) = 0.5 × (12/2) × (6 × 10^{-14}) × (0.8/125) × 10^{-2} × (21 × 10^3/50 × 10^{-12}) = 2.5 mW

Power penalty = −10 log(1−P_R) = 0.011 dB

Brillouin Stimulated Scattering Stimulated Brillouin scattering (SBS) also occurs when the high pump power causes nonlinearity in the fiber and causes the new lower frequency signal. The difference in Brillouin scattering is that the generated photon has much smaller bandwidth of few MHz, although the SBS threshold is much smaller than in the case of SRS. The exact value of the average threshold power depends on the modulation format and is typically ~5 mW. It can be increased to 10 mW or more by increasing the bandwidth of the optical carrier to be greater than 200 MHz through phase modulation. SBS does not produce inter-channel cross-talk in WDM systems because the Brillouin frequency shift of 10 GHz or less is much smaller than typical channel spacing of 50 GHz or more.

8.2.4 Impairment Awareness and Compensation in Optical Networks

In the previous section, we discussed the impairments in all-optical networks and estimated the system penalty each would cause. In the present section, we learn how the network is made aware of the impairments with monitoring and then try to compensate them.

At the system design level, the penalties are reduced by adjusting the transmitted power, length of the link, or data rate of the signal. This is the static design which takes into account the offline estimation of physical layer impairments at the time of designing the physical layer of the network. The analytical models of DWDM impairments, though are available, but have approximations involved in their derivation and hence, larger penalty margins are taken to determine an adequate OSNR. Therefore, one should have real-time feedback from the physical layer which can avoid an overuse of costly and power-consuming regenerators. Moreover, as many impairments are time-varying, in the dynamic networks a fixed penalty margin can result in network performance degradation. Therefore, in order to ensure impairment aware (IA) operation in dynamic optical networks, it is necessary to have reliable and cost-effective OPM with real-time access to physical layer parameters. If the network cannot efficiently adapt to dynamically changing impairments in the absence of real-time access to physical layer performance metrics, network performance can be adversely affected.

The physical impairments are captured by monitoring a number of performance parameters, such as OSNR and power levels. Performance monitoring in optical networks traditionally has been in the electrical domain for BER and other QoS measures, whereas OPM is the physical-layer monitoring of the signal quality in the optical domain. In 2004, ITU-T defined a list of OPM parameters that can be used for impairment-aware RWA [5]. In OPM real-time measurements of channel power levels, wavelength drift, and the OSNR are carried out. This involves WDM channel-layer monitoring, which gives the optical-domain characteristics essential for transport and management at the optical layer. OPM also involves monitoring the data protocol information and digital measurements, such as the BER, which can be used to infer properties of the analog optical signal. In WRN, both modeling and real-time OPM provide information on a per channel basis, which are matched to the needs of *lightpath* computation [6–8]. The quality of a *lightpath* estimation for aggregate physical impairments is expressed by one or more link-state parameters.

In WRN, path computation and wavelength selection is done dynamically based on real-time OPM with control and management planes support. The routes and wavelength assignment dynamically adapt to time-varying impairments and take decisions on the fly to reroute lightpaths, resulting in improved network performance [6]. The required signal quality that the lightpath must satisfy in terms of estimated BER from OSNR and latency, serve as thresholds for deciding whether particular link constraints are acceptable by the RWA algorithm. As discussed previously, all paths to be established within a domain are computed and established either by the centralized network management system (NMS) or distributed path computation element (PCE). The performance monitoring system updates the traffic engineering database (TED) information with the complete set of performance parameters within the entire domain, which is considered by the routing computation process. When the connection request arrives to the source network element (NE), first the RWA algorithm is executed, taking into consideration its own TED and the potential constraints required by the connection. The on-line RWA algorithm computes a route considering network performance objectives and the required end-to-end optical signal quality. If the requirements cannot be satisfied, the connection is blocked.

To explain some details of OPM and dynamic establishing of routes according to the RWA and the constraints, a centralized OPM model is shown in Figure 8.3. As shown in the figure, a PCE is reachable by all the NEs within the transparent data plane. PCE is aware of the complete network topology, resource availability, and physical parameters in a central repository of TED. The performance monitoring system dynamically updates the TED information at regular intervals with the complete set of performance parameters within the entire domain. Once the connection request is received by the NMS, the routing computation process is launched considering both the current TED information and the received set of requirements of the connection request. Then the NMS configures all the NE's involved in the path in a parallel

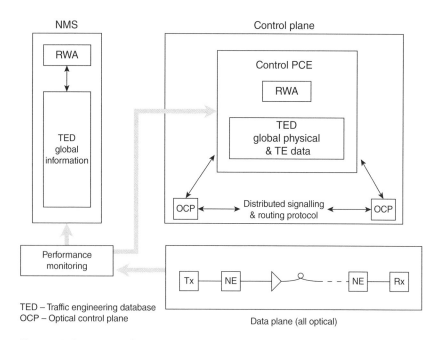

Figure 8.3 Dynamic performance monitoring system.

way using the network management interface to set up the lightpath. The centralized NMS-based system is aware of the complete, detailed, and periodical view of the whole network, including available resources, topology, physical impairments, through a central/global database. Therefore, all the path requests can be optimally served, while the potential effect on the signal quality of existing connections is taken into account.

8.3 Survivability in Optical Networks

The emergence of new multimedia-intensive online applications and services related to the medical field, cloud storage and computing, and financial services demands high connection availability and requires the underlying transport infrastructure to be resilient to all kinds of failures. In order for WDM optical networks to fulfill these resilience and connection availability requirements, it is necessary to specifically focus on schemes that are able to achieve maximum survivability under different network failure scenarios, without compromising on the important network parameters of bandwidth, latency, and connection blocking probability, etc. Optical networks carry terabits of traffic. Any optical cable cut or node or equipment failure due to human or natural disaster can cause communication disruption between millions of people and datacenters, causing large amount of important data to get lost. Hence, survivability against any failure for the network is extremely critical. We define *survivability* of a network as a network's capability to provide continuous service in the presence of any failures. In recent times, network availability of six nines (99.9999%) is being provided for certain services, which translates to 30 seconds of downtime per year. It means that fault management in these networks is extremely important because even a brief downtime can cause huge data loss, and also can result in a significant revenue loss for the service providers because of the violation of service-level agreements (SLAs) with clients.

Network failures are generally *link* and *node* failures. Link failure usually occurs because of cable cuts, while node failure is due to component/equipment failure at network nodes. Network failure can also be the result of a malfunction in the software residing in the network elements' controllers which are responsible for the control and management of the various networking operations. Even though failures cannot be avoided, quick failure detection, identification, and recovery make the network more robust and reliable and ultimately increase availability. Service recovery time is important because faster recovery means minimal loss in data during the outage and the vital services, such as banks, stock exchanges, airlines, or public safety systems that are supported in the network, are not affected. For this to be possible, the network's topology must have inherent survivability properties incorporated at the network design and planning stage. This will determine its ability to survive single or multiple link or equipment failures. Also, for a network to be a survivable redundancy is always necessary to ensure that a significant amount of information is not lost in the case of failure.

We start by outlining the basic concepts of survivability and with different protection and restoration schemes used in optical networks in Section 8.4. Survivability can be provided in different layers in network: logical layers and optical layer. The fault-management schemes in each layer have their own functionalities and characteristics. In Section 8.5 we discuss the survivability in the multilayer WDM optical networks. The protection techniques in the SONET and SDH networks are first briefly described, which are still used in many of the current protocols in the WDM networks. Next, we look at protection functions in the optical layer and discuss how protection functions in the different layers of the network can work together.

8.4 Protection and Restoration

There are several survivability schemes in optical networks for different topologies in the logical and optical layers. All these schemes can be distinguished by basically three aspects: (i) the first is the pro-active or re-active scheme. The proactive are the pre-planned protection schemes included at the design stage and the reactive are real-time dynamic protection schemes with restoration; (ii) second is the centralized or distributed backup route computation; (iii) the third aspect is whether the protection scheme is line, path, or ring-based. Protection schemes are referred to as preplanned systems, which provide an alternate protection path along with the working path precomputed at the design stage. Restoration is referred to as computation of recovery route in real time at the fly after the failure has occurred in the network. In both cases it is essential that once failure occurs, it should be quickly detected and identified so that before much data is lost the network is restored. For the network to achieve high availability, the network topology should be inherently more robust, have high survivable properties, and always have sufficient redundancy or spare capacity of resources [8, 9].

8.4.1 Protection and Restoration Schemes

In a protection scheme, the key characteristic is the in-advance or proactive reservation of network resources at network design time providing guarantees to protect against any possible future failures. The recovery speed from a failure is higher in protection schemes and can be guaranteed, but it all comes at the expense of low network capacity utilization. The protection schemes will have *working* paths and *protection* paths. Working paths carry traffic under normal operation and protection paths provide an alternate path to carry the traffic in case of failures. Working and protection paths are usually diverse routes so that both paths are not lost in case of a single failure. In contrast to a protection scheme, restoration is a reactive technique where there is no reserved path at the planning stage but it is computed in the real time when the failure occurs. The spare capacity available in the network is used to reroute traffic around the failure.

Both proactive and reactive restoration techniques can be either *link-based* or *path-based* in networks. Link-based protection employs local detouring around the fault of the total traffic on the link; meanwhile, path-based protection may traverse many intermediate nodes in a large network as it is an end-to-end detouring of a particular source-destination connection. The switching to the alternate path/link is done using automatic protection switching (APS) algorithm which is the most common predesigned protection schemes used for different topologies: point-to-point protection, ring or mesh topology. In *path* protection, a backup disjoint path is reserved in advance during the provisioning phase of path protection. The signaling and switching operations in path protection are far more complex than those in link protection. In the case of *link* protection, the unit being recovered is the *aggregate traffic* on the failed link, while in the case of *path* it is a connection from source to destination on a failed link that is recovered. For example, as shown in Figure 8.4a, for a five-node network (1-2-3-4-5-6), after a link fails between nodes 4 and 5, the total

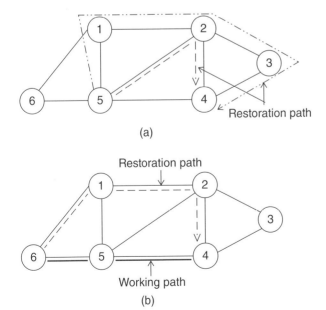

(a)

(b)

Figure 8.4 Example of (a) link protection, and (b) path protection.

affected traffic between the two nodes is rerouted through the backup paths 5–2–4 or 5-1-2-3-4. Here, the end nodes of the failed link (i.e. nodes 4 and 5) are responsible for recovery. In Figure 8.4b, if the link 4-5 breaks for the working path (6-5-4) of a connection, only the traffic on working path is switched by the diverse alternate path 6-1-2-4 unlike in the case of link protection path 5-2-4.The protection paths for every link failure are usually reserved at connection setup and should be disjoint with the failed link.

Protection schemes can be further classified as *dedicated* or *shared*. The backup protection path can have three architectures: $(1 + 1)$ or the dedicated backup reservation for a working connection. The source node transmits the information signal on both the working and protection links. The receiver at the destination node compares the two signals and chooses the better one. If one link fails, the destination node is still able to receive the signal on the operational link. In $(1 : 1)$ shared protection scheme, two working links share a protection link, but the source and destination nodes switch to the protection link only when a failure on the working link is detected. Under normal conditions, the protection link usually carries the low-priority traffic which can be taken off when the primary traffic is to be carried over it. Finally, in the case of $(1 : N)$ shared protection scheme, N working links share a single protection link, thereby providing protection against the failure of any one of the N working links. But, like in $(1 : 1)$ the traffic is switched from the protection link back to the working link after it has been repaired so that the protection link is available for any future working link failures. $(1 : N)$ shared protection results in more efficient usage of backup resources as compared with $(1 : 1)$ protection, but at the expense of slower recovery times.

Next, we have the ring topology. A ring is *two-connected* topology, hence it provides two separate paths between any pair of nodes that have no link or node shared except the source and destination nodes [10–12]. This makes a ring network inherently more resilient to failures. Unlike point-to-point APS in linear topology which handles link failures, the *self-healing ring* (SHR) protection technique is more flexible so as to deal with both link and node failures. The SHR incorporate protection mechanisms that automatically detect failures and reroute traffic away from failed links and nodes onto other routes rapidly. We will discuss further details of SHR with the SONET SHR systems.

The mesh topology is inherently more survivable because of higher node degree providing alternate routes in case of the link failure [13, 14]. Protection for such networks is usually more complicated than for point-to-point links or ring networks. Solutions for protection in mesh networks are found by forming *ring covers*, *cycle covers*, and *protection cycle (p-cycle)* [15, 16]. In ring cover, first a set of rings that covers every link in the mesh at least once is found and then the rings can be made into self-healing structures that can protect the network against a link cut or an equipment failure using the APS path/link, or SHR techniques. The mesh network then acts as a collection of SHRs, providing the advantages of simple control logic and speed in terms of failure recovery. However, a large spare capacity is required to get a full protection guarantee, but it does have fast failure recovery in milliseconds. The redundancy is at least 100%, but it can be more as few links may include multiple rings.

The p-cycle technique is a ring protection scheme which protects not only its own links but also any possible links connecting two non-adjacent ring nodes called the *chordal links*. By doing this, p-cycles reduce the redundancy required to protect a mesh network against link failure. There are two types of p-cycles namely *link p-cycles* and *node p-cycles*. Link p-cycles protect all WDM channels on a link whereas a node p-cycle protects all the connections traversing a node. p-cycles are the most efficient protection structures as for capacity minimization is concerned, but p-cycle planning is an NP-hard problem and is not scalable. The p-cycle approach is based on protection of individual channels on a link like the path protection, in contrast to rings and cycle double covers (CDCs). In CDC and ring cover, aggregate traffic on a link is recovered much like the link-based approaches. An advantage of the p-cycle approach over the ring cover and the CDC approaches is that it does not have to cover all links in the network with rings by omitting the chords that have their end nodes on the p-cycles but are not included in the p-cycles [16].

Another important parameter to be considered is the number of failures to be taken care of in a mesh network. Single link failures are most common in all-optical networks, but in large WDM networks there can be multiple failures at the same time. Therefore, in such networks it is important to protect against double link failures at least. For double link failures, two backup paths instead of one are provided, as the name indicates. However, extending to multiple backup paths increases backup resource usage, significantly reducing the capacity utilization of the network.

8.4.2 Restoration Schemes

In the case of restoration, protection resources are not reserved at the design stage of the network; rather, the restoration route in the network is identified once the failure occurs and is used dynamically to restore the affected services. This is typically more efficient than predesigned protection from the resource utilization point of view, but the restoration time is usually longer, and at times, total service recovery cannot be guaranteed because of not enough spare capacity being made available at the time of failure. Restoration techniques can also be *path* and *link* based. In *link* restoration, when a link failure occurs, start and end nodes of the failed link dynamically compute an alternate route around the failed link to reroute the traffic. In this case, source and destination of the failed path are not notified of the failure, and hence are not involved in the failure recovery procedure. In *path* restoration, source and destination nodes of the failed path/connection are notified about the failure and an end-to-end restoration is done dynamically by computing a new disjoint backup path to reroute the traffic.

In ring topology, intelligent switching devices, such as digital cross-connects (DCSs) in SONET are used for dynamic restoration, which can have centralized or distributed control systems. In the case of a mesh topology for large area network where there are limitations with preassigned redundant protection paths, dynamic restoration is more practical to provide restoration from single and double failures. There is a dynamic network control and management system (Chapter 7) which enables efficient connection provisioning and restoration once the failure is discovered. In the case of centralized restoration there is a central network controller to keep track of the current state of the network, identify failure and calculate the backup paths once the failure occurs. After the restoration path is computed, the controller sends signals to all relevant nodes to execute the connections needed to accommodate the new optical path. On the other hand, in distributed techniques local controllers or the elements agents at each node have all the network state information. The nodes are updated with the changing network state for establishing new routes and initiating the new connections to accommodate these new routes through signaling.

The centralized survivability approaches may be slower than the distributed ones; however, they are more efficient in terms of capacity utilization, as the backup paths are calculated based on complete information of the current state of the network in the centralized node. In the centralized provisioning, routing decisions are based on a global view of the network state information hence can make more optimal routing decisions. All these advantages, however, come at the price of a reduced robustness, i.e. the centralized entity becomes a single point of failure, and additional redundancy need to be taken to avoid service disruptions. The distributed system is more scalable and inherently less affected by failures, since path computation operations are not performed at a single node but at all the nodes. The distributed system has limitations in the case of larger networks with highly dynamic traffic because it is not able to maintain correct and up-to-date network state information at all the nodes. Hence optimal routing decisions may not be possible. The distributed approach, which is based on automatically switched optical network (ASON) and generalized multiprotocol label-switching (GMPLS) signaling algorithms, is discussed in Chapter 7.

8.5 Survivability in Multilayer WDM Optical Networks

In multilayered WDM optical networks, fault management can be done at the various layers. The fault-management schemes in each layer have their own functionalities and characteristics. A well-defined set of restoration techniques already exists in the higher layers, such as SONET, OTN, GMPLS, IP, etc., with several overhead bytes dedicated in the signal frame for these purposes. Faults can be more efficiently handled in the optical layer than in the higher layers. Each lightpath carries several sub-wavelength connections of the higher electronic layer, hence fewer entities, such as a lightpath or fiber itself need to be rerouted for quickly restoring all the traffic of these connections. Meanwhile, restoration in the higher layers is slow and requires intensive signaling. Nodes in the optical layer, which are transparent, can respond much faster as the routing protection techniques are independent of the protocols used in the higher layers. The purpose of performing restoration in the optical layer is therefore to decrease the outage time by exploiting fast rerouting of the failed connection. The restoration time is in the 50 ms range when automatic protection schemes are implement in the optical transport layer. But there are a few limitations as well, such as detection of all faults not being possible without regeneration. Also, optical layer survivability cannot protect against failures at higher layers, hence some survivability must be provided at higher client layers as well.

8.5.1 Survivability in the Electronic Logical Layer

The different logical layers of optical networks, viz. SONET/SDH, OTN, and others, have their well-developed protection and survivability mechanisms.

SONET standards specify an end-to-end availability of 99.99% in the network with failure recovery times shorter than 60 ms. APS is used in these systems for this fast failure recovery. SONET *linear* APS is the standard defined for $(1 + 1)$, $(1 : 1)$, or $(1 : N)$ protection architectures in point-to-point systems which may be unidirectional or bidirectional. Such techniques, however, have to provide the extra capacity requirements.

The ring topology of SONET formed with add/drop multiplexers (ADMs) can provide protection with APS shared resources within an acceptable recovery time of 60 ms. Because these SONET rings protect automatically against failures, hence are known as SHR. The SHR architectures have the ability to recover all traffic in the case of a cable cut and part of the traffic in the event of a node failure. SHR are able to provide this high survivability in the networks with the help of high-speed ADM and control mechanism.

As mentioned in Chapter 5 on metro networks, SHRs are unidirectional self-healing rings (USHR) and bidirectional self-healing rings (BSHR). The USHR can be either unidirectional path switched ring (UPSR) or unidirectional line switched ring (ULSR). UPSR is a $(1 + 1)$ dedicated protection scheme because the signals for every connection are transmitted on both rings as shown in Figure 8.5a. When a link or node fails and affects one of the ring, the ADM at each node decides which signal is still good and then chooses it. UPSR is the fastest SHR scheme because no switching of signals is needed. The UPSR architecture can survive only a single cable cut. Beyond the single cut other

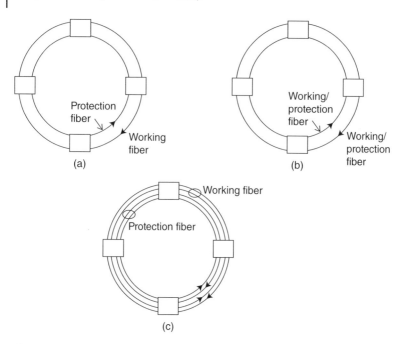

Figure 8.5 Self-healing rings: (a) UPSR, (b) 2F-BLSR, and (c) 4F-BLSR.

fault recovery techniques can be used to provide alternate routes for the connections at higher layers. Typically, UPSR architectures are used in the ingress node of a metro network where traffic is concentrated before it enters the core. On the other hand, ULSR has shared (1 : 1) or (1 : N) protection which is also called the *loopback* protection and both link and node failures can be taken care of in these schemes. The other SHR topology in the SONET ring is of that of the bidirectional ring. The typical architectures of bidirectional line switched rings (BLSRs) are two- and four-fiber line protection, 2F-BLSR and 4F-BLSR. In 2F-BLSR, shown in Figure 8.5b, half of the capacity on each ring is reserved for protection. When a failure occurs, the two nodes adjacent to the failed site will loop the affected traffic using the reserved capacity of the other ring. Two-fiber BLSRs have ring switching in case of failure of the link between the source and destination. In 4F-BLSR (Figure 8.5c), two fibers are dedicated as working fibers, so the capacity on them can be completely utilized. In a four-fiber BLSR, four fibers connect each adjacent pair of nodes, and the nodes are connected in a ring configuration. Upon failure, the nodes adjacent to the failure site simply loop the affected traffic from the working to the protection fibers. Typically, BLSR architectures are used in the core part of a metro network due to its efficiency in protection. Four fiber BLSRs are widely deployed in long-haul networks and two-fiber BLSRs in metro networks.

The total traffic supported on the ring is much higher by the BLSR architectures than over the UPSR architectures. The protection switching is done in the ADM node with time slot interchange. When a failure occurs, the working traffic from one fiber is looped back onto the opposite direction, carrying the working traffic in the opposite direction around the ring away from the failure. As in the case of (1 : 1) point-to-point APS, a signaling protocol using an APS channel carried in the SONET line overhead makes the affected nodes in the ring to do the line switching when failure occurs. The same procedure works in the case of a node failure. The line switching is performed at both sides of the failed node to redirect traffic away from that node.

In 4F-BLSR architecture a bidirectional connection between two nodes travels on separate fibers in the opposite directions through the same intermediate nodes. Similar to2F-BLSR case, traffic that would normally be carried on a working fiber in the broken link is looped back on protection fibers to direct it away from the failure. An additional advantage of the four-fiber architecture is that it can support both *ring* and *span* switching. Ring switching occurs when both the working and protection fibers are cut. In this case, the protection switches on both sides of the cut engage, and the traffic is routed away from the cut with the help of an APS signaling channel. Span switching occurs when only the working fiber fails. In that case, the traffic is switched from the working to the protection fiber in the same direction as the working traffic. This architecture is preferred by long-haul carriers. If a node fails, only the pass-through traffic can be restored. Alarm is used to account for the misdirected traffic in the case of node or multiple failures.

In pre-planned protection approaches for mesh topology protection in the SONET/SDH logical layer, ring protection techniques using ring covers are used. Along with the ring covers dynamic restoration approach with reconfigurability is also used to meet the routing demands on these rings which would otherwise require very high redundancy. In dynamic restoration when a failure occurs, connections are switched from the failed links and rerouted over others having spare capacity. Both centralized control and distributed control can be used for restoration process. The main disadvantage of DCS restoration techniques is that the reconfiguration time may not be fast enough to meet user demands for service availability. An advantage is that they require much less spare capacity than ring-based protection schemes. Typically, the time required for the restoration of traffic in a mesh SONET architecture using DCS equipment might be on the order of minutes if a centralized restoration algorithm is used and of the order of seconds if a distributed restoration algorithm is used.

8.5.2 Optical Layer Protection/Restoration

Providing survivability functionality at the optical layer has several advantages, such as fast recovery after failures, more efficient, transparency of protocols used in higher layers and others. The optical layer protection schemes are very similar to those in electrical layers. Similar to APS in SONET systems, in point-to-point WDM optical systems both dedicated $(1+1)$, and shared $(1:1)$ and $(1:N)$ optical protection are used, except that now switching is done in the optical domain either at the wavelength or fiber level. WDM SHR architectures also operate similar to SONET SHRs with advantage that due to the availability of multiple wavelengths in a single fiber, protection methods can be more flexible. Protection can be conveniently provided with few wavelengths dedicated for it in the fiber or even a dedicated fiber itself can be provided.

- *Predesigned protection techniques.* The predesigned protection techniques in WDM networks can also be link-based and path-based. The failure of a single fiber link causes the failure of all the channels on the link. Each working λ-channel has to have a protection wavelength path of the same capacity. The protection wavelength paths used for different working wavelengths on the same link may use different paths and/or different wavelengths. In the case of dedicated link protection a dedicated protection channel on a particular link is to be provided. In *shared link protection,* different protection paths share a wavelength on the overlapping portion if the corresponding working channels are on different links. The APS in the optical layer of mesh networks are shared optical link-based protection techniques that can potentially recover from failures within 50 ms by detouring traffic around a failed link or node. In the shared $(1:1)$ or $(1:N)$ case, the required spare capacity is less, protection paths are not active unless there is a fault, and they may be used for lower priority traffic but the failure recovery is slower and more complex [17].

In WDM systems, path-based protection refers to the reservation of a protection path and wavelength. In path-based protection a mechanism is required to notify the affected connection end nodes of the failure which is difficult to achieve, as many network-nodes get involved in achieving it, therefore, instead a disjoint protection path, a wavelength is usually reserved at connection setup for every link failure. Upon link failure, the wavelength paths reserved for this failure scenario are activated. But when the protection wavelength path is disjoint for every link of the working path, the same protection wavelength path can be used to restore the connection for any single link failure along the working path. Similar to link-based protection, path-based protection can be dedicated or shared. In dedicated path protection, the backup wavelength on the links of a protection path is reserved for a specific working connection. Therefore, two overlapping protection paths should have different wavelengths even if the working paths do not overlap. Dedicated path protection requires a large amount of extra capacity for protection purposes which is idle when there is no failure. But in shared path protection it is possible to utilize the capacity more efficiently by allowing the use of the same wavelength on a link for two different protection paths if the corresponding working paths are link-disjoint.

In WDM mesh networks rings and cycles are also used with protection switching schemes. Ring-covers, CDCs, and *p*-cycles are the commonly used techniques. In the WDM mesh, multiple logical rings are used to cover the whole physical mesh network and hence is referred to as *ring cover* scheme [16], as shown in Figure 8.6a. The six-node network is covered with three wavelength logical rings and six physical nodes. These logical rings of wavelengths behave like physical SHRs. When a connection is to be made between two nodes, the traffic can be routed through a single or multiple ring. These stacked rings lead to increase in geographical and capacity increase with multiple wavelengths in the fiber links.

CDC operates on a *link-protection* basis and the fundamental unit being protected here is a transmission link or switching node rather than an end-to-end path connection. We have a set of directed cycles which covers all the links in the network. Therefore if any link breaks we can have a cycle to recover the traffic on the failed link which can achieve APS in a mesh network. Figure 8.6b

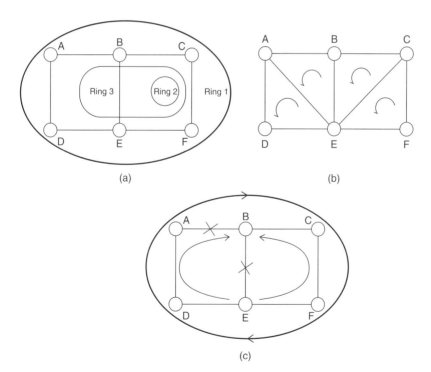

Figure 8.6 Example of protection in optical mesh networks: (a) ring cycles, (b) cycle double cover (CDC), (c) *p-cycle*.

shows one such case. The CDC networks have a pair of unidirectional working fibers and a pair of unidirectional protection fibers in each link. Half of the capacity is devoted for protection purpose, thus the system requires 100% redundancy. In the case of logical rings used for protection purpose, directed cycles are required to cover all the links in the network, so that if any link goes down, there is a cycle available to recover the traffic on the failed link, which can achieve APS in a mesh network [16].

Protection cycle (*p-cycle*) are also used in WDM mesh network for survivability. P-cycle have ring like recovery speed of 50 ms and the flexibility of mesh networks [18]. To achieve this, the spare capacity of the mesh network is pre-connected to form a closed structure analogous to ring connected all the nodes. The closed structure now can provide protection on the shared basis to all the on-cycle links as well as the straddling links (which are not covered by the ring) of the closed structure, which is the p-cycle. A straddling link is an off-cycle link having p-cycle nodes as endpoints. As the pre-connected spare capacity is a ring only two switching actions are needed at the end nodes of the failed link to switch the traffic on the back-up path in the event of failure. In the case of a straddling-link failure, the two node-disjoint paths around the p-cycle running between the two end nodes can be used to protect the failed link. As an example, in Figure 8.6c, A-B-C-F-E-D-A is the p-cycle covering all the six nodes of the mesh network. Out of the seven links besides straddling B-E link, the rest six are on-cycle links and get the p-cycle protection, as shown for the on-cycle A-B cut link. For the cut on straddling link B-E protection, there are two protection paths available: B-C-F-E and E-D-A-B. A p-cycle is more capacity-efficient than other ring protection schemes since the on-cycle capacity is used to protect both on-cycle link failures and the failures of straddling links. The implementation uses simple and reliable protection switches in each network node so that protection is accomplished without significant processing, transmission, or propagation delays. In the case of a link failure in the wavelength-routed mesh optical network, the protection switches at the ends of the failed link move to their protection state. This automatically reroutes the affected traffic around the fault over a path consisting of protection fibers which are organized in *p-cycles* topology computed off-line. The key to the successful operation of this protection technique is the creation of a family of p-cycles that covers the network in a particular way. The basic concept of p-cycles can be extended to protect against switching node failure as well. However, it turns out that node failure protection is a more complex issue than link failure protection.

- *Restoration schemes.* The dynamic reliability in optical network is achieved by using various survivability schemes. It needs to have an advanced network control and management system with good algorithms for fast restoration and survivability. Survivability in these networks involves all three *functional planes.* In the case of *data plane* protection, the configuration of protection is the responsibility of the management plane. However, the data plane has to inform the control plane about all failures of forwarding devices as well as their additions or removals. Protection in the case of *control plane* is most critical, hence it has to have high redundancy. The control plane creates both a working connection and a protection connection. Control plane restoration is based on rerouting of connections using spare capacity of the data plane. Such a rerouting operation takes place between the edges of the rerouting domain and is entirely contained within it [19]. There are two types of rerouting: *hard rerouting service* is a failure recovery mechanism triggered by a failure event and the *soft rerouting service* is associated with services like path optimization, network maintenance, etc., usually activated by the management plane. In hard rerouting, the original connection segment is released prior to creation of a new alternative segment and in soft rerouting the original connection is removed after creation of the rerouting connection. An important principle is that the existing connections in the transport plane should not be affected by failures in the control plane. However, new connection requests may not be processed by the failed control plane.

There are many factors that effect network survivability provisioning. A WDM optical network needs to have a complete updated network state information of network topology, the primary and backup paths and resources used by each path, available wavelengths over each fiber link, etc. Network state information needs to be exchanged or updated whenever the state of the network changes. As the amount of information that needs to be kept and updated is very large as the network size and speed grows, hence, it may not even be very accurate. There is also a tradeoff between the scalable provisioning of survivable services and the cost in terms of resource used, network state information maintenance, and network control complexity.

8.6 Summary

In the first part of the chapter significant physical impairments that may be relevant for transparent WDM networks have been addressed. Both linear and nonlinear impairments and their effect on performance have been given. Also, optical performance measurements techniques have been presented and dynamic impairment awareness solutions implemented at the management plane/control plane level have been given. Under high data rate, the impact of transmission impairments on a lightpath's quality can become very prominent, requiring appropriate techniques in both the physical layer and the network layer to mitigate the impairment effects on network performance. Such impairment-aware algorithms automatically consider the effects of impairments when setting up a lightpath.

The later part of the chapter has been devoted to studying survivability of multilayered WDM optical networks. After presenting some general classification criteria, the main protection and restoration strategies that can be adopted for the WDM network are presented. Specifically, discussed are the various protection and restoration schemes, the concepts of stacked rings, and protection cycles. Engineering the network for survivability is an important role in transport networks. Protection techniques are well established in SONET and SDH and include point-to-point, dedicated protection rings, and shared protection rings. These protection techniques are used in the optical layer as well. Optical layer protection is needed to protect the data services that are increasingly being transported directly on the optical layer without the SONET/SDH layer being present. It can also be more efficient with respect to reducing the protection bandwidth required and therefore is more cost-effective. Optical channel layer protection is needed if some channels are to be protected while others are not. Shared mesh protection in the optical layer can lead to more bandwidth efficiency and flexibility compared with traditional ring-based approaches.

Problems

8.1 (a) What are the major issues of signal degradation in a transparent network and can we define a transparent transmission length in the network?

(b) How could physical-layer devices affect the performance of the network and eventually the RWA algorithm?

8.2 (a) Compute the length of a point-to-point link with standard single-mode fiber at 1550 nm with bit rate of 2.5 and 10 Gbps for a directly modulated DFB laser with a spectral width of 0.1 nm and an externally modulated DFB laser with a spectral width of 0.01 nm for a power penalty of 1 dB. Assume RZ modulation and dispersion parameter of 17 ps/nm-km. (b) Repeat for non-zero dispersion shifted fiber (NZ-DSF) assuming a dispersion parameter of 5 ps/nm-km.

8.3 Chirping of optical pulses affects pulse broadening. Repeat the above problem with $C = -2$ to calculate the link length with chirped pulses.

8.4 (a) Give a reason to explain why in optical systems it is the third-order nonlinearities and not the second-order nonlinearities that typically affect the performance?

(b) Explain how the effects of dispersion and nonlinearities can be reduced.

8.5 Show that, for an unchirped Gaussian pulse travelling along the fiber, the pulse broadening effect of chromatic dispersion can be minimized by selecting an optimum pulse width of the propagating pulse to be:

$$T_0^{opt} = \sqrt{\beta_2 L}$$

where L is the length of the fiber link.

8.6 For a WDM amplified system with peak power limited laser output, calculate the extinction ratio for a power penalty of 10 dB. Next consider the case when the same system is also affected by the inter-channel cross-talk from two adjacent channels. How will the power penalty change now if the cross-talk level is 15 dB?

8.7 Consider the WDM link shown in Figure 8.1b. Each MUX and DEMUX introduces inter-channel cross-talk from the adjacent channel which is 30 dB below the desired channel. Compute the maximum cross-talk (i) after the DEMUX and (ii) after the MUX. What is the worst-case cross-talk penalty after the MUX? Assume the system is not an amplified system. (iii) If the cross-talk after each link stage is coherently added to the signal what will be the cross-talk after N such links?

8.8 Explain how stimulated Raman scattering can cause cross-talk in multi-channel lightwave systems. Derive Eq. (8.28) after approximating the Raman gain spectrum by a triangular profile.

8.9 Derive Eq. (8.24) by considering the nonlinear phase change induced by both self- and cross-phase modulation.

8.10 Determine and compare the maximum transmit power per channel for a WDM system operating over a standard single-mode fiber and a NZ-DSF to obtain the four-wave mixing limit for maximum allowable penalty due to FWM to be 1.5 dB. Assume that the channels are equally spaced with equal power in each, dispersion parameter for single-mode fiber and NZ-DSF to be 17 ps/nm-km and 3 ps/nm-km, respectively, link length to be 1000 km with optical amplifier every 80 km. Assuming the WDM system to have (i) four channels spaced 100 GHz apart, (ii) 16 channels spaced 100 GHz apart, and (iii) 32 channels spaced 50 GHz apart. Assume FWM resultant new wave power to be expressed as;

$$P_{ijk} = \left(\frac{\omega_{ijk} n}{2cA} \right)^2 P_i P_j P_k L^2$$

where, P_{ijk}, P_i, P_j, P_k are the powers in FWM resultant new compo-nent wave and each mixing wavelength, ω_{ijk} is the angular frequency of FWM component wave, n the nonlinear refractive index of the fiber (3.0×10^{-8} $\mu m^2/W$). A ($50 \mu m^2$) and L are the effective area and length of the fiber link, respectively.

8.11 In multi-layer WDM network different layers above the physical layer can provide restoration; give the advantages and disadvantage of it. Also, cer-tain interoperability issues arise when all the layers use their own protec-tion/restoration schemes. Discuss them and give the possible solutions.

8.12 A connection request 1–3 arrives in the network given in Figure 8.7. (i) Find a primary and a backup path for the request. (ii) Calculate the average protection-switching time in case of a failure of either of the two links on the primary path, given message processing time at each node is 5 μs, propagation delay on each link is 100 μs, reconfiguration time at any node is 60 μs, and time to detect a link failure is 100 μs.

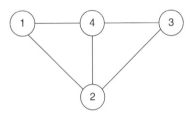

Figure 8.7 A four-node network topology.

8.13 Given the topology in Figure 8.8, where we have the path 1–2–3–4 already set up, and assuming that there is a failure on link 2–3, what will be the recovery procedure for the path restoration and link restoration?

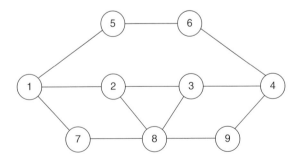

Figure 8.8 A nine-node network topology.

8.14 Write a computer program that can handle a variety of topologies, traffic patterns, and different routing/protection computation algorithms. It should:

(i) input different network topology graphs with a set of lightpaths and route the lightpaths using a shortest-path algorithm.

(ii) compute protection bandwidth in the network for dedicated wavelength protection using an algorithm to provide two disjoint shortest paths for each lightpath.

(iii) Compute protection bandwidth in the network for shared wavelength protection using an algorithm which determines the amount of protection capacity required on each link for each failure in the network. Determine the total protection capacity needed on the links with all other failures in the network.

References

1 Agrawal, G.P. (2012). *Nonlinear Fiber Optics*, 5e. San Diego, CA: Academic Press.

2 Suzuki, M. and Edagawa, N. (2003). Dispersion managed high-capacity ultra-long-haul transmission. *J. Lightwave Technol.* 21: 916–929.

3 Daikoku, M., Yoshikane, N., Otani, T., and Tanaka, H. (2006.). Optical 40 Gbps 3R regenerator with a combination of the SPM and XAM effects for all-optical networks. *J. Lightwave Technol.* 24: 1142–1148.

4 Buck, J.A. (1995). *Fundamentals of Optical Fibers*. New York: Wiley.

5 Huang, Y., Heritage, J.P., and Mukherjee, B. (2005). Connection provisioning with transmission impairment consideration in optical WDM networks with high-speed channels. *IEEE/OSA J. Lightwave Technol.* 23 (3): 982–993.

6 Pandya, R.J., Chandra, V., and Chadha, D. (2014). Impairment aware routing and wavelength assignment algorithms for optical WDM networks and experimental validation of impairment aware automatic light-path switching. *Opt. Switching Networking* 11, Part A,: 16–28.

7 Kilper, D.C. et al. (2004). Optical performance monitoring. *IEEE/OSA J. Lightwave Technol.* 22 (1): 294–304.

8 ITU-T Draft new Recom. G.697, *Optical Monitoring for DWDM Systems*, June 2004.

9 Sam Canhui, O. and Mukherjee, B. (2011). *Survivable Optical WDM Networks*. Springer.

10 Mouftah, H.T. and Ho, P.-H. (2003). *Optical Network Architecture and Survivability*. New York: Springer.

11 Lin, W.-P., Kao, M.S., and Chi, S. (2003). A DWDM/SCM self-healing architecture for broadband subscriber networks. *J. Lightwave Technol.* 21: 319–328.

12 Gerstel, O., Ramaswami, R., and Sasaki, G.H. (1998). Fault tolerant multiwavelength optical rings with limited wavelength conversion. *IEEE J. Sel. Areas Commun.* 16 (7): 1166–1178.

13 S. Ramamurthy and B. Mukherjee, *Survivable WDM Mesh Networks, Part II — Restoration, Proc. ICC*, 1999, pp. 2023–30.

14 Grover, W.D. (2004). *Mesh-Based Survivable Networks: Options and Strategies for Optical, MPLS, SONET, and ATM Networking*. Upper Saddle River, NJ: Prentice Hall.

15 S. Ramamurthy and B. Mukherjee, *Survivable WDM Mesh Networks, Part I — Protection, Proc. INFOCOM*, Mar. 1999, pp. 744–51.

16 W. D. Grover and D. Stamatelakis, *Cycle-Oriented Distributed Reconfiguration: Ring-like Speed with Mesh-Like Capacity for Self-Planning Network Restoration* In *Proceedings of the IEEE Int'l Conf. Commun. (ICC)*, Atlanta, GA, June 1998.

17 Maier, G., Pattavina, A., De Patre, S., and Martinelli, M. (2002). Optical network survivability: protection techniques in the WDM layer. *Photonic Network Commun.* 4 (3–4): 251–269.

18 Grover, W.D., Doucette, J., Kodian, A. et al. (2017.). Design of survivable networks based on p-cyclesChapter 16. In: *Handbook of Optimization in Telecommunications* (ed. M.G.C. Resende and P.M. Pardalos). New York: Springer International Publications.

19 Jajszczyk, A. and Rozycki, P. (2006). Recovery of the control plane after failures in ASON/GMPLS networks. *IEEE Network* 20 (1): 4–10.

9

Flexible Optical Networks

9.1 Introduction

With the exponential growth of multimedia services and with diverse models of Internet traffic, demand of dynamically varying small and large traffic across the network is increasing. Although current wavelength-division multiplexing (WDM)-based optical network architectures offer advantages of high-capacity transmission and reconfigurable wavelength switching, they also have drawbacks of rigid bandwidth and coarse granularity, which not only reduces dynamism in the network but is also not spectrum efficient. This dynamism can be brought about by changing network structure and a whole new network management strategy. The fixed grid can accommodate maximum up to 200 Gbps bitrate in the 50 GHz fixed International Telecommunications Union (ITU) grid with coherent and more sophisticated modulation schemes. This fixed wavelength grid will no longer work for 400 Gbps and above as the spectral width occupied by 400 Gbps at standard modulation formats is too broad to fit in the 50 GHz ITU grid, and when forced to fit by adopting a higher spectral-efficient modulation format, it would only allow much short transmission distances. Meanwhile, in the conventional fixed-grid network when sub-wavelength demands are sent over a full wavelength capacity, there is huge underutilization of scarce optical spectrum. Also, sub-wavelength services with time division multiplexing (TDM) grooming over the optical transport network use electrical switches, leading to high cost and energy consumption.

With the proliferation of data centers and cloud computing the optical networks will soon be required to support Tbps bit rates class transmission in the near future. This is not possible with present fixed DWDM optical transmission technology. Fixed DWDM suffers from the electrical bandwidth bottleneck limitation and physical impairments which become more severe as the transmission speed increases. Also, fixed and coarse granularity of current WDM technology will restrict spectral efficiency of optical networks due to

Optical WDM Networks: From Static to Elastic Networks, First Edition. Devi Chadha.
© 2019 John Wiley & Sons Ltd. Published 2019 by John Wiley & Sons Ltd.

stranded bandwidth provisioning, inefficient capacity utilization, and high cost. Hence what is required is a more data-rate flexible, agile, reconfigurable, and resource-efficient optical network. To meet future Internet traffic requirements, a novel elastic optical network (EON) architecture with flexible data rate and spectrum allocation, high resource efficiency, and low power consumption is required.

In this chapter, we present the flexible and scalable optical transport network architecture called EON. In EON, instead of fixed grid the fiber has flexible grid spacing. Thus, in the case of smaller traffic demands for sub-wavelength it alleviates the stranded bandwidth of the fiber by selecting a smaller grid size. When traffic demands are of higher bit rate, such as 400 Gbps or 1 Tbps, the grid required can be easily stretched to accommodate it in multiple slots with flexible grid spectrum. It thus provides support of various data rates, including possible future ones, in a highly spectrum-efficient manner. To add further to the flexibility, the required optical devices, such as flexible optical cross-connects (OXC)s and transponders, are envisaged for EON so that signal transmission can be possible to long distances with different multi-rate modulation schemes [1, 2].

In this chapter, we start with the description of a few of the coherent modulation schemes and the multicarrier modulation schemes (Sections 9.2 and 9.3, respectively) used in EON for better utilization of spectrum and adaptability. In Section 9.4 we give a general description of EON followed in Section 9.5 by the various network elements used in EON to make it flexible. Section 9.6 in brief gives the wavelength and spectrum assignment (RSA) algorithms used in these networks, and finally we close the chapter with requirements of network management and control in EONs in Section 9.7.

9.2 Coherent Modulation Schemes

With data-rate demands of more than 40 Gbps and long-haul networks over 1000 km, coherent modulations schemes have taken over the non-coherent intensity modulated direct detection schemes, as the non-coherent modulation schemes cannot provide the required bit error rate (BER) because of their very low signal to noise ratio (SNR) and spectral efficiency with direct detection. For data rates of 40 Gbps and above, intensity modulation with direct detection schemes cannot be used in the 50 GHz DWDM grid; instead we use either dual-polarization quadrature phase-shift keying (DP-QPSK) or M-quadrature amplitude modulation (M-QAM) spectrally efficient modulation schemes [3, 4]. The DP-QPSK and M-QAM code 4 bits per symbol and M bits per symbol, respectively. These reduced spectral width coded signals are so spectrally efficient that they can propagate high data rate through long distances. The spectral efficiency can be further improved by using coherent transmission and increasing transmission rates by coding more bits per symbol. As regard with generation of such high data-rate signals are concerned it can be done now with much less complexity with the advances we see in photonics. We can generate super-channels, in which multiple coherent optical carriers are combined to create a unified channel with a higher data rate in DWDM channels using large-scale photonic integrated circuits. With all the advantages of these high transmission rates there are a few limitations as well. The high bits/symbol schemes with dense signal constellation are sensitive to phase errors which are caused by the nonlinear phase noise and cross-phase modulation of fiber nonlinearities. Therefore, one will require more optical signal-to-noise ratio (OSNR) due to the Shannon's limit, which becomes difficult to achieve. Fortunately, for high data-rate networks with coherent detection, the phase information of the optical signal is preserved after electro-optic detection, therefore, the distortion due to chromatic and polarization mode dispersion in the signal can be compensated electronically in the receiver circuit with advanced DSP and adaptive equalizer [5]. Nevertheless, the coherent optical systems have the disadvantage of requiring much more complex electro-optics than the direct-detection schemes. This increases the complexity and system cost of coherent schemes. Having said that, to enable greater distance and low-latency connectivity with higher data rates, coherent detection is important. In the following sections, we give details of DP-QPSK and M-QAM coherent schemes which are now the standard for 100 Gbps and above transmission in optical networks.

9.2.1 Dual Polarization-Quadrature Phase Shift Keying (DP-QPSK)

Coherent DP-QPSK modulation scheme is typically used for long reach 40G/100G optical systems [6]. The basic functional block diagrams for an optical modulation scheme, with control of the amplitude of both in-phase (I) and quadrature-phase (Q) components of the modulated signal [7], are shown in Figure 9.1a. DP-QPSK codes 4 bits per symbol. The transmit side consists of a laser, Mach–Zehnder modulator (MZM) structures (two for each polarization), and four driver amplifiers. The modulated beams from two orthogonal polarizations are combined to obtain the DP-QPSK signal.

The coherent receiver requires mixing the received signal light with a tunable-laser local oscillator. On the receiver side, in Figure 9.1b the polarization beam splitters and optical phase hybrids are included in the receive structure to provide polarization and phase diversity. The four signals which are received are detected by the balanced photodiodes followed by the trans-impedance amplifiers. The signal is first quantized using quad analog-to-digital converters (ADCs). The adaptive equalizer in the DSP then provides equalization of chromatic dispersion, polarization mode dispersion, filtering

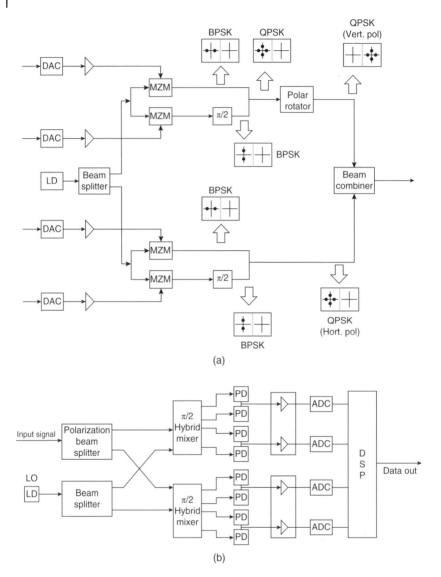

Figure 9.1 (a) Optical transmitter with DP-QPSK modulation. (b) Optical receiver for DP-QPSK demodulation.

distortion, etc. The carrier phase estimation, polarization, and I&Q de-multi-plexing are all achieved in the electronic domain using very fast ADCs and DSP. This approach alleviates the traditional problem with optical coherent technology since this design does not require a highly stable optical phase-locked loop.

9.2.2 M-Quadrature Amplitude Modulation (M-QAM)

Coherent M-QAM enables data rates many multiples higher than the electronics speed by coding multiple bits-per-symbol. With M-QAM there is the advantage that the bit rate transmitted can be traded for optical reach with programmable modulation. Higher-density M-QAM constellations has lower OSNR sensitivity and higher nonlinear distortion, resulting in reduced reach. The achievable spectral efficiency is related directly with number of coded symbols and the spectral width with the baud rate. For example, 400 Gbps data throughput can be supported by both DP-256QAM and DP-16QAM at 25 and 50 Gbaud, respectively, but 50 Gbaud occupies twice the spectral bandwidth. The spectral efficiency of DP-256QAM is obviously better but with reduced OSNR sensitivity. M-QAM is more sensitive to nonlinear phase noise and distortion, so fiber nonlinearities pose a major obstacle to transmission distance. As the symbols in QAM have more distance between them, the receiver can recover the signal better. Nevertheless, more power is needed to transmit the symbols. Figure 9.2 shows the 16-QAM transmitter and receiver circuits with coherent detection with ADC and DSP for long-distance transmission system, much the same as for DP-QPSK in Figure 9.1.

9.3 Multi-Carrier Modulation Schemes

The performance of high data-rate signals starts deteriorating significantly due to linear and nonlinear dispersion effects of the fiber for long-distance communication, which leads to very poor BER. Instead of single-carrier there are multi-carrier modulation (MCM) schemes available for high data rates above 100 Gbps transmission to take care of this dispersion effects in the links. In MCM schemes, the data stream to be transmitted is divided into a number of lower data-rate streams. Each of the lower data-rate streams in turn modulates an individual narrow bandwidth carrier. After receiving the transmitted signal, the receiver reassembles the overall data streams from those received on the individual carriers. MCM techniques are particularly important at high data-rate transmission when the channel or receiver bandwidths are smaller, causing dispersion. MCM systems are less susceptible to interference than single-carrier high data-rate systems; one, due to lower data rates on each subcarrier, and two, the interference may affect a small number of subcarriers, hence the effect of nonlinearities of the channel becomes limited.

There are several MCM systems proposed for optical fiber systems, such as coherent optical orthogonal frequency-division multiplexing (CO-OFDM) [8, 9], coherent wavelength-division multiplexing (CoWDM) [10], Nyquist-WDM, as well as dynamic optical arbitrary waveform generation (OAWG) [11, 12]. In all these MCM schemes, many low-speed subcarriers are generated to form

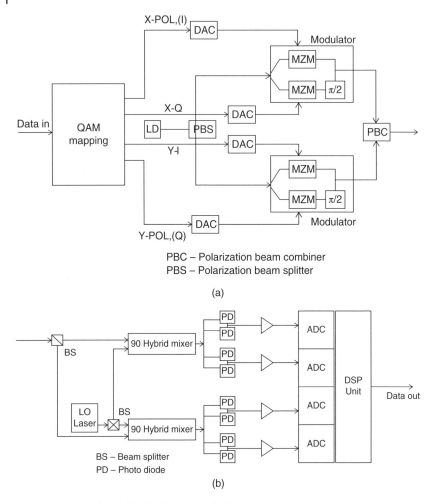

Figure 9.2 Optical 16-QAM. (a) Transmitter. (b) Receiver.

broadband data waveforms using lower-speed modulators, but they differ with respect to their operation principles and capabilities. In the case of CO-OFDM [9, 12, 13], orthogonality between the subcarrier is maintained with their frequencies spaced at multiple of inverse of the symbol rate. In the case of CoWDM, the orthogonality between subcarriers is maintained by setting the CoWDM subcarrier symbol rate equal to the subcarrier frequency spacing. Nyquist-WDM attempts to minimize the spectral utilization of each channel by reducing the spectral guard bands between WDM channels generated from independent lasers. Finally, in OAWG, the coherent combination of many spectral slices generated in parallel enables the creation of a continuous output spectrum. All the four schemes have their own advantages and drawbacks. They can adopt the various modulation formats and can generate Tbps super-channels. The transmitter structures in each are based on many modulators in parallel at low speeds modulating several optical subcarriers. In the case of CO-OFDM, both optical and electronic subcarriers can be used. On the receiver side, multiple receivers in parallel are required at low speeds with an equal number of subcarriers.

In the following section, we will be limiting our discussion on the CO-OFDM MCM scheme as OFDM has emerged as a leading modulation technique in the broadband RF wireless and in the optical domain as well. It is widely accepted commercially in almost every major communication standard. The rest of the schemes details can be found in [1, 13, 14]. We first explain a few basics of OFDM before we give details of optical OFDM systems.

9.3.1 Orthogonal Frequency Division Multiplexing

OFDM is a special class of multicarrier modulation that consists of transmitting a higher bit rate signal over several lower-rate *orthogonal* sub-channels. It is used in various standards, such as 802.11a/g Wi-Fi, 802.16 WiMAX, LTE (long-term evolution), DAB, and DVB (digital audio and video broadcasting), and DSL (digital subscriber loop) around the world. The fundamental principle of OFDM is to overlap multiple-channel spectra within limited bandwidth without interference, taking into consideration both filter and channel characteristics. In Figure 9.3a, different subcarriers (f_1–f_4) that form one OFDM symbol are depicted. Each subcarrier has a $|sin(x)/x|^2$ spectrum, and at the maxima of individual subcarrier the other subcarriers have a minima. But all the OFDM subcarriers have side lobes over a frequency range which includes many other subcarriers. Though this causes issues of sensitivity to frequency offset and phase noise, but it is very robust against channel dispersion because the symbol is divided into narrow sub-bands.

An OFDM multicarrier modulated signal in time domain (Figure 9.3b) can be represented as:

$$s(t) = \sum_{m=-\infty}^{\infty} \sum_{k=1}^{N} c_{km} s_k (t - mT_s) \qquad (9.1)$$

where:

$$s_k (t) = \Pi(t) e^{j2\pi f_k t} \qquad (9.2)$$

$$\Pi(t) = \begin{bmatrix} 1, \ldots\ldots\ldots\ldots (0 < t \leq T_S) \\ 0, \ldots\ldots\ldots\ldots (t \leq 0, t > T_S) \end{bmatrix} \qquad (9.3)$$

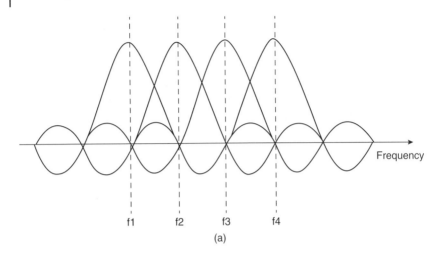

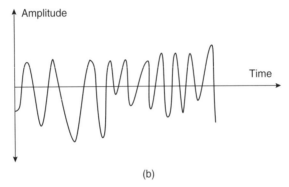

Figure 9.3 (a) OFDM signal spectrum. (b) OFDM signal time variation.

and c_{km} is the mth information symbol at the kth subcarrier, s_k is the signal for the kth subcarrier with $\Pi(t)$ as the symbol shaping function, N is the number of subcarriers, f_k is the frequency of the subcarrier, and T_s is the symbol period. The optimum detector for each subcarrier uses a matched filter or a correlator, herefore, the detected information symbol c'_{km} at the output of the correlator is given by:

$$c'_{km} = \int_0^{T_s} r(t - mT_s) s_k^* dt = \int_0^{T_s} r(t - mT_s) e^{-j2\pi f_k t} dt \qquad (9.4)$$

where r(t) is the received time-domain signal. The orthogonality between the subcarriers is achieved by the correlation between any two subcarriers, given by:

$$\delta_{kn} = \frac{1}{T_s}\int_0^{T_s} s_k s_n^* dt = \frac{1}{T_s}\int_0^{T_s} e^{j2\pi(f_k-f_n)t} dt = e^{j2\pi(f_k-f_n)T_s}\frac{\left(\sin\pi(f_k-f_n)T_s\right)}{\pi(f_k-f_n)T_s} \quad (9.5)$$

If the OFDM MCM signal has the following condition between the subcarriers satisfied:

$$f_k - f_n = i\frac{1}{T_s} \quad (9.6)$$

then the two subcarriers are orthogonal to each other, where f_k and f_n are the subcarrier set where i is an integer. The orthogonal subcarrier sets with the above condition satisfied can be recovered with the matched filters without inter-carrier interference.

Unlike the conventional single carrier modulation which has a broad and extended spectrum, the OFDM spectrum is a rectangular one with most of the energy confined inside the rectangular spectrum. Hence the occupied spectrum is much narrower compared with conventional single carrier modulation. OFDM modulation and demodulation can be implemented using computationally less complex and efficient fast Fourier transform and inverse fast Fourier transform (FFT/IFFT) functions, which are normally used in OFDM systems to implement OFDM modulation and demodulation. The discrete value of the transmitted OFDM signal s(t) is a N-point IFFT of the information symbol c_k, and the received information symbol c'_k is a N-point FFT of the received sampled signal r(t).

9.3.1.1 Cyclic Prefix in OFDM

Inter-symbol interference (ISI) between neighboring OFDM symbols can happen for long distance communication. This is due to the relative delay spread between different subcarrier symbols as the carrier passes through the dispersive channels. Also, the orthogonality condition for the subcarriers can get lost during transmission resulting in an inter-carrier-interference (ICI) penalty.

In order to mitigate dispersive channel effects, such as chromatic dispersion (CD) and polarization dispersion (PD), an enabling technique for OFDM is the insertion of cyclic prefix [8, 9] in the guard band between the symbols. Cyclic prefix (CP) is used to resolve the channel dispersion-induced ISI and ICI. In most OFDM systems, a CP is added to the start of each time domain OFDM symbol before transmission. In other words, a number of samples from the end of the symbol are appended to the start of the symbol. Figure 9.4 shows insertion of the CP into the guard interval, ΔG. Although the CP introduces some redundancy and reduces the overall data rate, it adds equalization in the OFDM signal.

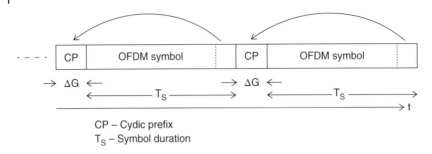

CP – Cydic prefix
T_S – Symbol duration

Figure 9.4 Cyclic prefix in OFDM.

The condition for ISI-free OFDM transmission is given by:

$$t_d < \Delta G \tag{9.7}$$

where t_d is the relative delay spread of the subcarriers.

9.3.1.2 Peak-to-Average Power Ratio for OFDM

The OFDM technology has several advantages in transmission systems, such as high spectrum efficiency, high data-rate transmission with resilience to ISI and ICI, and an energy-efficient operation by adaptive modulation and dynamically switching on/off specific subcarriers when needed according to customer bandwidth requirement and also the channel conditions.

Nevertheless, OFDM has a drawback of high peak-to-average-power ratio (PAPR). PAPR normally occurs due to the power amplifiers at the transmitter end. PAPR occurs due to high input power causing the amplifier gain to saturate. Fortunately, in the optical systems, EDFA is used as post-amplifier at the transmitter end which ideally has linear characteristics regardless of its input signal power due to its slow response time of the order of millisecond. Nevertheless, PAPR poses a challenge for optical fiber communications due to the nonlinearities in the optical fiber.

The PAPR of the OFDM signal in the communication system is defined as the ratio between the maximum peak power to the average power of the transmitted OFDM signal as:

$$PAPR = \frac{\max\left\{|s(t)|^2\right\}}{E\left\{|s(t)|^2\right\}}, \ldots \ldots t \in [0, T_s] \tag{9.8}$$

The theoretical limit of PAPR depends on the number of subcarriers and can be written in dB as:

$$PAPR = 10\log_{10} N \tag{9.9}$$

where N is the number of subcarriers. Therefore, in the limiting case PAPR depends only on the number of subcarriers. The more common technique used to measure PAPR is the complementary cumulative density function (CCDF), which is the probability of PAPR being greater than the threshold value, expressed as:

$$CCDF = \Pr\left(PAPR > PAPR0\right) = 1 - \Pr\left(PAPR \leq PAPR0\right) \qquad (9.10)$$

where PAPR0 is the threshold of PAPR desired of the system. Equation (9.10) is valid only for a large number of subcarriers ($N \geq 64$). The theoretical limit of threshold is very high (for example, for N = 256, PAPR = 24 dB), but practically PAPR should be much smaller, of the order of 3–5 dB. Several techniques have been proposed to reduce PAPR [15].

The robustness against channel dispersion and its ease of transmitting high data signal, OFDM using IFFT/FFT techniques makes it a suitable advanced modulation format for optical communications systems. However, there are a few drawbacks of high PAPR, frequency and phase noise leading to ICI, etc. Once these impairments are mitigated, OFDM becomes very attractive to be used in optical communications, RF communications and others.

9.3.2 Optical OFDM

The use of OFDM in optical fiber communications brings spectral efficiency and tolerance to impairments, such as CD and PD in the optical fiber system providing high data-rate transmission across dispersive optical media. Optical OFDM can be broadly categorized in two parts: one in which subcarrier frequencies are in the RF domain which are then up-converted in the optical domain, and the other in which we have the subcarriers themselves in the optical domain. In this all-optical approach, an optical OFDM signal is directly generated in the optical domain through modulation of multiple optical subcarriers, without electrical IFFT processing [8]. The main advantage of the optical approach is that the electronics of the ADC/DAC are eliminated. Different approaches can be used to generate OFDM subcarriers in the optical domain.

With the RF subcarrier, the optical OFDM is further said to be of two types: first, the direct-detection optical OFDM (DDO-OFDM) which has a simple realization based on low-cost optical components, and second, the coherent detection optical OFDM (CDO-OFDM) with higher spectral efficiency and receiver sensitivity. In the DDO-OFDM, data is mapped onto individual RF subcarrier symbols and converted to the baseband OFDM signal through inverse fast Fourier transformation (IFFT) and digital-to-analog conversion (DAC). Then this baseband OFDM signal is up-converted to the optical domain using a wavelength-tunable laser and an optical I/Q modulator. In the

CDO-OFDM the difference is that on the receiver side the signals are detected coherently.

QPSK and QAM modulation formats are commonly used modulation formats before IFFT is carried out. The block diagram of a basic OFDM transmitter is shown in Figure 9.5a. It is composed of a serial to parallel converter, a symbol mapper, an IFFT of N points, a block to insert a CP followed by a parallel to serial converter. With symbol mapping, a certain number of bits represent each symbol. The IFFT block generates an OFDM symbol with N orthogonal subcarriers. Finally, the resulting signal is serialized, digital-to-analog converted and it is sent through the channel.

In the receiver, Figure 9.5b, the inverse process takes place. First, after the detection of the signal, the data is serial to parallel converted in order to remove

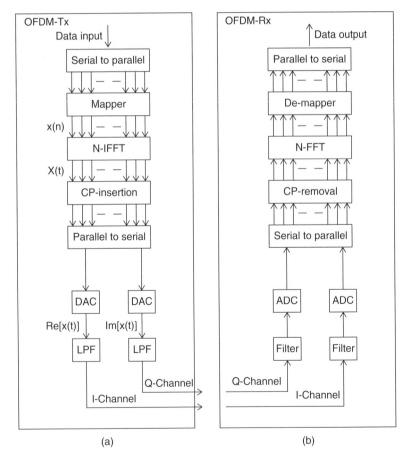

(a)   (b)

Figure 9.5 OFDM. (a) Transmitter. (b) Receiver.

the CP in the following step. If the receiver FFT window is aligned with the start of the main symbol period of the first arriving signal and the delay spread introduced in the system by the channel is smaller than the CP, then no ICI or ISI occurs. Then the FFT is implemented and finally the resulting signal is de-mapped, serialized, and analog-to-digital converted in order to recover the original bit stream.

9.3.2.1 Direct-Detection Optical OFDM

In direct-detection optical OFDM systems the information is carried on the optical intensity. DDO-OFDM is a low-cost approach and can be used in a broader range of applications, such as long-haul transmission, multi-mode fiber, and short-reach single-mode fiber transmission. DDO systems, as it is shown in Figure 9.6a, the electrical OFDM signal directly or externally modulate a laser at the transmitter to create the intensity modulated optical signal. A typically used external modulator is the MZM. In the receiver, the signal is recovered using a photodiode. As only real data can be modulated using a single modulator, it is necessary to manipulate the OFDM symbol at the transmitter end. Once the OFDM symbol is created (Figure 9.5a), their real and

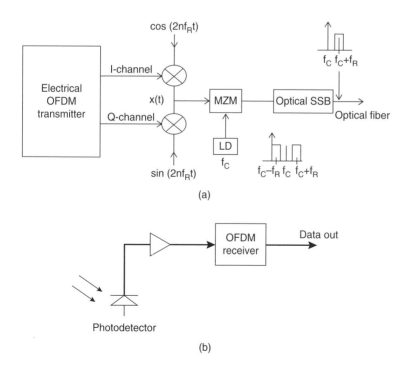

Figure 9.6 DDO-OFDM. (a) Transmitter. (b) Receiver.

imaginary parts are converted to the analog domain separately as shown in Figure 9.6a. The real part of the signal is multiplied by cosine and imaginary part by sine of the RF signal. Adding the two signals, a real valued OFDM signal is obtained to work in the IM/DD systems. Once we have real OFDM symbols they modulate the optical signal in order to transmit onto the fiber. The MZM creates a double side-band spectrum with respect to the optical carrier. So, in order to ensure that the OFDM subcarriers are represented only once by the optical frequencies, and to avoid CD fading, single-side-band modulation can be adopted. In short reach applications, double side band can also be used. On the receiver end, as shown in Figure 9.6b, the data is detected with a photodetector, it is electrically amplified, analogically converted and de-mapped with the OFDM demodulator in order to recover the original bit stream.

There can be some other options to obtain unipolar signal in optical OFDM systems. One is called the DC biased O-OFDM and other Asymmetric Clipped O-OFDM (ACO-OFDM). In DC-biased optical OFDM, a DC bias is added to the signal, however because of the large PAPR of OFDM, even with a large bias some negative peaks of the signal will be clipped resulting some distortion. With DC biased O-OFDM, we are able to transmit more information with the same bandwidth, implying higher spectral efficiency, but at the same time clipping noise affects the transmission. In the ACO-OFDM the bipolar OFDM signal is clipped at the zero level and the negative going signals are removed. If only the odd frequency OFDM subcarriers are non-zero at the IFFT input, all of the clipping noise falls on the even subcarriers, and the data carrying odd subcarriers are not impaired. ACO-OFDM requires a lower average optical power for a given BER and data rate than DCO-OFDM.

9.3.2.2 Coherent-Detection Optical OFDM

The coherent CDO-OFDM systems achieve better performance in receiver sensitivity, spectral efficiency, and robustness against PD but has higher complexity in the transceiver design than the direct detection systems. The superior performance of CDO-OFDM makes it an excellent candidate for long-haul transmission systems.

The CDO-OFDM transmitter basically has RF OFDM signal transmitter, much as in the direct detection system, and the RF to optical up-converter. Within RF to optical up-converter, the baseband OFDM signal is up-shifted onto the optical domain using an optical I/Q modulator, which is comprised of two MZMs with a 90° optical phase shifter. Then the optically modulated signal is transmitted onto the fiber channel. A coherent receiver with a local laser diode oscillator is used to down-convert the data to the RF domain. The OFDM optical receiver uses two pairs of balanced receivers and an optical 90° hybrid to perform optical I/Q detection and finally data is demodulated and sent to the detector and decoder to obtain the baseband signal.

9.3.2.3 All-Optical OFDM

In the last section we studied the optical OFDM systems with an electronic DSP based FFT/IFFT, where the generated digital OFDM samples were converted into analog signals through a DAC and later used to modulate an optical carrier. On the receiver side, optical signals are detected and digitally sampled by an ADC. Such OFDM systems have the limitation of speed due to the electronic processing in the ADC, DAC, and FFT/IFFT modules.

To overcome the speed limitation of electronics, all-optical OFDM are used where in the OFDM transmitters multiple optical subcarriers are generated. The subcarriers are separated with optical devices and each subcarrier is modulated at the baud rate close to the subcarrier spectral spacing. The spectra of the modulated optical subcarrier will overlap with each other, forming one optical OFDM channel band. On the receiver side, optical FFT circuits achieve the FFT function without O-E or E-O conversion. These optical IFFT/FFT circuits comprise of an integrated passive optical device providing appropriate temporal delays and phase shifts. Both IFFT and FFT functions can be realized all-optically with AWGs. This approach provides a fully integrated and scalable optical FFT/IFFT solution on a single chip with much less structural complexity. The AWG as FFT/IFFT circuits is especially suitable for optical OFDM super-channel system consisting of a large number of optical subcarriers.

An all-optical OFDM system setup with AWGs is shown in Figure 9.7a. The AWG at the transmitter end functions as a regular WDM-DEMUX and is used to separate all the subcarriers generated by a mode locked laser. Each subcarrier afterwards is modulated by the data symbol. The modulated subcarriers are combined together through a coupler to form an OFDM signal [1, 16]. The AWG at the transmitter end is used as an IFFT filter, and the one at the receiver is utilized as an FFT filter. The mode-locked laser generates a pulse train with a repetition rate of optical subcarriers. These generated multiple synchronous pulse trains are modulated individually by the data at each subcarrier. When these modulated data streams pass through the N × 1 multiplexer, an IFFT operation is performed. In creating the optical OFDM signal, the orthogonal condition is satisfied through proper pulse shaping and phase locking the optical subcarrier to orthogonal frequency, and the symbol rate of each optical subcarrier is kept equal to the optical subcarrier spacing.

At the receiver end, the received signal is injected to another AWG conducting the FFT processing. It does the de-multiplexing, implementing the phase delays and careful arrangement of time delays to each subcarrier, as depicted in Figure 9.7b. The demultiplexed signals at different subcarriers are presented at different AWG outputs and can be collected after optical sampling. Subsequently after the detection the data is recovered by DSP where the functions of clock recovery, equalization, carrier phase estimation, and recovery are performed.

With the intrinsic flexibility and scalability characteristics of optical OFDM technology, a novel EON architecture, discussed in the next section,

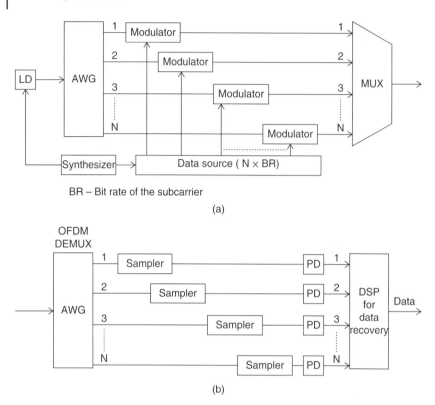

BR – Bit rate of the subcarrier

(a)

(b)

Figure 9.7 (a) Block diagram of the all-optical OFDM transmitter configuration. (b) Block diagram of the all-optical OFDM receiver configuration.

possessing the capability to manage signals with different data rate and variable bandwidth can be built to meet the requirements of future programmable optical networks

9.4 Elastic Optical Network

Figure 9.8 shows an existing ITU fixed grid and flexible grid structure in the optical fiber spectrum. In the present fixed grid WDM networks, as shown in the Figure 9.8a, have two limitations: one, it is not able to allocate resources with granularity finer than a wavelength of 10 Gbps because of which there is reduced utilization of the deployed network capacity when sub-wavelength traffic has to be transmitted over the fixed grid without TDM grooming; and two, it is not able to accommodate very high data-rate traffic required for enterprises, data centers, and other customers in a cost-effective manner. The

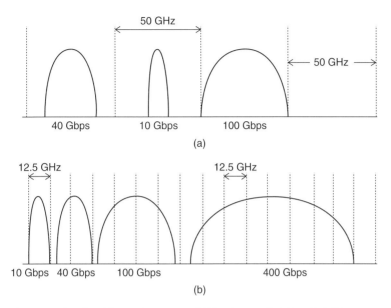

Figure 9.8 Fixed and flex-grid spectrum. (a) Fixed spectral widths of 50 GHz accommodating 10 Gbps, 40 Gbps, and 100 Gbps traffic. (b) Flex-grid with 12.5 GHz slots, 10 Gbps, 40 Gbps, 100 Gbps, and 400 Gbps data rate accommodated in flexible number of slots.

fixed grid cannot support bit rates of 400 Gbps and 1 Tbps at standard modulation formats because the bandwidth required for these bit rates is much more than one 50 GHz grid boundary. Though the fixed grid systems are being used to carry upto 200 Gbps bit rate traffic over the optical spectrum with sophisticated modulation schemes, but with demands over 400 Gbps, the 50 GHz fixed grid WDM system cannot achieve a significant reach. Also, with the changing models of traffic content there is huge heterogeneity in the pattern of the demands from the client networks. Therefore, it is desirable that the optical spectrum of the fiber, the network elements; the transceivers, OXC, and others, are all able to maximize spectral efficiency by adapting to the actual changing data rate and network conditions for each given traffic demand varying from sub-wavelengths in Gbps to well over 400 Gbps. In Figure 9.8b, as the grid is flexible, we observe traffic with any bit rate can be accommodated within multiple grid width of 12.5 GHz with better spectrum utilisation.

To address the challenge of the present day needs of the traffic requirement one needs flexible and adaptive networks with flexible grid spectrum of the fiber and equipped with transceivers and network elements that can adapt to the actual traffic needs. A network with combination of adaptive transceivers, flexible grid, and intelligent network and client nodes is called an EON [1–3]. With EON the service providers will be able to frequently change the network according to the needs of the clients with ease. With this kind of network flexibility, redesigning and adding of required features in the network blocks will be possible with minimal changes. For example, if the transceivers are able to transparently transmit signals with selectable modulation formats, adaptive rates, they will be able to allocate the capacity into one or several independent optical flows and transmit toward one or multiple destinations. Similarly, the OXC have flexible dynamic optical functions to be able to connect in any directions and wavelength by employing signal processing and routing [17–23].

The EONs thus alleviate the limitations of fixed-grid networks. It provides the support of various data rates, including the possible future ones in a highly spectrum-efficient manner. To summarize, migration from fixed-grid networks to EONs provides several benefits, such as the following:

- *Increased spectral efficiency.* By enabling optimization of the spectral allocation for each request, the flexible ITU channel grid increases *spectral efficiency* and the network capacity significantly.
- *Sub- to super-channel accommodation of client signals.* Client signal bit rates of 1 Tbps and beyond can be accommodated in flex-grid architectures with possibility also for sub-wavelength traffic.
- *Dynamic reconfiguration.* EONs enable dynamic reconfiguration of the network by using flex-grid optical channels, bandwidth variable transmitters, OXC which are independent of wavelength, port direction, and number as the basic building blocks.
- *Economical.* A reduction in terms of capital expenditure is achieved by equipment and device integration.

9.5   Elastic Optical Network Elements

To make the EON flexible, several network elements with requisite flexibility are required for generation, switching, and routing of the lightpaths. These elements are discussed in brief below.

9.5.1   Flexible-Grid Fiber

The conventional ITU grid which is based on 193.1 THz supports fixed channel spacing of 25, 50, and 100 GHz. In WDM optical networks, a fixed central frequency is assigned to an optical channel depending on the channel spacing, starting from 193.1 THz for the C-band. The standard ITU grid has built-in guard bands between each optical channel which waste up to 25% of fiber capacity.

For flexible grid, the allowed frequency slots have a nominal central frequency in THz defined by $193.1 + n \times 0.00625$ (n is a positive or negative integer including 0). A *slot* width is defined by $(12.5\,\text{GHz} \times m)$, with m as a positive integer, (ITU Recommendation G.694.1). Depending on the demand, a subset of the possible slot widths and central frequency can be selected. The flexible channel plan is a 12.5 GHz grid pattern. In a flex-grid optical fiber, two consecutive *slices* of 6.25 GHz each define a *slot* and the *central frequency* (cf) defines where the assigned spectrum is centered and thus it allows positioning the slots anywhere within the whole optical spectrum. Figure 9.9 shows the flex-grid optical spectrum with the details of slice and slot as explained above.

The flex-grid supports mixed channel sizes in increments of (m × 12.5 GHz) and can easily accommodate 100 Gbps, 200 Gbps, 400 Gbps, and 1 Tbps optical rates services. As mentioned earlier, WDM networks currently transmit 10 Gbps, 40 Gbps, and 100 Gbps optical signal on a single optical carrier that fits within a standard 50 GHz channel. In order to transmit higher data rates, say above 100 Gbps, several optical subcarriers will be required to accommodate these rates. For example, to carry 400 Gbps signal in the case of fixed grid, 8 WDM carriers of 50 GHz spacing will be required if the subcarriers were transported on independent 50 GHz channels. In the case of flex-grid, the higher data-rate signals will be transmitted as a *super-channel* which is one entity and is provisioned, transmitted, and switched across the network as a single block, as shown in Figure 9.8b.

There is significant spectrum saving with flex-grid transmission by having minimal guard bands and flexible width for each demand. Figure 9.10 compares the spectrum utilized by five mixed bit rates' demands of 100 Gbps and 400 Gbps with two modulation schemes of 16 QAM and QPSK in the case of fixed grid and flex-grid and shows significant spectrum saving advantage with flex-grid.

There can be significant spectrum efficiency increase with advance modulation techniques giving even larger benefits for EONs. The efficiency improvement will always be there, but it can be substantially different depending on the exact traffic scenario. Table 9.1 shows the spectral efficiency gain in a few of the cases for flex-grid over the fixed grid for QPSK and QAM schemes.

The modulation techniques standardized in the optical fiber transmission are the DP-QPSK using four subcarriers and DP-16QAM with two subcarriers. Each modulation scheme is optimized for different network applications and

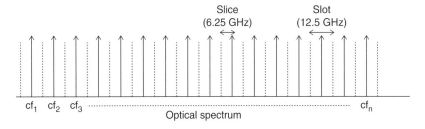

Figure 9.9 Flex-grid optical spectrum.

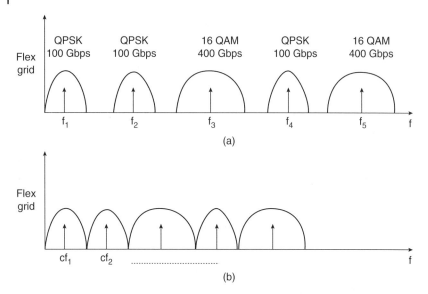

Figure 9.10 Total spectrum utilization for five demands of mixed bit rates of 100 Gbps and 400 Gbps for two modulation schemes of 16-QAM and QPSK with (a) fixed grid, and (b) flex-grid.

Table 9.1 Efficiency improvement for flexible spectrum assuming a 50 GHz grid for fixed DWDM and 10 GHz channel guard-band for EONs [1].

Bit Rate (Gbps)	Modulation format	Fixed-grid (50 GHz channel spacing)	Elastic BW with guard-band in GHz	Increase in spectral efficiency with EON
100	DP-QPSK	1 channel	37.5 + 10	5%
100	DP-16QAM	1 channel	25 + 10	43%
400	DP-QPSK	4 channels, each with 100 Gbps	75 + 10	135%
400	DP-16QAM	2 channels, each with 200 Gbps	75 + 10	17%

reach. For example, the 4×100 Gbps DP-QPSK approach is better suited to long-haul networks because of lower BER with better SNR, while the 2×200 Gbps DP-16QAM is used for metro distances with higher data rate for smaller reach for 400 Gbps.

9.5.2   Bandwidth Variable Transponder

To provide the flexibility required in EONs, their architecture requires special equipment with flexibility and programmability in comparison with the generic WDM optical network topology. The transponders used for EON are the bandwidth-variable transponders (BVTs) to transmit or receive data signal with adjustable rates or spectral bandwidth, and modulation formats. These BVTs are capable of dynamically tuning data rates or optical reach according to the network state, typically by adjusting modulation formats, number of sub-carriers in the case of OFDM, or baud rates on flexible elastic optical paths (EOPs). However, the single traffic flow BVT has a limitation when the flow is of low bit rate, then part of its capacity is wasted. This issue can be solved with elastic bit rate, multi-flow, sliceable bandwidth variable transponder (SBVT) approach as shown in Figure 9.11. When a SBVT is used to generate a low-bit rate channel, its idle capacity can be exploited for transmitting other independent data flows. The SBVT architecture [17, 19–24] was introduced in order to support sliceability, multiple bit rates, multiple modulation formats, and adaptive code rates flows. This enables the optical flows to be aggregated or sliced based on the traffic needs saving the fiber spectrum and other resources.

A functional block diagram of a sliceable OFDM BVT is shown in Figure 9.12. It consists of a source for generating multiple equally spaced optical subcarriers, a client demands flow distribution module, a set of multiple flexible subcarrier module for modulation and an optical multiplexer. The subcarriers are generated by either a single multi-wavelength optical source or multiple lasers sources, one per subcarrier. Different modulators in the OFDM transmitter

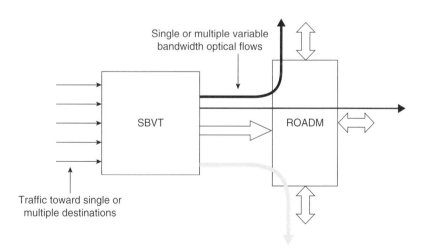

Figure 9.11 Concept of sliceable bandwidth variable transponder.

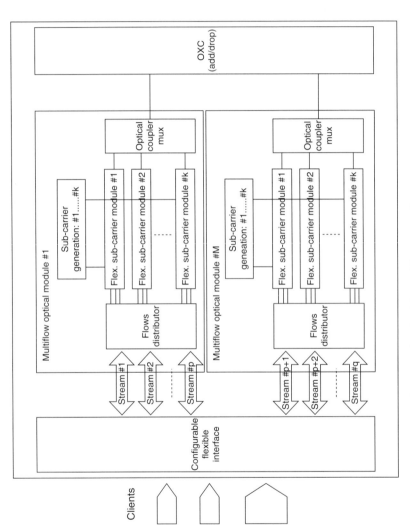

Figure 9.12 Architecture of sliceable bandwidth variable transponder.

generate many low bandwidth subcarriers to generate high speed data signal. In this configuration each flexible subcarrier module is utilized as a single carrier transponder that can generate various modulated signals, such as 16-QAM or QPSK. The idea is by *slicing* a single BVT several *virtual transceivers* are formed that serve separate EOP. With this there is a complete capacity utilization of 400 Gbps or 1 Tbps BVT by combining many EOPs of variable bit rates of 100, 150 Gbps, etc. If the demand is for large multiple channels they can be bundled into a single super-channel and carried as a single EOP and the demand will be switched as one entity in OXC and managed as a single entity in management systems. Subcarriers can also be sliced and directed to specific output ports according to the traffic needs. Thus, the number of managed entities and size of OXC get reduced with limited number of add and drop independent EOPs.

9.5.3 Flexible Spectrum Selective Switches

In EONs the bandwidth variable WSS(BV-WSS) are used to switch optical routing paths having different bandwidth. Flexible wavelength spectrum selective switches (flex WSSs) can switch variable spectral bands (see Figure 9.13) and are one of the key enabling technologies for the EON. The recent WSS technologies allow switching arbitrary spectrum slices in 3.125~6.25 GHz steps, enabling elastic reconfigurable optical add drop multiplexers (ROADM)s. These devices are based on one of several technologies, optical MEMS, LCoS, or silica planar lightwave circuits (PLCs) which can be employed as switching elements to realize an OXC with flexible bandwidth and center frequency. LCoS components are used to control the phase of light at each pixel to produce a programmable grating and beam deflection. The channel bandwidth is software configurable by selecting different numbers of pixels [18, 25].

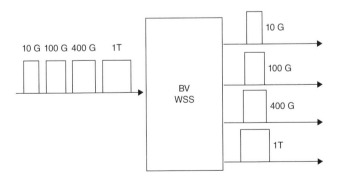

Figure 9.13 Bandwidth variable wavelength selective switch.

9.5.4 Reconfigurable Optical Add Drop Multiplexers

The ROADMs in fixed WDM system have fixed add/drop pairs, with a fixed relationship between the transponder and the direction of transmission, but have no other flexibility. In the flex-grid ROADMs, there is much more flexibility. To enable full flexibility in EONs the CDC [1, 2, 25] (colorless, directionless, and contention-less) ROADMs are used. *Colorless* technology in ROADMs means that any wavelength can be assigned to any port, instead of fixed wavelength to port relationship. This means that any wavelength can be dropped to any one of several client interfaces. *Directionless* technology further eliminates the fixed direction to port relationship, allowing any signal on any one of several ports to be dropped to any one of several client interfaces. Finally, *contention-less* technology eliminates the possibility of wavelength blocking within the ROADM due to the same wavelength being utilized on several trunks. For providing flexible grid-spacing there is the CDC grid-less (CDCG) ROADM architectures shown in Figure 9.14. The BV-WSS send variable sizes data rates which can then be switched to any port without any contention inside the switch now.

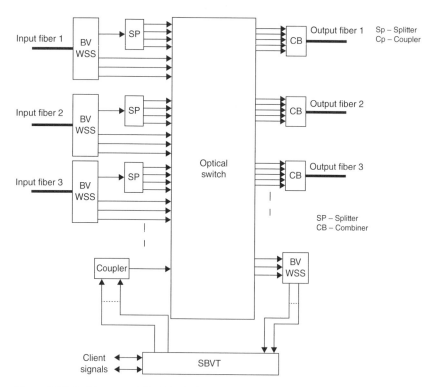

Figure 9.14 Architecture of CDCG (colorless, directionless, contentionless, and grid-less) ROADM.

Great deal of flexibility is added to ROADMs with grid-less channel spacing for EONs. It is possible to reassign or re-route any wavelength automatically from a network management system. All this flexibility in the CDCG ROADMs though adds more complexity to the architecture and adds to the cost.

9.6   Routing and Spectrum Assignment Algorithms

The problem of determining a route and wavelength for an optical channel in conventional transparent wavelength-routed optical networks was of the routing and wavelength assignment (RWA) problem with same wavelength maintained along the path. In the case of flex-grid networks, the *lightpath* or the EOP has to be assigned a route and an appropriate spectrum instead. The constraints in this case are the spectrum *contiguity* and *continuity* along the complete EOP route from ingress to egress. Therefore, contiguous spectral allocation for an EOP in EONs introduces another constraint to wavelength-continuity in terms of the available spectrum on each fiber link along the route. This is a called the *routing* and *spectrum* assignment (RSA) problem [26–29]. The RWA algorithms of WDM networks are not directly applicable to the RSA problems, so different algorithms with new constraints have to be developed.

The constraints on EOP of RSA are for its spectrum contiguity and continuity along the path, so what are the spectrum continuity and contiguity constraints? Equivalent to single wavelength constraint of WDM RWA for a connection over all the links in the lightpath of traditional WDM networks, spectrum continuity constraint is in the case of RSA on all the links on an EOP. A connection which requires capacity of several slots has to be assigned a number of contiguous slots to form a super-channel. These contiguous slots have to be routed and switched as a single identity from source to destination on all the links, is the contiguity constraint.

The first issue in planning RSA for EON requires path routing and the requirement of contiguous spectrum allocation. In EON with the flexibility of using adaptive modulation schemes, such as M-QAM and DP-QPSK, one may plan for adaptive modulation levels based on transmission distance or data-rate requirement. The choice of the modulation level also can consider the required QoT (quality of transmission) of the connection. Thus, for given length of an optical path, one can find the higher modulation level over the shorter optical paths and save spectrum or vice versa. Also, the other parameters related to transmitter/receiver characteristics, interference and nonlinear physical layer impairments, can also affect the QoT and thus eventually the modulation level choice. The minimum spectrum spectral resources required by a demand is determined by the required data rate and optical reach. In static case it is not all that complicated to determine the route with minimum spectral allocation. But in a dynamic environment it is quite complicated to determine the solution. In dynamic traffic, the transmission rate of a connection changes with time and there is requirement to provide non-overlapping spectrum to all the dynamic connections for their requested rates. Therefore, there are two issues during spectrum allocation one is the spectrum *fragmentation* which happens due to demands expanding and contracting dynamically leaving insufficient stranded spectrum. The other issue is when at a specific time any connection compresses, the unused spectrum could be shared and used for best-effort traffic, but this spectrum has to be de-allocated when the initially provisioned connection requires it. In dynamic traffic environment connections must expand or contract without affecting the existing traffic but due to fragmentation of the spectrum it is not always possible.

The spectrum required by an optical channel is defined in terms of number of spectrum slots with each slot of fixed width so that the full width of connection can be flexibly adjusted by changing the number of slots. Once the above choices of modulation schemes, other QoT parameters are taken into account and accordingly the number of slots selected, the signal transmitted over the optical path is routed through bandwidth variable OXCs toward the receiver. In this routing process, only the spectrum domain is essential as every BV-OXC on the route allocates a cross-connection with the corresponding spectrum to create an appropriate-sized end-to-end optical path. The BV-OXC then configures its spectral switching window in a contiguous manner according to the spectral width of the incoming optical signal.

9.6.1 Static and Dynamic RSA

9.6.1.1 Static ILP and Heuristic RSA Solutions

The static solution of RSA deals with the routing and resource allocation during the network planning stage, where an a-priori traffic matrix is given in terms of capacity needed, and the routing and spectrum assignment operations are performed offline [30–32].

Similar to the RWA problem, for optimum solution, the static RSA problem can be formulated as an Integer Linear Programming (ILP) problem through a combined routing and spectrum allocation. Much like the RWA, the objective of the ILP is to minimize the utilized spectrum, with the constraints of spectrum continuity along the path and contiguous slots assignment for each connection with no spectrum overlapping among different connections. The RSA problem can also be decomposed into routing and spectrum allocation subproblems and solved separately and sequentially in order to reduce the complexity of the combined RSA ILP. The solutions may not guarantee to be an optimum solution but are near-optimum solutions.

To solve the RSA problem efficiently for larger networks, several heuristic algorithms have been proposed to serve connection requests in static or dynamic scenarios. The heuristic algorithms can be designed as two-step or one-step approaches.

Two-Step Approach The RSA problem can be partitioned into routing and spectrum assignment sub-problems and solved sequentially. For routing, pre-calculated k shortest paths are used or some other metrics, such as load balanced routing to minimize the spectrum usage in the network, similar to the routing in the fixed WDM networks. Once the routing path is decided then spectrum allocation can be done using one of the following schemes:

- *First-fit.* Similar to the first-fit policy for wavelength assignment in WDM networks, in the case of flex-grid, first all spectrum slots are numbered. Then with the pre-calculated k shortest paths, starting from the shortest route to the longest one among the paths, the first-fit algorithm searches for the required consecutive slots in ascending order of the spectrum slot index and selects the first found route and slots for the connection request.
- *Last-fit.* All spectrum slots are numbered. Like the first-fit with the pre-calculated k shortest paths, the last-fit algorithm searches for the required consecutive slots in descending order of the spectrum slot index and selects the first found route and slots for the connection request.
- *Lowest starting slot.* This is void filling algorithm, which utilizes the stranded spectrum of size greater than the requested slots [33]. This algorithm searches for the first consecutive slots feasible for the new request in ascending order of the slot index and selects the path with the lowest starting slot among the set of candidate paths. The algorithm has better spectrum utilization because of its void-filling capability.

Finally, ordering of the traffic demands for the given traffic matrix is of key importance as different orderings may result in different spectrum utilization. Several ordering policies are proposed, such as *highest-spectrum demand-first* ordering, which orders the connection demands in decreasing order of their requested bandwidth and serves the connection with the highest bandwidth first, and *longest-path-first* ordering, which arranges connection demands in descending order of the number of links along their shortest path's use and serves the connection that has the longest path first, etc.

One-Step Approach The commonly used one-step approach is the modified Dijkstra's shortest path, implemented by checking the available spectrum in the Dijkstra's shortest-path algorithm. Another proposed algorithm is spectrum-constraint path vector searching, which builds a path-vector tree with spectrum constraint to search the global optimal route.

Distance-Adaptive Spectrum Allocation and Modulation For efficient utilization of the spectrum in EON, unlike the WDM system, all optical paths are not assigned the same spectrum width. The spectrum is assigned considering the transmission distance of each path. The distance-adaptive spectrum allocation concept can be introduced in EON, based on adaptive modulation and bit-rate variable transponders. For longer distances, a low-level modulation format with wider spectrum are allocated and a high-level modulation format with narrower spectrum to shorter distance paths using different modulation formats. For example, for the same data rate, 16-QAM carries twice the number of bits per symbol of QPSK, and consequently requires half the spectrum bandwidth, while its OSNR tolerance is lower than QPSK, meaning a shorter distance reach. The distance-adaptive spectrum allocation scheme enables spectrum efficiency improvement, as the assigned spectral bandwidth can be saved for shorter paths by increasing the number of modulated bits per symbol [34].

9.6.1.2 RSA for Time-Varying Traffic

There are several policies which can be adopted for time-varying traffic RSA. One can do the network planning of the time-varying traffic models by spectrum reservation and resource sharing among different connections. If connections have complementary transmission rates in time then they could be served by shared spectrum slots. An alternative approach is to assign each connection a guaranteed transmission rate, and a probabilistic model for exceeding this rate, where spectrum reservation and sharing between connections could be performed. Another scheme could be flexible spectrum zoning for different data-rates traffic [35–37].

One of the important considerations of RSA in EON is to take into consideration fragmentation of the optical spectrum while assigning the spectrum to different demands. Spectrum fragmentation refers to the existence of non-aligned, isolated, and small-sized blocks of spectral segments in EONs due to setting up and tearing down connections frequently in dynamic traffic scenario. Spectrum fragmentation can be caused by two factors during RSA: one, when allocating a block of contiguous spectrum to a request, we may cut the available spectrum on fiber links into small segments, and the second factor is when RSA leads to the situation that the available spectrum on a certain link is misaligned in the spectrum domain with those on the neighbor links. Spectrum fragmentation leads to low bandwidth utilization and high blocking probability.

To alleviate spectrum fragmentation, one can either use preemptive or proactive algorithms. The preemptive RSA algorithms are designed with the objective to minimize spectrum fragmentation during setting up of demands. These are the fragmentation-aware RSA algorithms. The objective of fragmentation-aware RSA is to minimize the above mentioned two factors. The proactive algorithms with spectrum defragmentation involve the rerouting and retuning of some existing connections on a fly. The spectrum defragmentation consolidates the available spectrum fragments and vacates spectrum resources over the path for future requests.

Network planning of the time-varying traffic models can be done by spectrum reservation and resource sharing among different connections. [35–39]

9.6.1.3 Network Defragmentation RSA

Under a dynamic traffic scenario, connection setup and release may lead to fragmentation throughout the network, by separating the available spectrum into small non-contiguous spectrum bands. This problem will increase the blocking probability of incoming connection requests, as there may be insufficient contiguous spectrum available. Also, with time the allocated optical routes and spectrum might not be optimal in terms of spectral utilization. Therefore, it is desirable for network operators to periodically reconfigure the optical paths and spectrum which is referred as network defragmentation [40–45]. The required mechanism will remove stranded fragments of spectrum between EOPs and redistribute them to allow for further growth.

Defragmentation in an EON requires the optical path to be reconfigured as a whole because of the continuity constraint over the links. Since defragmentation is performed in a live network, the key requirement is that it should be hitless so that existing services are not impacted. The defragmentation algorithm should give a step-by-step transition sequence to migrate the connections and minimize the number of moved connections to reduce service disruption. The defragmentation algorithm should incorporate metrics, such as blocking rate, resource utilization, network capacity, etc. An ILP formulation and a few heuristic algorithms, namely, the Greedy-Defragmentation algorithm and the Shortest-Path-Defragmentation algorithm were also provided for large-scale networks, with the objective of maximizing the consolidation of the spectrum as well as minimizing the number of service interruptions. There are several proposed algorithms to select contiguous *spectrum slots* from an unused spectral resource pool to minimize spectrum fragmentation and to retain as much as possible contiguous spectrum for future utilization [46–49].

In the past few years, considerable advancement been made in EON. Some field trial of EON transmission has demonstrated over 620 km distance with 10G/40G/100G/555G with defragmentation [50], and an experimental setup of EONs for data center applications has also been successfully demonstrated [51].

9.7 Network Control and Management

The functions of network management plane are path provisioning, performance monitoring and fault detection etc. In the case of elastic networks, high flexibility in dynamic bandwidth provisioning and bandwidth adjustment are the very important issues for advanced network control and management system to provide. The control plane signaling and routing protocols have to be extended to support the spectrum slot specification, distribution of the available spectrum information, bandwidth adjustment, recovery operations, etc. The control plane protocol extensions for traffic engineering (TE), survivability, etc. will have to be incorporated in the RSA [52, 53].

There are a few architectural choices for the EON control plane, which include the generalized multiprotocol label-switching (GMPLS) and path computation element (PCE), which we discussed for the DWDM system in Chapter 7. There is also the emerging software-defined networking (SDN) control which can be used [53]. This is given in detail in the next chapter.

9.8 Summary

The rapid growth of emerging applications with high dynamic Internet traffic requires a high-capacity, cost-effective, agile, reconfigurable, and energy-efficient optical network architecture for the future. To achieve these, a novel EON architecture with immense flexibility and scalability on spectrum allocation, data-rate accommodation, and reconfigurability is envisaged with prospects to build for any future applications.

The chapter gives the different modulation and OFDM multi-carrier schemes necessary to provide the required flexibility in the EONs. Bandwidth-efficient advance modulation M-QAM and DP-QPSK schemes with optical OFDM, which are the promising technologies for high-speed transmission and adaptability based on its subcarrier multiplexing technology, are discussed. Also presented in detail are the flexible-grid optical spectrum for transmission of traffic from sub-lambda to terabits rate, SBVT, and flexible switching nodes to have complete transparency and flexibility. Along with the above data layer structure, suitable dynamic routing and spectrum assignment algorithms, network planning, and traffic engineering are taken up.

Although a lot of effort has been made to solve these problems, there are still many outstanding issues in most of the above-mentioned areas and more research is needed to realize the full potential of this novel EON architecture.

References

1 Ori Gerstel, Masahiko Jinno, Andrew Lord, S. J. Ben Yoo, 'Elastic Optical Networking: A New Dawn for the Optical Layer? IEEE Communications Magazine, February 2012 , S12-S20
2 MasahikoJinno, Hidehiko Takara, Bartlomiej Kozicki, Yukio Tsukishima, Yoshiaki Sone, and Shinji Matsuoka, Spectrum-Efficient and Scalable Elastic Optical Path Network: Architecture, Benefits, and Enabling Technologies, IEEE Communications Magazine, February 2009 , 66-73

3 Couch, L.W. II (2007). *Digital and Analog Communication Systems*, 7e. Upper Saddle River, NJ: Prentice Hall.

4 Zacharapoulos, I., Tzanakaki, A., Parcharidou, D., and Tomkos, I. (2005). Optimization study of advanced modulation formats for 10 Gbs metropolitan networks. *J. Lightwave Technol.* 23: 321–329.

5 Savory, S.J. (2008). Digital filters fir coherent optical receivers. *Opt. Express* 16 (2): 804–817.

6 Zhou, X. and Xie, C. (2016). *Enabling Technologies for High Spectral-Efficiency Coherent Optical Communication Networks*. NJ: Wiley.

7 Multilevel Modulation Formats Push Capacities Beyond 100 Gbits/sec," Shubhashish, Data, and Crawford, *Laser Focus World, February, 2012*, pp. 58–63 https://www.laserfocusworld.com/.../ multilevel-modulation-formats-push-capacities-be…

8 Shieh, W. et al. (2008). Coherent optical OFDM: theory and design. *Opt. Express* 16: 841–859.

9 Armstrong, J. (2009). OFDM for optical communications. *J. Lightwave Technol.* 27 (3): 189–204.

10 Frascella, P. et al. (2010). Unrepeatered field transmission of 2 Tbit/s multi-banded coherent WDM over 124 km of installed SMF. *Opt. Express* 18: 24745–24752.

11 Geisler, D.J. et al. (2011). Bandwidth scalable, coherent transmitter based on the parallel synthesis of multiple spectral slices using optical arbitrary waveform generation. *Opt. Express* 19: 8242–8253.

12 Hanzo, L., Munster, M., Choi, B.J., and Keller, T. (2003). *OFDM and MC-CDMA for Broadband Multi-User Communications, WLANs and Broadcasting*. New York: Wiley.

13 Hara, S. and Prasad, R. (2003). *Multicarrier Techniques for 4G Mobile Communications*. Boston: Artech House.

14 John, A. C. Bingham. "Multicarrier Modulation for Data Transmission: An Idea Whose Time Has Come", IEEE Communications Magazine, May 1990, pp. 5–14.

15 Han, S.H. and Lee, J.H. (2005). An overview of peak-to-average power ratio reduction techniques for multicarrier transmission. *Wireless Commun. IEEE* 12 (2): 56–65.

16 Lee, K., Thai, C.D.T., and Rhee, J.-K. (2008). All optical discrete Fourier transform processor for 100 Gbps OFDM transmission. *Opt. Express* 16 (6).

17 Sambo, N., Castoldi, P., D'Errico, A. et al. (2015). Next generation sliceable bandwidth variable transponders. *IEEE Commun. Mag.* 53 (2): 163–171.

18 Napoli, A., Bohn, M., Rafique, D. et al. (2015). Next generation elastic optical networks: the vision of the European research project IDEALIST. *IEEE Commun. Mag.* 53 (2): 152–162.

19 H. Takara, T. Goh, K. Shibahara, K. Yonenaga, S. Kawai, and M. Jinno, "Experimental demonstration of 400 Gb/s multi-flow, multi-rate, multi-reach optical transmitter for efficient elastic spectral routing," *in Proc. ECOC,* Sep. 2011.

20 J. Fabrega, M. SvalutoMoreolo, F. Vflchez, B. Rofoee, Y. Ou, N. Amaya, G. Zervas, D. Simeonidou, Y. Yoshida, and K. Kitayama, "Experimental demonstration of elastic optical networking utilizing time-sliceable bitrate variable OFDM transceiver," in *Tech. Dig. OFC/NFOEC,* Mar. 2014.

21 Zhang, J., Ji, Y., Song, M. et al. (2015). Dynamic traffic grooming in sliceable bandwidth-variable transponder-enabled elastic optical networks. *J. Lightwave Technol.* 33 (1): 183–191.

22 Sambo, N., D'Errico, A., Porzi, C. et al. (2014). Sliceable transponder architecture including multiwavelength source. *J. Opt. Commun. Networking* 6 (7): 590–600.

23 Zhang, J., Zhao, Y., Yu, X. et al. (2015). Energy-efficient traffic grooming in sliceable transponder-equipped IP-over-elastic optical networks [invited]. *J. Opt. Commun. Networking* 7 (1): A142–A152.

24 Jinno, M., Takara, H., Sone, Y. et al. (2012). Multiflow optical transponder for efficient multilayer optical networking. *IEEE Commun. Mag.* 50 (5): 56–65.

25 S. Poole, S. Frisken, M. Roelens, and C. Cameron, "Bandwidth flexible ROADMs as network elements," *in Tech. Dig. OFC/NFOEC,* 2011.

26 Klinkowski, M. and Walkowiak, K. (2011). Routing and spectrum assignment in spectrum sliced elastic optical path network. *IEEE Commun. Lett.* 15 (8): 884–886.

27 Velasco, L., Castro, A., Ruiz, M., and Junyent, G. (2014). Solving routing and spectrum allocation related optimization problems: from off-line to in-operation flexgrid network planning. *J. Lightwave Technol.* 32 (16): 2780–2795.

28 Shi, W., Zhu, Z., Zhang, M., and Ansari, N. (2013). On the effect of bandwidth fragmentation on blocking probability in elastic optical networks. *IEEE Trans. Commun.* 61 (7): 2970–2978.

29 Olszewski, I. (2017). Improved dynamic routing algorithms in elastic optical networks. *Photon. Network Commun.* 34: 323–333.

30 Christodoulopoulos, K., Tomkos, I., and Varvarigos, E. (2011). Elastic bandwidth allocation in flexible OFDM-based optical networks. *J. Lightwave Technol.* 29 (9): 1354–1366.

31 Jinno, M., Kozicki, B., Takara, H. et al. (2010). Distance-adaptive spectrum resource allocation in spectrum-sliced elastic optical path network. *IEEE Commun. Mag.* 48 (8): 138–145.

32 N. Sambo, G. Meloni, F. Cugini, A. D'Errico, L. Potì, P. Iovanna, and P. Castoldi, "Routing, code, and spectrum assignment (RCSA) in elastic optical networks," in *Tech. Dig. OFC/NFOEC,* 2015.

33 López, V. and Velasco, eds, L. (2016.). Routing and spectrum allocation. In: *Elastic Optical Networks Architectures, Technologies, and Control* (ed. L. Velasco, M. Ruiz, K. Christodoulopoulos, et al.), 55–81. Springer.

34 Abkenar, F.S. and Rahbar, A.G. (2017). Study and analysis of routing and spectrum allocation (RSA) and routing, modulation and spectrum allocation (RMSA) algorithms in elastic optical networks (EONs). *Opt. Switching Networking* 23: 5–39.

35 X. Wan, L.Wang, N. Hua, H. Zhang, and X. Zheng , 'Dynamic Routing and Spectrum Assignment in Flexible Optical Path Networks' *In Proc. of the Optical Fiber Communication Conference (OFC 2011)*, Anaheim (USA), JWA55, March 2011.

36 André C. S. Donza, Carlos R. L. Francês, João C. W. A. Spectrum Allocation Policies in Fragmentation Aware and Balanced Load Routing for Elastic Optical Networks, *International Conference on Digital Society Costa*, ICDS 2015. https://www.thinkmind.org/download.php?articleid=icds_2015_2_40_10126

37 T. Takagi, H. Hasegawa, K. Sato, Y. Sone, B. Kozicki, A. Hirano, and M. Jinno, "Dynamic Routing and Frequency Slot Assignment for Elastic Optical Path Networks that Adopt Distance Adaptive Modulation," *Proc., OFC/NFOEC 2011*, Paper OTuI7.

38 Tessinari, Rodrigo S., et al. "Zone based spectrum assignment in elastic optical networks: a fairness approach." IEEE *Opto-Electronics and Communications Conference (OECC)*, 2015.

39 M. Dallaglio, A. Giorgetti, N. Sambo, and P. Castoldi, "Impact of S-BVTS based on multi-wavelength source during provisioning and restoration in elastic optical networks," in Proc. of *European Conference on Optical Communication*(ECOC) , 2014.

40 Wan, X., Hua, N., and Zheng, X. (2012). Dynamic routing and spectrum assignment in spectrum-flexible transparent optical networks. *J. Opt. Commun. Networking* 4 (8): 603–613.

41 Beyranvand, H. and Salehi, J. (2013). A quality-of-transmission aware dynamic routing and spectrum assignment scheme for future elastic optical networks. *J. Lightwave Technol.* 31 (18): 3043–3054.

42 André C. S. Donza, Carlos R. L. Francês, J. W. A. Costa, 'Spectrum Allocation Policies in Fragmentation Aware and Balanced Load Routing for Elastic Optical Networks', *International Conference on Digital Society*, pp. 46–51, 2015

43 Yin, Y., Zhang, H., Zhang, M. et al. (2013). Spectral and spatial 2D fragmentation-aware routing and spectrum assignment algorithms in elastic optical networks [invited]. *J. Opt. Commun. Networking* 5 (10): A100–A106.

44 M. Zhang et al., "Bandwidth defragmentation in dynamic elastic optical networks with minimum traffic disruptions," in Proc. of ICC 2013, pp. 3894–3898, 2013.

45 Lu, W. and Zhu, Z. (2013). Dynamic service provisioning of advance reservation requests in elastic optical networks. *J. Lightwave Technol.* 31: 1621–1627.

46 Khodashenas, P., Comellas, J., Spadaro, S. et al. (2014). Using spectrum fragmentation to better allocate time varying connections in elastic optical networks. *J. Opt. Commun. Networking* 6 (5): 433–440.

47 Castro, A., Velasco, L., Ruiz, M. et al. (2012). Dynamic routing and spectrum (re)allocation in future flexgrid optical networks. *Comp. Networks* 56 (12): 2869–2883.

48 Pages, A., Perello, J., Spadaro, S., and Comellas, J. (2014). Optimal route, spectrum, and modulation level assignment in split spectrum-enabled dynamic elastic optical networks. *J. Opt. Commun. Networking* 6 (2): 114–126.

49 Velasco, L. and Ruiz, M. (2017). *Provisioning, Recovery, and In-Operation Planning in Elastic Optical Networks*. Wiley.

50 N. Amaya et al., "Gridless Optical Networking Field Trial: Flexible Spectrum Switching, Defragmentation and Transport of 10G/40G/100G/555G over 620-km Field Fiber," in Proc. of *European Conference on Optical Communication* (ECOC '11), Geneva, Switzerland, 2011.

51 Zhang, J., Yang, H., Zhao, Y. et al. (2013). Experimental demonstration of elastic optical networks based on enhanced software defined networking (eSDN) for data center application. *Opt. Express* 21 (22): 26990–27002.

52 Lopez, V., Jimenez, R., de Dios, O.G., and Fernandez, J.P. (2018). Control plane architectures for elastic optical networks. *IEEE/OSA J. Opt. Commun. Networking* 10 (2).

53 R. Muñoz, R. Casellas, R. Martínez, and R. Vilalta, "Control plane solutions for dynamic and adaptive Flexi Grid optical networks," in Proceedings of *European Conference on Optical Communication* (ECOC 2013), paper We.3.E.1, 2013

10

Software-Defined Optical Networks

10.1 Introduction

In the last decade there has been significantly increased utilization of Internet-based applications such as streaming, multimedia, social media, sharing, and other emerging dynamic high-bandwidth applications, including data center (DC) networking, cloud computing, etc. Global internet traffic will reach 3.3 zetabytes/year at the end of 2021, up from 1.2 zetabytes/year at the end of 2016 [1], and with Gartner indicating the market for Internet of Things (IoTs) devices to reach nearly 21 billion connected devices by 2020 [2], will push the traffic further. Besides the capacity requirement, the dynamic nature of this traffic, bandwidth on demand (BoD), etc. needs agility in the next-generation future optical communication systems, as they will finally carry this traffic in the backbone, metro, and even access part of the network.

The current networks are being substantially redesigned. For example, in the case of cloud computing, the services provided are dependent on the cloud's physical infrastructure comprising of not only computing, storage, IT resources, but also significantly the network interconnecting these infrastructures together and with the users. Furthermore, in order for users to utilize cloud-computing services, data centers (DC) platforms, they need to be integrated with the operator's network infrastructures. Therefore, transport optical network and optical switching technologies have to include BoD, latency, agility, and programming of the network for the different traffic flows of various applications in traffic engineering (TE).

In previous chapters we have studied the vastly laid DWDM networks. These networks are reasonably advanced but their networking functions are not agile and they have difficulty in meeting present communication trends. The client need is not only for large bandwidth and highly dynamic networks, they also require BoD, access on demand to software and storage, infrastructure for cloud computing, and *big data* applications. At the same time, with more computing and storage resources placed remotely in the cloud, efficient access

Optical WDM Networks: From Static to Elastic Networks, First Edition. Devi Chadha.
© 2019 John Wiley & Sons Ltd. Published 2019 by John Wiley & Sons Ltd.

to these resources via a network is becoming critical to fulfill today's computing needs. There is a paradigm shift in the new communication trends and they call for an evolution in existing networking design. Having discussed the applications and technologies which are bringing this change, the dynamic control plane (CP) with generalized multiprotocol label-switching (GMPLS) and the elastic optical networks in the previous chapters, now we discuss the concept of software-defined optical networks which is signaling a whole new shift in the optical networking paradigm.

As the emerging applications are highly dynamic, managing such traffic requires a programmable flexible network with bandwidth management. But the current network cannot be programmed easily due to the fact that control of the devices is integrated with the individual network element (NE), which makes it very difficult to program and then change network functions dynamically. We can overcome this challenge with a software-defined optical networking (SDON) which can have a programmable control and data plane (DP) that can manage to change different network functions and protocols according to the applications. A software-defined network can thus be defined as the network in which the connections can be dynamically provisioned and reconfigured and the network can have complete topological flexibility and scalability. These features, in general, can be achieved in the presently prevailing any software-defined networking (SDN) [3, 4] circuits, which has a framework of automatic and dynamic management for multiple NEs. SDN makes networks programmable and less expensive to build, operate, and also upgrade. There is virtualization of different NEs and services with their characterization in terms of software. This provides greater flexibility and dynamism in different types of communications networks with the addition of new and innovative features. Enabling such features is also possible in the optical networks with the recent advances in flexible and programmable device technologies, as discussed in Chapter 9, such as S-BVTs (sliceable bandwidth variable transceivers), CDC-ROADMs (colorless, directionless and contention-less reconfigurable optical add drop multiplexers), elastic wavelength grid of the fiber, coupled with emergence of SDN paradigms, such as *OpenFlow* [5], etc., to be discussed later. The physical infrastructure, NEs are virtualized in software and can be used by the applications and network services directly. To fully exploit the hardware agility in the NEs, network operators need to have the ability to efficiently, reliably, and dynamically reconfigure both the wavelength and route of optical channels through the network entirely through software management without any manual intervention. This allows the user to define, address, and manipulate many new features on the logical maps of the network, creating multiple virtual networks (VNs) independent of underlying transport technology and network protocols.

First in Section 10.2 we describe the general architecture of a software-defined network before discussing SDON in Section 10.3, for better appreciation of its features. The functions of the three planes and those of the different layers within them are explained. The data layer and network function virtualizations and other features have been discussed and how these have been done for SDON have been introduced along with the ongoing work in the field.

10.2   Software-Defined Networking

Software-defined networking (SDN), as defined by the Internet Engineering Task Force (IETF) [4–7], is a networking paradigm which enables the programmability of networks. This is achieved by using white boxes network hardware made from off-the-shelf devices which does several standard packet-processing functions and decouple the software that controls the network from the devices that implement it.

The SDN achieves a high potential of flexibility and responsiveness because it fundamentally has (i) a complete separation of the *control plane* from the *forwarding plane* – control functionality is removed from NEs, they are just simple packet forwarding elements; (ii) has a *logically* centralized control, and has the network intelligence and control centralized in *software-based* SDN controllers that keep a holistic view of the network. The SDN controller maintains a global network map, which then makes it simpler to change routing and switching strategically for different applications, quality of service (QoS), balance traffic load, etc., and then decide on global optimization of the network resources; (iii) has network function virtualization (NFV) of the network equipment. The network equipment thus consists of commercial-off-the-shelf servers and storage devices. As there is no requirement for function-specific hardware, therefore, the equipment cost will be much less. Network functions are not embedded firmware of various network devices, but are software running on commodity hardware in a cloud, eliminating the dependency on dedicated vendor-specific hardware.

A simplified layered architecture of SDN is shown in Figure 10.1. This illustrates the different layers in the three planes of data, control, and application of the SDN. The data plane consists of infrastructure and south bound interface (SBI) layers, also referred to as the data-controller plane interface (D-CPI); the control plane consists of the network hypervisor which runs virtual machines (VM), network operating system (NOS) and the north bound interface (NBI), also known as the application-controller plane interface (A-CPI) layers; and finally, the management plane consists of network applications. Each layer operates independently, allowing multiple solutions to coexist within each layer.

SDN has three abstraction features to operate on: (i) forwarding, (ii) state distribution, and (iii) specification. The architecture and the abstractions of SDN are shown in Figure 10.2. Any data-forwarding function desired by the

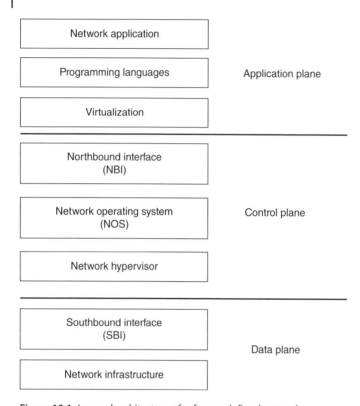

Network application	
Programming languages	Application plane
Virtualization	
Northbound interface (NBI)	
Network operating system (NOS)	Control plane
Network hypervisor	
Southbound interface (SBI)	Data plane
Network infrastructure	

Figure 10.1 Layered architecture of software-defined networks.

control plane is to be carried out by the *forwarding abstraction* while hiding details of the underlying hardware. *OpenFlow* [8] is one of the popular abstraction communication interfaces defined between the control and forwarding planes, which works like a device driver in an operating system of a computer system. Next, SDN has distributed control, but with help of the *state distribution abstraction* it shields the control mechanisms from the difficulties faced with distributed control state. The state distribution abstraction allows the access through a *logically centralized* control state. This is done by splitting global consensus-based distributed algorithms into a distributed database system and a centralized algorithm. The centralized algorithm is the *global network view abstraction* implemented by the NOS, thus making it a centralized graph algorithm instead of a distributed protocol. NOS is a distributed system that creates and maintains the global network view (GNV). It installs the control commands on the forwarding devices and collects the status information of the forwarding layer in data plane to offer a GNV to the network applications. Finally, the *specification abstraction* allows a network application to express the desired network behavior without being involved in implementing

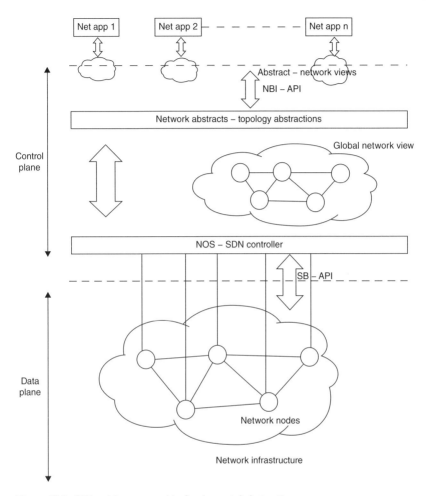

Figure 10.2 SDN architecture and its fundamental abstractions.

the behavior itself. This can be achieved through virtualization solutions and network programming languages. With the help of both of these, the applications express the abstract configurations that is based on a simplified abstract model of the physical configuration of the GNV expressed by the SDN controller. Abstraction models have just enough details to specify goals and does not have to provide information needed to implement these goals.

10.2.1 Functions of SDN Layers

Next, we discuss the functions of different layers in the three planes of SDN in detail.

10.2.1.1 Infrastructure Layer

The infrastructure layer in the data plane (DP) has the data-forwarding devices either in virtual or actual hardware which are interconnected through wireless channels or wired cables. The functional characteristics of NEs are known through the SBI to the control plane (CP), unlike the case of legacy network devices, discussed in Chapter 2, where the control mechanisms are embedded within an NE and have the decision-making capabilities to perform network actions, such as switching or routing. Additionally, as discussed, these forwarding actions in the NEs are autonomously established based on self-evaluated topology information that is often obtained through proprietary vendor-specific algorithms. In contrast, SDN decouples the autonomous control functions, such as forwarding and neighbor discovery algorithms of the network nodes, and moves these control functions out of the infrastructure to a centrally controlled *logical controller*. In doing so, the NEs act only as dumb switches which act upon the instructions of the controller. This decoupling reduces the NE complexity and improves reconfigurability.

There are two main functions of this layer. First, it is responsible for collecting network status, storing them temporally in local devices, and sending them to controllers. The network status includes information of network topology, traffic statistics, and network usages. Second, the infrastructure layer does the processing of packets based on rules provided by a controller.

The forwarding decisions in the data plane are flow-based instead of destination-based. A flow is defined by a set of packet field values acting as a match criterion and set of actions. In SDN, all packets of a flow receive identical service policies at the forwarding devices. Different types of network devices, including routers, switches, firewalls, etc. have the same flow abstraction giving the required flexibility.

10.2.1.2 South Bound Interface

A logical interface that interconnects the SDN controller and the NEs operating on the infrastructure layer is referred to as SBI. SBI is the communication protocol between the forwarding devices and control plane elements which formalizes the way they interact. With SBI each NE in the data plane, such as switches, hosts, or links, etc., is abstracted and represented in its generic form. Through this abstraction, the distributed core of the NOS can maintain the state of the NEs without having to know the specifics of the element, making the NOS core protocol device agnostic. This brings the flexibility in the network of adding new networking technology, accepting interoperability of networks, and allowing the deployment of vendor-agnostic network devices. Thus, the SBI enables the NOS to control or manage multiple diverse devices, even if they use different protocols such as *OpenFlow* or *NetConf*, etc. The main reason for the multiple protocol label switching (MPLS)-based control plane not being commercially deployed in Internet Protocol (IP) networks to control the devices was due to its distributed nature and high complexity, whereas the logically centralized SDN networks are built on top of open and standard interfaces (e.g. *OpenFlow*), which ensures configuration and communication compatibility and interoperability among different data and control plane devices.

OpenFlow [8, 9] protocol is the most commonly used SBI. It is an open standard which is vendor and technology agnostic that allows separation of the data and control plane. It is based on flow switching with the capability to execute software or user-defined flow-based routing, control, and management in the controller. *OpenFlow* can program the entire data plane and can allow virtualization of IP routers, Ethernet switches, wavelength space switches (WSS), optical cross-connect (OXC), etc. Through the appropriate programming of these NEs the network can be logically partitioned into separate virtual networks. This allows each virtual network to use its own set of protocols and policies to support different services.

In SDN *OpenFlow* architecture, there are two main elements: the *OpenFlow* controller and the forwarding devices or hardware, as shown in Figure 10.3. The controller is a software stack running on a commodity hardware platform, and the network devices, as discussed earlier, are the hardware/software elements for packet forwarding. The *OpenFlow* logical switch consists of flow tables and a group table which perform packet forwarding, and the *OpenFlow* channels connect to an external controller. The switch communicates with the controller and in turn the controller manages the switch via the *OpenFlow* switch protocol. Using the OpenFlow switch protocol, the controller can add,

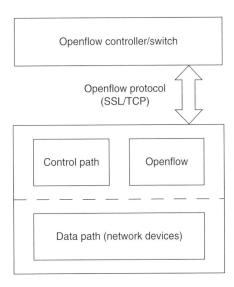

Figure 10.3 *OpenFlow* architecture.

update, and delete flow entries in flow tables, either proactively or reactively due to response from the packets. The *OpenFlow* protocol provides messages to NOS sent by forwarding devices when there is change in their state. The controller collects the flow statistics generated by the forwarding devices. The hardware details of the NEs are abstracted and the device features and capabilities are expressed as concrete device class. For example, in case of SDON it can provide the necessary specifications to represent an optical device class. The *OpticalFlow* API (application plane interface) maps this abstracted information into the *OpticalFlow* protocol and its extensions which are supported by vendor devices.

An *OpenFlow* enabled forwarding device is based on a pipeline of flow tables. Each entry of a flow table has three parts: (i) a matching rule, (ii) actions to be executed on matching packets, and (iii) counters that keep statistics of matching packets. Inside an *OpenFlow* device, a path through a sequence of flow tables defines how packets should be handled. When a new packet arrives, the lookup process starts in the first table entry and ends either with a match in one of the rules of the pipeline, or with a miss when no rule is found for that packet. A flowrule can be defined by combining different matching fields, as illustrated in Figure 10.4. If there is no default rule, the packet will be discarded. Possible actions by the device include; forwarding the packet to outgoing port, encapsulating it and forwarding it to the controller, either dropping it or sending it to the normal processing pipeline. Flow entries may forward to a port. This is usually a physical port, but it may also be a logical port defined by the switch. This high-level and simplified model derived from *OpenFlow* is at present used in the design of SDN data plane devices.

There are many off-the-shelf software switches now available for SDN solutions by several vendors for data centers and virtualized network infrastructures [10]. Various kinds of *OpenFlow*-enabled devices ranging from Gigabit Ethernet (GbE) switches [11, 12] to high-density switch chassis with up to 100 GbE connectivity for edge-to-core applications are moving into the commercial space, as *OpenFlow* provides a common specification to implement OpenFlow-enabled forwarding devices and for the communication channel between data and control plane devices, e.g. switches and controllers.

10.2.1.3 Network Hypervisors

The network hypervisor layer is used to create virtual networks (VNs) to implement new services by abstracting the underlying physical network and launching the required functionality virtually as a software running on the server. In the traditional networks, when any new feature or functionality is to be added, the network operator has to change or install physical hardware. Instead, in the case of SDN with the NFV a network operator can obtain the required new feature immediately at any required place. In SDN architecture, with the non-propriety hardware in the data plane and the various network

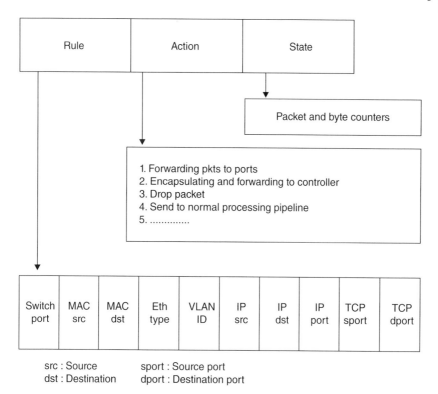

Figure 10.4 Flow table of *OpenFlow*.

functions of these networks, such as routing, content delivery, firewall, load balancing of server, etc., they are deployed by NFV as software on standard non-propriety server hardware. The advantage of the NFV is that it can provide each user to have its own virtual resources, and can allocate resources on-demand from a shared physical infrastructure at a relatively low cost. Also these *virtual machines* can be easily migrated from one physical server to another and can be created and/or destroyed on-demand, enabling to provision any elastic services with the flexible and easy management.

10.2.1.4 Network Operating System

The network intelligence of the SDN network is in the NOS or the controller. It provides a logically centralized control for network management and solving networking problems. The functions of NOS are providing abstractions to facilitate the programming of forwarding devices, providing essential services for the generic functionalities and common APIs to the developers and operators to manage the network. The NOS software platform runs on commodity server technology, unlike the closed proprietary NOSs of the legacy networks.

To efficiently manage the network, the NOS gets the information of the SDN infrastructures, such as flow statistics, topology information, neighbor relations, and link status through SBI. The NOS then provides advanced capabilities, such as virtualization, application scheduling, and database management, while supporting the SDN operations. The logical design of the controller has two counter-directional information flows, as illustrated in Figure 10.5. In the downward information flow, the controller translates the application policy of the management plane into packet forwarding rules as the network status. This process ensures validity and consistency of the forwarding rules. For the upward information flow, the controller synchronizes the network status collected from the infrastructure monitoring for the networking decision making. With the above architecture of the controllers, their logical design can be decoupled into four building process components: high-level language, rule update process, network status collection process, and network status synchronization process. These are explained further below.

1) *High-level language.* One of the key controller functions is to translate application requirements from the application layer into packet forwarding rules, which forms the communication protocol between the application layer and the control layer. There can be several high-level programming languages which can be used for this purpose. These high-level languages should be able to easily describe the network management strategies and their requirements. There are the earlier C++, Java and Python and other

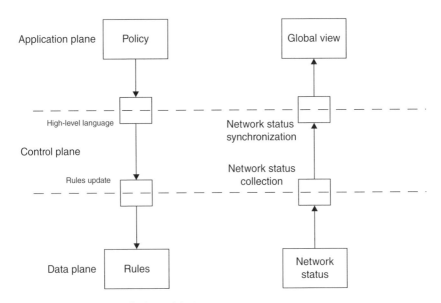

Figure 10.5 SDN controller logical design.

more specific for SDN applications, such as Nettle, Maple, and several others [13, 14].

2) *Rules update.* An SDN controller is responsible for generating packet forwarding rules and then installing them into network devices for operation. At the same time, forwarding rules in network devices need to be updated because of configuration changes and dynamic control.

3) *Network status collection.* In the upward flow the controllers collect network status to build a global view of an entire network and provide it to the application layer. The main network status is the traffic statistics, such as traffic type, duration time, packet number, data size, and bandwidth share of a flow. For network status collection, each network device collects and stores local traffic statistics within its own storage. These local traffic statistics may be retrieved by controllers, or proactively reported to controllers.

4) *Network status synchronization.* Finally, the network status synchronization is important at the centralized controller. Maintaining a consistent updated global view among all the controllers in the distributed global network is essential to ensure proper network operations. Inconsistent or stale states in the distributed control, may result in the application layer making incorrect decisions, which then leads to inappropriate or suboptimal operations of the network.

Common NOS software is deployed and runs on each of the server in the cluster. This distributed core nature of NOS provides scalability, agility, high availability, and brings the carrier grade performance features to the SDN control plane. The deployment symmetry is an important design consideration as it enables rapid recovery in the event of a particular NOS server failure. The distributed core NOS instances work together, so that to the rest of the network and applications it appears as a single platform. Applications and network devices do not have to know whether they are with single instance or with multiple instances of NOS. The distributed core manages a variety of operational states, such as application intents, topology database, and flow tables, etc. The NBI provides the API for developing applications. In summary, the distributed core is the key architectural feature of the NOS that brings carrier grade features to the SDN control plane.

There are several benefits of this logical centralization of the controller. First, the centralization of the control logic in a controller with global knowledge of the network-state simplifies the development of more sophisticated networking functions, services, and applications. The applications can take advantage of the GNV information leading to a more consistent and effective policy decisions. It is much simpler and less error-prone to modify network policies through high-level languages and software components of the controller. The applications can take actions, such as reconfigure forwarding devices from any part of the network, and it becomes easier to program these applications since the abstractions provided by the control platform and the network programming languages can be shared.

10.2.1.5 North Bound Interfaces Layer

The top layer, NBIs of the control plane in Figure 10.1, handles transactions with the application layer, i.e. receiving policies described in high-level languages from SDN applications and presenting the GNV to them. There are two northbound abstractions: one is the *intent framework* (IF) and the other is the GNV. The IF abstraction allows an application to request a service from network without having to know details of how the service will be performed. They can simply specify their intent policy statement or connectivity requirement, such as setting up a path between two switches with a required bandwidth, etc. After taking the requests from all applications, IF figures out which ones can/cannot be accommodated, resolves conflicts between applications, and the objectives are then compiled into instructions that are sent to the network devices. The process is done under the control of policies specified by the NOS after resolving conflicts. The GNV provides the application with a view of the network: the hosts, switches, links, bandwidth, etc. An application can program this GNV through APIs, which creates the particular application to calculate with the network view available. Technically, the northbound abstractions and APIs insulate applications from details of the network which in any case is not required by the particular application.

Though, there is a common southbound interface (*OpenFlow*), which is standardized, whereas, a common NBI is yet to be evolve. Once it is made available, the abstraction would allow network applications to explore the full potential of SDN. They will not depend on specific implementations but will promote application portability and interoperability among the different control platforms. The northbound interface is mostly in software and not a hardware implementation, as in the case of the southbound APIs. Also, each network application can be quite different from the other, so having a standard API can be quite complex in this case.

10.2.1.6 Network Application Layer

The network application layer comprises different network applications and services that utilize the control plane to realize network functions over the physical or virtual infrastructure. The different network applications include the routine functions of routing, load balancing, reliability, QoS, security policy implementation. The other applications are network virtualization, mobility management in wireless networks, network data analytics, or specialized functions requiring processing in large data centers, etc. NOS through the NBI presents the required part of abstracted global network status information and permits the particular application to adaptively control the network by utilizing the SDN platform. Equipped with this information, SDN applications can implement strategies to manipulate the underlying physical networks using a *high-level language* provided by the control layer.

To summarize, the idea of SDN is to decouple the control plane from the data plane and allow flexible and efficient operation and management of the network via software programs. Network intelligence is taken out of switching devices and placed on controllers. Thus, in the data plane the switches and routers are simple programmable devices which perform packet forwarding based on rules installed by the controllers in the CP. They are externally controlled by software and not by on-board intelligence. With the global network map as input, the SDN controllers form the *abstract forwarding model* for forwarding function. Under this new paradigm, innovative solutions for specific purposes e.g. network security, survivability, energy efficiency, and network virtualization can be rapidly implemented in the form of software and deployed in networks with real traffic by the applications.

10.3 Software-Defined Optical Networking

We are aware of the importance of optical networks in our present communication networks due to their high transmission capacities and security. At the same time they are quasi-static and are not amenable to fast changes in its topology and network characteristics due to its circuit switching nature and pose challenges in making them flexibly controllable and agile. Nevertheless, software-defined optical networks can gain the flexibility and programmability of SDN control for supporting networking applications with an underlying optical network infrastructure.

Optical networks have a few typical features specific to them when compared with IP. One, optical networks are vast, have presence from access, metropolitan area networks (MANs) to wide area networks (WANs), therefore, they need to have heterogeneous devices ranging from the end user nodes and local area networks (LANs), optical network units (ONUs), and optical line terminals (OLTs) to the edge routers of metro network nodes and then on to the transport network infrastructures from large OXCs to bandwidth variable transceivers (BVTs), etc. These different devices often coming from different vendors may require manual configuration and maintenance of optical networks. Second, in optical networks, besides optical fiber there may be many diverse communication technologies used with different propagation features, such as, wireless and cables. The other aspect regarding optical network is the diverse nature of traffic it has to support, from voice to IP multimedia data packets and lightpaths. Furthermore, IP networks are packet-switched while optical transport networks are basically space and time division multiplexed or wavelength division multiplexed circuit-switched. The circuit-switched networks are managed tightly with high reliability and redundancy, but the IP networks have very few management capabilities. The multilayered optical network, therefore, always had independent control mechanism to support individual technology. GMPLS though has been borrowed from IP, but has a unified control plane for both IP and optical transport plane. Another fundamental difference between IP and optical networks is the data plane traffic flow resources; in IP, packets are the only flow resource, but in optical there are many flows; wavelengths, fiber, and time-slots which are circuit switched. These various flows are quite different from the control perspective when compared with packets in IP. With all these diversities it becomes difficult to control and manage the optical network with flexibility and programmability as would be desired.

Nevertheless, with the added features and advantages discussed in the previous section on SDN, efforts are being made to combine the IP also over multilayer optical networks to form the software-defined optical networks [15–18]. The automatically switched optical network (ASON) and automatically switched transport network (ASTN), standardized by ITU-T in mid-2000, introduced control plane with distributed path computation on top of the optical network. In the last section, we saw that the logical centralized CP in SDN for path computation and other TE features adds several advantages with the SDN controller. What we need is, therefore, to have a commonly-operated converged multilayered optical network with a common flow-switched forwarding plane in SDON. In order to do so, there is a need to unify the management of all layers of the multilayered optical network and have a converged data plane totally independent of the converged control plane.

With flex-grid networks' programmable components, discussed in the previous chapter, the optical transport layer in SDON can be abstracted to a set of shared, common resources, that can be programmed and used dynamically on demand. In the traditional optical network, the control plane and the data plane resided in the same hardware. In SDN, the controller is separated from the hardware and is centralized, or runs over cloud servers. Optical SDN takes these concepts from the packet world and applies them to the underlying optical layer. It brings many advantages, such as reduced latency for dynamic traffic, more economy, and flexibility in the network. Extending SDN to the transport optical layer with centralized orchestration of all resources will result in end-to-end optimization, multilayer data flows and much larger network utilization.

10.3.1 SDON Architecture

In order to evolve optical software-defined networks having the programmability of SDN, several challenges need to be overcome. These issues are basically due to the various switching granularities, the rigidity of optical devices and various heterogeneous transport technologies which the optical network supports. What needs to be changed or included in the optical networks in this direction is as the following:

- One needs to have a generalized unified transport plane and switching granularity for different optical transport technologies, such as, fixed DWDM, flexi DWDM, space, etc. along with electronic packet switching technology.
- Design and implementation of an abstraction mechanism that can hide the heterogeneous optical transport layer technology details and have a generalized abstracted switching entity.
- Have a common *data plane-abstraction* that would give a common model for multilayer control. This will eliminate the vendor islands and propriety interfaces running multi-vendor networks and replace them with flexible and standardized interfaces.
- Have a common *control-abstraction* that will have all network-functions and services automated across packets and circuits providing fast deployment and development. It will also provide the network operator the flexibility of selecting the best mix of technologies for their service needs.
- The TE issues should be accounted for bandwidth allocation and traffic mapping in optical networks with multilayer hybrid transport technologies.

Thus, we need a common unified control architecture for the abstractions for common flow for packet switched traffic and circuit switching traffic in space, wavelength, or TDM. Abstraction for the common map for applications and services in SDON. We in the following paragraphs, underline the details of upcoming developments in the three planes: data, control, and application of the SDON system.

10.3.1.1 SDON Data Plane
The data plane, much as discussed in the SDN, consists of the common infrastructure layer, the abstraction layer and the SBI interface.

- *Infrastructure layer.* The infrastructure blocks of SDON need to be flexible and programmable. They need to be centrally controlled for any SDN implementation. Therefore, the fundamental requirement for SDN-enabled optical network is the ability of every infra-element; optical, photonics, and electronic, to be software programmable. Electronics and photonics are much easily amenable to changes with software but optical resources have been traditionally treated as dumb, fixed pipes. However, with recent advancement in optical technologies, optical transmission resources are increasingly becoming software-controlled and programmable [16, 19–24]. The different blocks which have to be amenable to programming are the fiber transmission, transceivers, switches, etc. The transceivers can permit SDN control of the optical signal transmission characteristics, such as, bit rate, modulation format, etc. on the lightpaths. Also, the SDON controlled optical switching, including the switching elements and the all switching paradigms; packet and circuit switching need to be programmable. The optical signal quality needs to be monitored as this information is required for controlling the transceivers, amplifiers, and the switching elements. Along with these modules, the basic optical wavelength channels need to be flexible. In conclusion, to bring dynamicity and flexibility to optical networking, significant developments in the following areas are required in these networks.

- *Flexible wavelength grid.* ITU-T standards for the fixed grid WDM had defined a rigid grid of 100 GHz and 50 GHz frequency spacing. The WDM wavelength grid specification from ITU-T G.694.1, has defined a flexible grid with flexible center frequency of WDM channels spaced at 12.5 GHz but with the ability to define an aggregate super-channel spectral width of (N × 12.5) GHz with N as positive integer. These super-channels flexibly accommodate any combination of optical carriers, modulations, and data rates. In addition, the (N × 12.5) GHz spectral width of the super-channel can be tuned, enabling rapid changes of the operating specifications [25, 26] as discussed in last chapter.
- *Software-defined port speeds, protocols, and wavelengths.* The different physical ports that are attached to the network devices must be software defined in terms of protocol, speed, as well as, wavelength selection. This allows new services to be provisioned or changed on an optical network via software without any prior knowledge of the service type. Towards this direction, tunable lasers which enable flexible wavelength selection are available to provide network flexibility as desired. Also, the other related technologies and software infrastructures are now being defined to allow protocol and wavelength tunability and become part of an overall software-defined flexible network design.
- *Advanced modulation and detection schemes.* For speeds up to 10 Gbps, optical transmitters used a simple on-off keying (OOK) encoding but for higher data rates it has the limitation of longer reach. In order to increase the spectral efficiency and reduce the per-channel electronics speed within the optical transmitters and receivers, more advanced modulations were developed [27]. These modulation schemes use both amplitude and phase parameters of the light signal. In particular, coherent detection and advanced digital signal processing have become the standard for high speed optical signal detection for 100 Gbps and above. For 100 Gbps and above transmission dual-polarization quadrature phase-shift keying (DP-QPSK) and M-quadrature amplitude modulation (M-QAM) are the standard transmission modulation schemes. The transmitter modulation scheme can be selected optimally with trade-off between bandwidth and reach for a specific network span. The properties of these transponder, such as bit rate, optical reach, bandwidth requirement need to be adjusted based on the link length and physical properties of the

channel. Therefore, software programmable transponders with the modulation scheme defined in the software are possible.

Taking into consideration the above details, we now discuss the different software-defined infrastructure building blocks of SDON.

5) Transceivers

Software-defined optical transceivers are optical transmitters and receivers that can be flexibly configured by SDN to transmit or receive a wide range of optical signals [22, 23]. Software-defined optical transceivers can have variable modulation and bit rate of the transmitted optical signal. These transceivers have BVTs generating single or multiple signal flows.

i) *Single-flow bandwidth variable transponder.* Single-flow BVTs have fully programmable transmission characteristics and a single flow with software-controlled transmission signal modulation, bandwidth, and rate. For example, a flexible transceiver may have a transmitter based on flexible Mach–Zehnder modulators (MZMs) and a corresponding flexible receiver for SDN control. By adjusting the MZM direct current bias voltages and amplitudes of drive signals, the amplitude and phase of the generated optical signal can be varied. Thus, modulation formats ranging from M-PSK to M-QAM can be generated. The amplitudes and bias voltages of the drive signals can be signaled through an SBI from the control plane to achieve the different modulation formats and rates. The corresponding flexible receiver may have polarization filter that may feed parallel photodetectors. The detected electronic analog signal will be converted to digital by an analog-to-digital converter (ADC). The outputs of the parallel ADCs can then be processed with DSP techniques to automatically detect the transmitted modulated signal.

ii) *Sliceable multi-flow bandwidth variable transceivers.* For better utilization of network bandwidth with multiple rate data flows in the network, parallel optical signals from tunable or multiple discrete lasers are used to design multi-flow transponders, as discussed in the last chapter. These multi-flow transceivers can generate multiple parallel optical signal flows and allow for the independent SDN control of multiple traffic flows which can have DSP based or encoder based programmable transponder designs. The sliceable BVT has a SDN controlled adaptive DSP of multiple parallel orthogonal frequency-division multiplexing (OFDM) signal subcarriers. Each subcarrier is fed by a DSP module that configures the modulation format, bit rate, and the power level of the carrier by adaptable gain coefficient. The output of the DSP module is then passed through digital to analog conversion that drives laser sources and the combined flow can be sliced into multiple distinct subflows for distinct destinations. A virtualizable multi-flow BVT based on

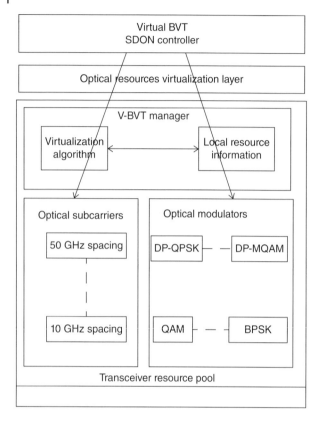

Figure 10.6 Virtualizable multi-flow bandwidth variable transceiver.

a combination of an optical subcarriers pool with an independent optical modulators pool have been developed with the emphasis on implementing Virtual Optical Networks (VONs) at the transceiver level with programmable channel spacing, center frequency, modulation type, etc. [28], as shown in Figure 10.6.

6) SDN-controlled switching elements

   i) *SDN controlled reconfigurable add-drop multiplexer (ROADM).* ROADMs in the available optical networks have limited configurability in terms of the switching channels, ports, etc. as they have pre-configured routes. New ROADM node designs which are *colorless, directionless, contentionless, and grid-less (CDCG),* allow changes of wavelength channels and routes, have no contention in the switch and have spectrum size selected through a management control plane [26, 29]. CDCG ROADM operation means that any wavelength can be routed on any port in any direction without causing resource contention. CDCG ROADM designs consist of an express bank for interconnecting the input and output ports coming from links or devices, and an add-drop bank that connects the express bank with the local receivers for dropping wavelength channels or transmitters for adding wavelength channels. The OXC backplane allows for highly flexible add/drop configurations implemented through SDN control.

ii) *Open transport switch (OTS).* The OTS has been proposed as an *OpenFlow*-enabled optical virtual switch. The OTS design abstracts the details of the underlying physical switching layer which can be with packet or circuit switched or a virtual switch element. Three agent modules – discovery, control, and data plane – interface with the physical switching hardware. These modules are controlled from an SDN controller through extended *OpenFlow* messages.

iii) *Logical xBar.* The *OpenFlow* switch for optical data plane have been developed with logical *xBar* programmable switch. Multiple small elementary xBars [29, 30] can be recursively merged to form a single large switch with a single forwarding table. The forwarding plane is envisioned to be managed by the labels based on SDN and GMPLS. This concept of *xBar* is considered to be the building block for forming large networks.

iv) *Optical white box.* There have been efforts toward designing an optical white box switch which is completely software programmable [18, 20, 31]. It combines a programmable optical and electronic backplane with programmable switching node elements. Through the backplane flexible arbitrary connections can be made between the switch node elements. The switch node elements are the programmable interfaces that can build SDN-controlled BVTs, programmable switching elements and DSP elements. The protocol agnostic switching elements can support both wavelength channel and time-slot switching in the optical backplane, as well as, programmable switches with a high-speed packet processor in the electronic backplane. The DSP elements support both the network processing and the signal processing for executing a wide range of network functions.

10.3.1.2   SDON Control Plane

The unified control plane architecture of the SDON, much like the SDN, will comprise of abstraction for common flow and common map and a controller running the NOS [31–34]. The control plane being a server plane to the application layer has to facilitate the applications through the data plane consisting of various software-defined infrastructure devices. In order to achieve the above role, it provides the common-map view and the common-flow abstraction for the packets, optical wavelength, time-slots, and other types of flows of the lower infra layer with the switch-API. The main benefits of the common-flow abstraction is that it will no longer require independent distributed control planes for switching the different flows of the different layers and domains in the optical network. Instead with an external logically centralized controller it will define packet and circuit flows flexibly and map them to each other, de-layering the multilayer network by treating packets and circuits as part of flows and reducing the complexity in control planes. Second, the common-table abstraction with the use of the switch-API makes the solution independent of vendor-specific hardware control solutions. In the following paragraphs the work in this direction is explained.

Data Layer Abstraction and OpenFlow In the case of SDON, the control layer has to control the data layer consisting of optical network infrastructure elements, optical spectrum and the multiple optical network layers and their operational aspects. What needs to be done is to have unified NEs and switching granularity that can be generalized for different optical transport technologies, such as fixed and flexi DWDM, sub-wavelength switching, etc. as shown in Figure 10.7 in the data plane. The data layer abstractions need to hide the transport layer technology details and realize the generalized switching. SDON also has to account for the physical layer specific features of different optical transport technologies such as power, impairments, switching constraints, etc. and the cross-technology constraints for bandwidth allocation and traffic mapping in networks comprising heterogeneous technological domains e.g. packet over several hybrid optical transport technologies. These features are represented by the different intra and inter domain flow tables and data base characteristics blocks in Figure 10.7.

With resource or hardware abstraction the technological details of underlying heterogeneous network resources are masked, enabling a programmable interface for hardware state configuration. The commonly used *OpenFlow* protocol therefore needs extensions to include controlling the optical transmission and switching components. The abstraction mechanism is realized by an extended *OpenFlow* controller and the *OpenFlow* protocol to include the optical network components and flows. By the OF controller and protocol, we can generalize the *flow switching* concept for the underlying heterogeneous optical technologies, integration with various switching paradigms and number of switched domains. *OpenFlow*-enabled switch in the controller is represented by one or more flow tables, and each table entry consisting of match fields, counters, and a set of associated actions. In SDON this switch besides the flow table on packet domains switching also has to address the optical domain switching, considering the Synchronous Optical NETwork (SONET)/synchronous digital hierarchy (SDH), OXCs, and Ethernet/TDM

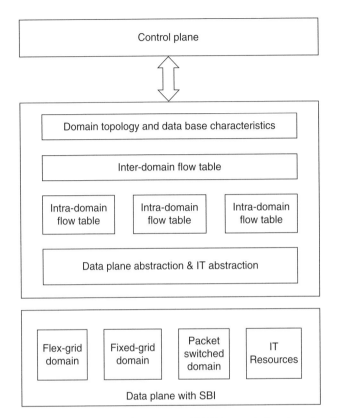

Figure 10.7 Data layer abstractions of SDON control plane based on *OpenFlow*.

convergence as circuit switched technologies. Additionally, what is required for the non-SDN switching device is a virtual *OpenFlow* switch with virtual interfaces that correspond to physical switching ports for their abstractions. When a flow entry is added between two virtual ports in the virtual *OpenFlow* switch, the abstraction layer uses these non-SDN switching devices' management interface to add the flow entry between the two corresponding physical ports [34–38]. This is also called the retrofitting of the optical devices.

In optical switched networks, the data information flows over the fiber, wavelength, or in time-slots as a *circuit flow* model. In IP network the information flow is as the *datagram* model, rather than flow of packets of end-to-end communication. But data packets can also be treated as flows with some *logical association* with the overheads in the packet headers. To define the common-flow abstraction for both the traffics, we define the information data forwarding as flows. In the case of circuit-switched transport network the circuits made with fiber, wavelength, or time-slot itself are the logical association for the connection of an incoming wavelength, time-slot, or fiber input-port to an outgoing port. In the case of datagram services we can classify the packets by their logical association for the particular connection with packet header. This soft state of packets is retained in switches to remember the nature of the flows which are passing through them. Now the same set of actions is performed on all packets that have the same logical association, which is the flow definition. Thus, the fundamental data-plane unit to control in transport or packet networks is *the flow* and not the individual circuits or packets.

The common-flow abstraction is therefore a common-forwarding table abstraction [33], where we abstract all kinds of packets and circuit switch hardware, by presenting them as forwarding tables for direct manipulation by a switch-API. In other words, switches are no longer viewed as Ethernet switches, IP routers, L4 switches, MPLS label switched routers (LSRs), SONET switches, OTN switches, ROADMS, or multilayer switches- they are just tables. These tables support the flow identifiers irrespective of the traditional layer of networking (L0-L4) or combination of them. As shown in Figure 10.8, the vendor and layer specific switching with Ethernet, IP router or SONET, OTN switches, common flow abstraction is based on a data-abstraction by switch *flow tables*, manipulated by a *switch-API*. In the case of packets, the flow tables take the form of lookup-tables and for circuit-switching they take the cross-connect tables form, and together with the common switch-API they abstract all the different layers and vendor specific hardware and interfaces.

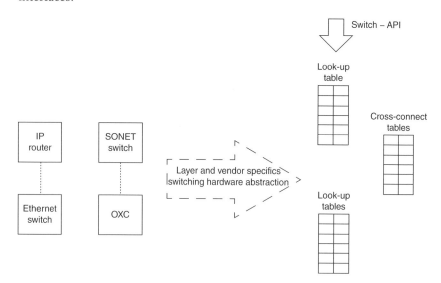

Figure 10.8 Common-flow abstraction in optical networks.

SDON Control of Optical Infrastructure Components

- Controlling Optical Transceivers with *OpenFlow*
Recent generations of sliceable bandwidth variable transponder (SBVT) utilize DSP techniques, discussed in Chapter 9, that allow many parameters of the transceiver to be software controlled. Parameters such as bit rate, modulation scheme, symbol rate, and wavelength can be added to the *OpenFlow* cross-connect table entries to support these programmable features of software-defined optical transceivers. With software-defined control, modulation technique and forward error correction (FEC) code for each optical subcarrier of the super-channel transponder and the optical amplifier power level can be controlled via *OpenFlow* [33]. Also, the transponder subcarriers can be treated as *OpenFlow* switch ports that can be configured through the *OpenFlow* protocol via port modification messages. By doing so would allow the SDN controller to adaptively modify amplifiers to compensate for channel impairments while minimizing energy consumption.

- *Controlling Optical Switches with OpenFlow*
OpenFlow can be used for the control of optical circuit, packet, as well as burst switching. Circuit switching can be enabled by *OpenFlow* by adding new circuit switching flow table entries [35]. These WSS cross-connect tables are configured via *OpenFlow* messages inside the circuit switches. The cross-connect table can have entries consisting of the fields to identify the input and output ports, and entries for circuit switching in space, fixed-grid wavelength and time, and also support flexible wavelength grid optical switching [34–36]. The computation of forwarding tables can be off-loaded to an SDN controller which with its global view provides a more effective means for resolving any contention leading to packet loss and select the path with the best available resources among multiple available paths between two nodes [37]. Paths can be computed periodically or on demand to account for changes in traffic conditions. Optical packet switching can be implemented with a semiconductor optical amplifier switch. *OpenFlow* flow tables can also be used to configure optical burst switching devices.

Common Map Abstraction The common map abstraction, which is created and kept updated with network state by the control-plane, allows network applications to be implemented in a centralized manner. The global map has the database of the network topology of both packet and circuit switching nodes. The switching capabilities of these switches are represented by the forwarding tables and this database is created and presented to control applications and kept up to date by the map-abstraction.

The several network control functions, such as routing, access control, mobility, TE, recovery, BoD, etc., are provided either by packet networks or by the circuit networks, and sometimes by both. These control function programs can be best formulated when they operate in a centralized way with a global view of the network. By centralizing the optical network control in the SDON controller, the SDN networking paradigm creates a unified view of the entire optical network. With map-abstraction, a network-API can be used to write programs that introduce new control-functionality as the need arises. The network API include the networking tasks of discovery, configuring switches, controlling their forwarding behavior and monitoring network state. The network API can present calls for all three tasks together to network applications. With its full visibility, the common map allows these new features to be supported taking advantage of both packets and circuits in as indicated in Figure 10.9.

Common-map abstraction has several benefits. First, it offers client applications ability to perform joint-optimization of network functions and services across both packet and circuit switching technologies with a global view of the network. Second, the common-map abstraction makes inserting of any new control functions into the network easy and extensible as the distributed control is abstracted away with a centralized control which makes implementing individual control functions simpler. Also, it allows the network programmer to write program in different ways, treating the packet and circuit flows in the same layer or in different layers having separate topologies but still be commonly controlled as shown in Figure 10.9. It can bring innovation by offering a network API to program for controlling network behavior.

10.3.1.3 SDON Application Layer

Network application, such as routing, QoS, load balancing, access control, security, energy efficiency, and failure recovery, etc. of the optical application

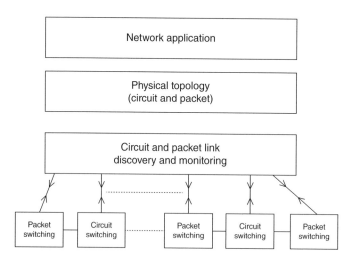

Figure 10.9 Common-map abstraction of SDON control.

layer, interacts with NOS to implement network services through the NBI. The controllers in turn with global map abstraction provides the required services of the data plane with the help of SBI. In the case of SDON the different applications for traffic and network management mechanisms are supported through the *OpenFlow* protocol and the central SDN controller. The centralized knowledge of the traffic and network is utilized to allocate or configure resources, such as data-center resources, application bandwidths, security, and topology configurations or routes. SDON QoS needs to optimize the interactions of the controller with the network applications and data plane to quickly and correctly react to changing user demands and network conditions, so as to assure consistent QoS [39, 40]. There are further a number of applications in the areas of security, failure recovery, and energy efficiency being carried out for SDON.

With the vision of SDON in place, significant work is being carried out in the forwarding and the control planes. Network abstractions and the APIs for the transport as well as the access part of the network are also making rapid progress [41–43].

10.4 Summary

Software-defined networking promises a simplification and unification of network management for optical networks which will allow programmability, flexibility, and automation of operational tasks. In the first part of the chapter an introduction with sufficient details of the architecture and functioning of the SDN has been given. In the latter half, SDON infrastructure plane and control framework for the wide range of optical network transmission approaches and network structures are given. The common-flow abstraction of the data plane lets us deal both packets and circuits commonly as flows, thereby providing a common paradigm for flexible and vendor-agnostic control across multilayered optical networks, network virtualization, by which flexible operation of multiple VONs can be installed over a given physical optical network infrastructure. The work in the area of SDON is in the developing stage and it foresees a huge potential in networked high-performance computing, intra- and inter-data center networks connectivity, and other applications.

References

1 http://www.cisco.com/c/en/us/solutions/collateral/service-provider/visual-networking-index-vni/mobile-white-paper-c11-520862.html.
2 http://www.informationweek.com/mobile/mobile-devices/gartner-21-billion-iot-devices-to-invade-by-2020/d/d-id/1323081.

3 Open Networking Foundation, "SDN Architecture Overview, Version 1.1, ONF TR-504," Palo Alto, CA, USA, Nov. 2014.

4 https://tools.ietf.org/html/rfc7426 RFC 7426 - Software-Defined Networking (SDN): Layers and Architecture Terminology.

5 Hu, F., Hao, Q., and Bao, K. (2014). A survey on software-defined network and OpenFlow: from concept to implementation. *IEEE Commun. Surv. Tutorials* 16 (4): 2181–2206.

6 Kreutz, D., Ramos, F.M.V., Verissimo, P. et al. (2015). Software-defined networking: a comprehensive survey. *Proc. IEEE* 103 (1): 14–76.

7 Nunes, B., Mendonca, M., Nguyen, X. et al. (2014). A survey of software-defined networking: past, present, and future of programmable networks. *IEEE Commun. Surv. Tutorials* 16 (3): 1617–1634.

8 Lara, A., Kolasani, A., and Ramamurthy, B. (2014). Network innovation using OpenFlow: a survey. *IEEE Commun. Surv. Tutorials* 16 (1): 493–512.

9 Liu, L., Tsuritani, T., and Morita, I. (2012). From GMPLS to PCE/GMPLS to OpenFlow: how much benefit can we get from the technical evolution of control plane in optical networks? In: *Proc. 14th ICTON*, 1–4.

10 https://northboundnetworks.com/products/zodiac-fx.

11 http:// www.pica8.com/product.

12 https://noviflow.com/products/noviswitch.

13 Foster, N. et al. (2010). Frenetic: a high-level language for OpenFlow networks. In: *Proc. Workshop PRESTO*, 6:1–6:6.

14 Voellmy, A. and Hudak, P. (2009). Nettle: a language for configuring routing networks. In: *Proc. IFIP TC 2 Working Conf. DSL*, 211–235.

15 Thyagaturu, A., Mercian, A., McGarry, M.P. et al. (2016). Software defined optical networks (SDONs): a comprehensive survey. *IEEE Commun. Surv. Tutorials* 18 (4): 1–47.

16 Steven Gringeri, Nabil Bitar, and Tiejun J. Xia, Extending Software Defined Network Principles to Include Optical Transport, *IEEE Communications Magazine*, vol. 51, no. 3, pp. 32-40, March 2013.

17 Bhaumik, P. et al. (2014). Software-defined optical networks (SDONs): a survey. *Photonic Network Commun.* 28 (1): 4–18.

18 Channegowda, M., Nejabati, R., and Simeonidou, D. (2013). Software-defined optical networks technology and infrastructure: enabling software-defined optical network operations [invited]. *IEEE/OSA J. Opt. Commun. Networking* 5 (10): A274–A282.

19 Wang, Y. and Cao, X. (2012). Multi-granular optical switching: a classified overview for the past and future. *IEEE Commun. Surv. Tutorials* 14 (3): 698–713.

20 Marom, D.M., Colbourne, P.D., D'Errico, A. et al. (2017). Survey of photonic switching architectures and technologies in support of spatially and spectrally flexible optical networking [invited]. *J. Opt. Commun. Networking* 9 (1): 1–26.

21 Collings, B., 'New Devices Enabling Software-Defined Optical Networks', IEEE Commun. Mag., 51(3), 66-71, 2013.

22 Moreolo, M.S. et al. (2016). SDN-enabled sliceable BVT based on multicarrier technology for multiflow rate/distance and grid adaptation. *J. Lightwave Technol.* 34 (6): 1516–1522.

23 Geisler, D.J. et al. (2011). Bandwidth scalable, coherent transmitter based on the parallel synthesis of multiple spectral slices using optical arbitrary waveform generation. *Opt. Express* 19: 8242–8253.

24 Channegowda, M., Nejabati, R., and Simeonidou, D. (2013). Software-defined optical networks technology and infrastructure: [invited]. *J. Opt. Commun. Networking* 5 (10): A274–A282.

25 M. Jinno et al., "Distance-Adaptive Spectrum Resource Allocation In Spectrum-sliced Elastic Optical Path Network," *IEEE Commun. Mag.*, 2010, pp. 138–45.

26 M. Jinno et al., "Spectrum-Efficient and Scalable Elastic Optical Path Network: Architecture, Benefits, and Enabling Technologies," *IEEE Commun. Mag.*, vol. 47, 2009, pp. 66–73.

27 Winzer, P.J. and Essiambre, R.-J. (2006). Advanced optical modulation formats. *Proc. IEEE* 94 (5): 952–985.

28 Ou, Y. et al. (2016). Demonstration of virtualizeable and software-defined optical transceiver. *IEEE/OSA J. Lightwave Technol.* 34 (8): 1916–1924.

29 Strasser, T.A. and Wagener, J.L. (2010). Wavelength-selective switches for ROADM applications. *IEEE J. Sel. Top. Quantum Electron.* 16: 1150–1157.

30 Azodolmolky, S. et al. (2011). Integrated OpenFlow-GMPLS control plane: an overlay model for software defined packet over optical networks. *Opt. Express* 19 (26): B421–B428.

31 Liu, L. et al. (2013). Field trial of an OpenFlow-based unified control plane for multilayer multigranularity optical switching networks. *IEEE/OSA J. Lightwave Technol.* 31 (4): 506–514.

32 Liu, L., Muñoz, R., Casellas, R. et al. (2013). OpenSlice: an OpenFlow-based control plane for spectrum sliced elastic optical path networks. *Opt. Express* 21 (4): 4194–4204.

33 Chan, V.W. (2012). Optical flow switching networks. *Proc. IEEE* 100 (5): 1079–1091.

34 Liu, L. et al. (2012). First field trial of an OpenFlow-based unified control plane for multi-layer multi-granularity optical networks. In: *Proc. OFC/NFOEC*, 1–3.

35 Channegowda, M., Nejabati, R., Fard, M.R. et al. (2013). Experimental demonstration of an OpenFlow based software-defined optical network employing packet, fixed and flexible DWDM grid technologies on an international multi-domain testbed. *Opt. Express* 21 (5): 5487–5498.

36 Das, S. et al. (2010). Packet and circuit network convergence with OpenFlow. In: *Pro. OFC/NFOEC*, 1–3.

37 Liu, L., Tsuritani, T., Morita, I. et al. (2011). Experimental validation and performance evaluation of OpenFlow-based wavelength path control in transparent optical networks. *Opt. Express* 19 (27): 26578–26593.

38 Liu L. et al., "Experimental demonstration of an OpenFlow/PCE integrated control plane for IP over translucent WSON with the assistance of a per-request-based dynamic topology server," in *Proc. Eu. Conf. and Exhibition on Optical Commun. (ECOC)*, 2012, p. Tu.1.D.3.

39 Li, K., Casellas, R., Tsuritani, T. et al. (2013). Experimental demonstration of an OpenFlow/PCE integrated control plane for IP over translucent WSON with the assistance of a per-request-based dynamic topology server," OSA Publishing. *Optics Express* 21 (4): 4183–4193.

40 Li, K., Guo, W., Zhang, W. et al. (2014). QoE-based bandwidth allocation with SDN in FTTH networks. In: *IEEE Network Operations and Management Symposium (NOMS)*. Poland: IEEE.

41 Cao, X., Yoshikane, N., Popescu, I. et al. (2017). Software-defined optical networks and network abstraction with functional service design [invited]. *J. Opt. Commun. Networking* 9 (4): C65–C75.

42 Szyrkowiec, T., Autenrieth, A., and Kellerer, W. (2017). Optical network models and their application to software-defined network management. *Int. J. Opt.* 2017: 5150219. https://doi.org/10.1155/2017/5150219.

43 Kitayama, K., Hiramatsu, A., Fukui, M. et al. (2014). Photonic network vision 2020—toward smart photonic cloud [invited]. *J. Lightwave Technol.* 32 (16): 2760–2770.

Index

An *f* or a **t** following a page number denotes a figure or a table, respectively.